T0271292

CAMBRIDGE TRACTS IN MATHEMATICS

GENERAL EDITORS

J. BERTOIN, B. BOLLOBÁS, W. FULTON, B. KRA,
I. MOERDIJK, C. PRAEGER, P. SARNAK,
B. SIMON, B. TOTARO

CAMBRIDGE TRACTS IN MATHEMATICS

GENERAL EDITORS

J. BERTOIN, B. BOLLOBÁS, W. FULTON, B. KRA, I. MOERDIJK,
C. PRAEGER, P. SARNAK, B. SIMON, B. TOTARO

A complete list of books in the series can be found at www.cambridge.org/mathematics
Recent titles include the following:

Variations on a Theme of Borel

An Essay on the Role of the Fundamental Group in Rigidity

SHMUEL WEINBERGER

University of Chicago

CAMBRIDGE
UNIVERSITY PRESS

CAMBRIDGE
UNIVERSITY PRESS

University Printing House, Cambridge CB2 8BS, United Kingdom

One Liberty Plaza, 20th Floor, New York, NY 10006, USA

477 Williamstown Road, Port Melbourne, VIC 3207, Australia

314–321, 3rd Floor, Plot 3, Splendor Forum, Jasola District Centre,
New Delhi – 110025, India

103 Penang Road, #05–06/07, Visioncrest Commercial, Singapore 238467

Cambridge University Press is part of the University of Cambridge.

It furthers the University's mission by disseminating knowledge in the pursuit of
education, learning, and research at the highest international levels of excellence.

www.cambridge.org
Information on this title: www.cambridge.org/9781107142596
DOI: 10.1017/9781316529645

© Shmuel Weinberger 2023

First published 2023

A catalogue record for this publication is available from the British Library.

Library of Congress Cataloging-in-Publication Data
Names: Weinberger, Shmuel, author.
Title: Variations on a theme of Borel / Shmuel Weinberger, University of Chicago.
Description: Cambridge, United Kingdom ; New York, NY : Cambridge University Press,
2021. | Series: Cambridge tracts in mathematics ; 213 | Includes bibliographical references.
Identifiers: LCCN 2022024936 (print) | LCCN 2022024937 (ebook) | ISBN
9781107142596 (hardback) | ISBN 9781316507490 (ebook)
Subjects: LCSH: Rigidity (Geometry) | Manifolds (Mathematics) | Three-manifolds
(Topology) | Graph theory. | BISAC: MATHEMATICS / Geometry / General |
MATHEMATICS / Geometry / General
Classification: LCC QA640.77 .W45 2021 (print) | LCC QA640.77 (ebook) | DDC
514/.34–dc23/eng20220823
LC record available at https://lccn.loc.gov/2022024936
LC ebook record available at https://lccn.loc.gov/2022024937

ISBN 978-1-107-14259-6 Hardback

To Devorah, Baruch, and Esther

In memory of my parents, Rabbi Philip and Hannah Weinberger

and

In memory of Armand Borel and Bill Thurston, with awed acknowledgement of the profound ways their varied visions have enriched our understanding of geometry and topology

Contents

Preface

This essay is a work of historical fiction – the "What if Eleanor Roosevelt could fly?" kind.[1] The Borel conjecture is a central problem in topology: it asserts the topological rigidity of aspherical manifolds (definitions below!). Borel made his conjecture in a letter to Serre some 69 years ago,[2] after learning of some work of Mostow on the rigidity of solvmanifolds.

We shall reimagine Borel's conjecture as being made after Mostow had proved the more famous rigidity theorem that bears his name – the rigidity of hyperbolic manifolds of dimension at least three – as the geometric rigidity of hyperbolic manifolds is stronger than what is true of solvmanifolds, and the geometric picture is clearer.

I will consider various related problems in a completely ahistorical order. My motive in all this is to highlight and explain various ideas, especially recurring ideas, that illuminate our (or at least my own) current understanding of this area.

Based on the analogy between geometry and topology imagined by Borel, one can make many other conjectures: variations on Borel's theme. Many, but perhaps not all, of these variants are false and one cannot blame them on Borel. (On several occasions he described feeling lucky that he ducked the bullet and had not conjectured smooth rigidity – a phenomenon indistinguishable to the mathematics of the time from the statement that he *did* conjecture.)

However, even the false variants are false for good reasons, and studying these can be quite fun (and edifying); all of the problems we consider enrich our understanding of the geometric and analytic properties of manifolds. *Verum ex erroris.*

The tale I shall tell moves between topology and geometry, Lie groups,

[1] See *Saturday Night Live*, season 4, episode 4.
[2] May 2, 1953.

arithmetic, and operator theory, algebraic K-theory, and topics in Banach space geometry that are also of interest in theoretical computer science. The goal is to develop an appreciation for this landscape – not to explain the most recent or important results on the conjecture itself.[3]

The extent of the canvas that forms the natural backdrop to this problem is both a joy and a challenge. I cannot explain all the detail or even sketch all action going on about this canvas, but I will try to tell some good stories[4] – simplifying enough to explain the key ideas, and providing references as best I can to papers that have the missing parts, trying to do a bit more than that when the results have not appeared elsewhere, but hopefully not overdoing it[5] and making anything unnecessarily complicated. The goal is to give a feeling for what we understand rather than to give the most precise or complete statements – a moving target that, even if hit at the moment of writing, quickly turns into a miss.

While there is some overlap between this book and various other surveys, almost always their treatments are superior. In particular, I recommend the varied surveys (Farrell and Jones, 1991b; Ferry *et al.*, 1995; Gromov, 1996; Farrell, 2002; Farrell *et al.*, 2002; Valette, 2002; Roe, 2003; Higson and Guentner, 2004; Kreck and Lück, 2005; Lück, 2022) My hope is that the current treatment will at the very least be useful to my own students as a response to their FAQs and that the brevity of the discussion will be stimulating to some.

The astute reader should be able to figure out what's in this book from its table of contents, and the knowledgeable reader will be able to figure out what's missing.

This book grew out of two lecture series given in 2013, the "Frontiers of Mathematics" lectures at Texas A&M University, and a mini-course two weeks later at "Noncommutative Geometry and Operator Algebras XIII" at Vanderbilt University, followed by another lecture series in Bloomington in 2014. It probably had its genesis in a lecture series I gave in memory of Borel at ETH Zürich in 2005, although much of the material presented here reflects developments that have occurred since then. I reworked the exposition somewhat in the succeeding years, and finally gave up at the point when I felt that my edits were ruining whatever sense of freshness and excitement the original showed. Given the choice between two evils, I chose the one that involved less work for me. I would like to thank my audiences in all these venues for their suggestions, questions, and interest.

Even more, I am indebted to my collaborators, Arthur Bartels, Jean Bel-

---

[3] Although the book would feel incomplete without some discussion of this.
[4] More O'Henry than Homer.
[5] I told myself that I didn't want this to be more than 250 pages long!

lissard, Jonathan Block, John Bryant, Sylvain Cappell, Stanley Chang, Jim Davis, Mike Davis, Sasha Dranishnikov, Benson Farb, Michael Farber, Steve Ferry, Erik Guentner, Nigel Higson, Bruce Hughes, Tadeusz Januszkiewicz, Alex Lubotzky, Wolfgang Lück, Washington Mio, Alex Nabutovsky, Jonathan Rosenberg, Julius Shaneson, Larry Taylor, Semail Ulgen-Yildirim, Zhizhang Xie, Min Yan, and Guoliang Yu, for teaching me so much and sharing in the joy of discovery of both theorems and counterexamples. I also owe a debt to my collaborators Erik Pederson, John Roe and Bruce Williams whom I no longer can thank. Chapters 6 and 7 owe a lot to unpublished joint work with Cappell and with Cappell and Yan, and conversations with John Klein. I also owe a large debt to my colleagues at Chicago, Danny Calegari, Frank Calegari, Kevin Corlette, Matt Emerton, Alex Eskin, Benson Farb, Bob Kottwitz, Andre Neves, Leonid Polterovich, Mel Rothenberg, Amie Wilkinson, David Witte-Morris, and Bob Zimmer, and at Hebrew University, especially Hillel Furstenberg, Gil Kalai, David Kazhdan, Nati Linial, Alex Lubotzky, Shachar Mozes, Ilya Rips, Zlil Sela, and Benjy Weiss, who created such wonderful intellectual environments for discussing geometric problems, especially involving groups or graphs. I believe that all of these people will be able to see reflections of our conversations below, as will many friends and coworkers whose names I have not mentioned. Comments I received from Bena Tshishiku, from David Tranah, and from anonymous referees at Cambridge University Press were invaluable in the revision process.

Finally, and most importantly, I need to thank my family, Devorah, Baruch, and Esther, for many things that are more important than their encouragement of my work and putting up with all that goes with the modern academic life.

1

Introduction

1.1 Introduction to Geometric Rigidity

Our story begins with the Bieberbach theorems about the structure of compact flat manifolds (i.e. compact Riemannian manifolds whose sectional curvatures are everywhere 0, i.e. that are locally isometric to $\mathbb{R}^n$). The universal cover of such a manifold, M, is Euclidean space, and therefore its fundamental group π is a discrete subgroup of $\text{Iso}(\mathbb{R}^n)$. There is a (split) exact sequence

$$1 \to \mathbb{R}^n \to \text{Iso}(\mathbb{R}^n) \to O(n) \to 1$$

so that π has a rotational part, and a translation subgroup. (Thus $\text{Iso}(\mathbb{R}^n)$ is a semidirect product of the linear = orthogonal group, and the group of translations, where the former acts on the latter in the obvious way.)

Bieberbach showed that the rotational part of π is always finite, so that π has a subgroup of finite index that is pure translation, and simple considerations then guarantee that this is rank n, i.e. that M is finitely covered by a torus, i.e. by $\mathbb{R}^n/\Lambda$ for some lattice $\Lambda \cong \mathbb{Z}^n$.

We shall first assume that this is a 1-fold cover for simplicity:[1] the structure of the manifold M we started with is then understood as a structure on a torus, and by an analysis of its isometries.

The space of tori, though, is very interesting and quite nontrivial already. (Indeed the $n = 2$ case gives rise to the beautiful theory of modular forms (Serre, (1973).) Let us normalize by demanding that $\text{vol}(M) = 1$, and furthermore let us pick the isomorphism $\Lambda \to \mathbb{Z}^n$ (which is tantamount to giving a homotopy equivalence $M \to \mathbb{T}^n$). There is a unique linear map in $\text{GL}_n(\mathbb{R})$ taking $\Lambda \to \mathbb{Z}^n$. Notice that the translation group is conjugate to the standard action (as a group action of $\mathbb{Z}^n$) iff (if and only if) this matrix is orthogonal. Thus, the space of

---

[1] Although this is but one of a superexponentially growing number of possibilities as n increases.

"polarized flat tori of volume 1" is the same as $SL_n(\mathbb{R})/SO(n)$, a contractible manifold – e.g. by the Gram–Schmidt process.[2]

At this point, we can pick up the theory for general flat manifolds if we want: the finite holonomy group (the group of rotations we ignored before) acts on the space of flat tori, and whose fixed point set is the space of flat structures on the given manifold (with volume equal to 1/#holonomy). The fixed set of a compact group acting on a complete simply connected non-positively curved manifold is another such space, by a theorem of Hadamard provided it is nonempty and connected. It is nonempty (in general, this is Cartan's fixed-point theorem: a fixed point can be given as the unique "median" of any orbit – the point which makes the largest distance to any point of the orbit finite) in our case, because we assumed there was a flat manifold, and connected, because a geodesic connecting two fixed points to each other would be fixed and therefore lie in the fixed set. Anyway, we then see that there is a unique such manifold as a smooth manifold, and that any two are conjugate in the affine group.

Mostow (1968) showed in a celebrated paper that for constant negative curvature manifolds, the rigidity is much stronger. Perhaps the first hint of this comes from the Gauss–Bonnet theorem: In this case it says that:

Proposition 1.1 *If M is a closed manifold*[3] *of constant curvature -1, i.e. if M is a closed hyperbolic manifold of even dimension, then*

$$\chi(M^{2n}) = 2(-1)^n \, \text{vol}\,(M)/\omega_{2n},$$

where ω_{2n} is the volume of the sphere (of radius 1).

To foreshadow other developments, we note that if $\text{vol}\,(M) < \infty$, then M has finite topological type (i.e. is the interior of a compact manifold with boundary) so that both sides of the equation make sense, and in fact the equation holds.

As a consequence of the Gauss–Bonnet theorem, we see that in the hyperbolic case, unlike the flat case, the fundamental group determines the volume.[4] Perhaps even more straightforwardly, flat manifolds have a nonrigidity because of homotheties, but hyperbolic manifolds have a scale because of their nonvanishing curvature.

Mostow's theorem then gives what seems like the ultimate strengthening

---

[2] In the spirit of later developments, we should say that $SL_n(\mathbb{R})/SO(n)$ is a complete simply connected manifold of non-positive curvature – as is any semisimple Lie group modulo its maximal compact subgroup – and is thus, by Hadamard's theorem, diffeomorphic to Euclidean space.

[3] Recall that a closed manifold is a compact manifold without boundary.

[4] At least in even dimensions. Mostow rigidity implies that this is true in all dimensions; a cohomological explanation for this is provided by Gromov's theory of bounded cohomology (Gromov, 1982).

of this line of thought. The contractible manifold occurring in the flat case degenerates (if the dimension is greater than >2) to a point!

Theorem 1.2 *Suppose that M and M' are closed hyperbolic manifolds of dimension d > 2, then any isomorphism h: $\pi_1(M) \to \pi_1(M')$ is induced by a unique isometry between M and M'.*

As a minor point, strictly speaking, an induced map on fundamental groups requires the map to preserve base points, but the isometry will almost surely not (as it's unique, it either does or does not). Consequently, we should actually assume that one has a conjugacy class of homomorphisms of the fundamental group, or use groupoids, or some similar device.

We note that this is not true in dimension 2; for a surface of genus g, the space of marked[5] hyperbolic structures is called Teichmuller space, and is topologically $\mathbb{R}^{6g-6}$.

Mostow's theorem is a beautiful and perhaps initially surprising result. However, it can feel a bit sterile if one doesn't know examples of hyperbolic manifolds and indeed it is not so easy to construct hyperbolic manifolds in dimension >2 (in dimension 2 they can be built easily using tessellations of the hyperbolic plane). Even after knowing some constructions, how are you going to find two not obviously isometric hyperbolic manifolds that have isomorphic fundamental groups?

However, the uniqueness statement in Mostow's theorem gives us quite nontrivial information even when $M = M'$. Any self-isomorphism of π must be realized by a self-isometry, giving the following conclusion:

Corollary 1.3 *If π is the fundamental group of a compact hyperbolic manifold M, then $\mathrm{Iso}(M) \cong \mathrm{Out}(\pi)$, where $\mathrm{Iso}(M)$ is the isometry group of M, and $\mathrm{Out}(\pi)$ is the group of outer automorphisms of π: it is the quotient of the automorphisms $\mathrm{Aut}(\pi)$ by $\mathrm{Inn}(\pi)$, the normal subgroup of inner automorphisms of π.*

$\mathrm{Out}(\pi)$ is the set of components of the self-homotopy equivalences of M to itself: it is not $\mathrm{Aut}(\pi)$ because we do not insist that maps and homotopies preserve base points.

The isometry group of a compact manifold is always a compact Lie group (Myers–Steenrod), so we learn that, in the hyperbolic case, this group is always finite. Then we then deduce the purely algebraic fact that $\mathrm{Out}(\pi)$ must be finite – if the dimension of the hyperbolic manifold >2.

In dimension 2, the first conclusion holds (as we will discuss later), but the second does not. $\mathrm{Out}(\pi)$ is the celebrated mapping class group, an object of

---

[5] That is, ones where we are given an identification of the fundamental group or, equivalently, a homotopy class of a map to a standard surface.

fundamental importance in low-dimensional topology and in algebraic geometry. Elements of infinite order in Out(π) can never be realized by isometries of a compact Riemannian manifold.

The conclusions of Mostow's theorem can be greatly generalized. First of all, hyperbolic space can be generalized to be any locally symmetric manifold with no Euclidean factors and no hyperbolic plane factors: in other words, as Mostow showed in subsequent work, it applies to G/K, if G is a semisimple Lie group (i.e. a Lie group with no connected normal solvable subgroups) and K its maximal compact subgroup. We will discuss these in much greater length in Chapter 2.

In addition, according to Prasad (1973), all of these rigidity theorems hold for noncompact finite volume hyperbolic manifolds (and locally symmetric manifolds).

Amazingly enough, there are many additional extensions of these theorems, not thought of as uniqueness theorems *per se*. We will discuss some of the important work of Margulis on (the aptly called) superrigidity in Chapter 2.

1.2 The Borel Conjecture

The striking results of §1.1 show that for various "geometric structures" (let's say that this means a given choice of a local model for germ neighborhoods of points), the space of given marked structured manifolds is either a point, or the algebraic topologist's "point": a contractible space.[6]

Although a contractible space isn't as good as a point, for some purposes it's quite good. For example, that it is connected is already a type of uniqueness statement. In the situation where one has a structure on this space with non-positive curvature, one can geometrically make conclusions that are stronger than follow from the algebraic topology alone. For instance, the non-positive curvature on the space of flat tori enables one to prove Bieberbach's theorem that any torsion-free group that is virtually free abelian (of rank k) is the fundamental group of a compact aspherical manifold (of dimension k). (Exercise or see the footnotes.)

Borel suggested that the topological conclusion that the hyperbolic manifolds were homeomorphic[7] could be traced to a purely topological hypothesis:

---

[6] There should be a kind of mathematician for whom a point is a non-positively curved space – someone informed by both algebraic and geometric intuitions.

[7] Borel actually made his conjecture on the basis of an earlier result of Mostow (1954) on solvmanifolds, where the conclusion was "isomorphic," i.e. diffeomorphic. Borel expressed relief that he hadn't conjectured diffeomorphic in light of this result.

Conjecture 1.4 *If* $h: M' \to M$ *is a homotopy equivalence between closed aspherical manifolds, then h is homotopic to a homeomorphism.*

Recall that a space is aspherical if its universal cover is contractible. It is a $K(\pi, 1)$ in the language of the algebraic topologists, meaning its homotopy groups π_i vanish for $i > 1$. This can be tested by checking whether the universal cover has vanishing reduced integral homology (by the Hurewicz isomorphism theorem). A homotopy (class of) equivalence(s) between aspherical spaces is essentially the same thing as a (conjugacy class of) isomorphisms between their fundamental groups.

If M is non-positively curved or of the form $K\backslash G/\Gamma$ (where G is a real Lie group and K its maximal compact), then it satisfies the hypothesis of the Borel conjecture. In these cases, the conjecture is an astounding theorem of Farrell and Jones.[8]

One can also try to reverse this mode of thought, and ask whether the moduli space of non-positively curved structures on a closed topological aspherical manifold is contractible (if it is nonempty!). Farrell and Jones have shown that the answer to this is negative as well: the space isn't even connected. But, I am running ahead of the story.

Borel is suggesting here that aspherical is the topological analogue of "locally symmetric of noncompact type" or of "non-positively curved." In Chapter 2 we will discuss various constructions of aspherical manifolds – although in Borel's time there were no examples that were very far away from the lattice setting.

Of course, in the topological setting, one cannot expect the homeomorphism to be unique. However, it might seem reasonable to believe that the space of homeomorphisms is contractible, i.e. the analogue of a point. Unfortunately, this is not true and we will later discuss the reason for this; it is an indirect consequence of the conjecture that there is a type of uniqueness: uniqueness up to pseudo-isotopy.

Definition 1.5 Two homeomorphisms $f, g: M \to N$ are pseudo-isotopic if there is a homeomorphism $M \times [0, 1] \to N \times [0, 1]$ that restricts to $f \cup g$ on the boundary $M \times \{0, 1\} \to N \times \{0, 1\}$.

For high-dimensional closed manifolds, one knows due to the work of Cerf and of Hatcher and Wagoner (see Hatcher and Wagoner, 1973) that pseudo-isotopies between homeomorphisms are isotopic to isotopies iff the manifolds are simply connected. This work shows that always there's typically an infinite number of isotopy classes of homeomorphisms in the given homotopy class.

---

[8] The important point being that M' is *not* assumed to be a space of this sort (for then, the relevant result is part of differential geometric rigidity). We will explain some of the ideas of this result in Chapter 8.

The reasonable optimist might therefore choose to append "unique up to pseudo-isotopy" to the statement of the Borel conjecture. As we will discuss in Chapter 3, this both follows from the Borel conjecture in general, and is part of the "correct" natural extension to manifolds with boundary.

Uniqueness up to pseudo-isotopy is not as strong as uniqueness, and it will need some study. If one had uniqueness in families, one could immediately learn things about bundles. The weaker type of uniqueness has implication for "block bundles"[9] and has more relevance to the topological category than the bundle result would have (in other words, this is a feature, not a bug).

As we noted in the geometric setting, uniqueness would also immediately have implications regarding the symmetries of aspherical manifolds. Borel himself proved some of these, and we will discuss them in Chapter 7. For example, if M is an aspherical manifold whose fundamental group has trivial center, then the only connected compact Lie group that can act continuously on it is trivial.[10] We saw that it implied that any finite subgroup of $\text{Out}(\pi)$ was realized by a group action – at least when π is centerless;[11] this statement is called the Nielsen realization problem.[12]

It also would imply certain uniqueness statements about group actions – or if you like, it would imply "equivariant Borel conjectures." We will see, by contrast (in Chapter 6), that these conjectures are false – for several different reasons.

Another variant of the Borel conjecture goes like this: Given a group π, the Borel conjecture asserts the uniqueness of the aspherical topological manifold whose fundamental group is π. Shouldn't there be an existence theorem to go with such a uniqueness one? Wall (1979) has conjectured that the correct condition is that π should satisfy Poincaré duality.[13] We will discuss some of the evidence for Wall's conjecture – most comes from the Borel conjecture –

---

[9] And even more to "approximate fibrations," which it would surely be taking us far afield to introduce at this point. Let us leave it as saying that if one tried to extend the Borel philosophy to some singular settings, and took seriously the idea that one is looking for topologically invariant notions rather than modeling closely the topological analogue of the smooth category, then one would be led to "pseudo"s.

It is worth noting that Mostow's work on hyperbolic manifolds is based on extending the map of universal covers to certain ideal ∂s. These extensions, as is critical to Mostow's work, are naturally continuous and not smooth. These ideas of Mostow from the late 1960s are fundamental to almost all of the work on the Borel conjecture since the early 1980s.

[10] Equivalently, every continuous circle action on M is trivial.

[11] When π is not centerless, the isometries tend not to be unique, and the realization is false for certain nilmanifolds, an example of Raymond and Scott (1977).

[12] The original Nielsen problem was for surfaces and was proven true first by Kerckhoff (1983) using geometrical properties of Teichmuller space. By now, there are a number of proofs.

[13] For a group to satisfy Poincaré duality it means that its $K(\pi, 1)$ satisfies Poincaré duality. In Wall's conjecture, one means Poincaré duality with arbitrary coefficient systems, to the same extent that one has such Poincaré duality for manifolds. This is equivalent to there being a

and we'll also discuss variants of Wall's conjecture where one weakens the type of Poincaré duality the group satisfies.

Yet another way of thinking about Borel's philosophy is the following. If knowing the group means knowing the manifold, then every topological property of manifolds has to be reflected in its fundamental group. Thus one can conjecture that an aspherical manifold is a nontrivial product iff its fundamental group is.[14] Similarly one can hope that a manifold "fibers" over another if there is a suitable exact sequence of groups. We will discuss these kinds of problems later.

If one were a wild optimist,[15] one could easily go very far and conjecture that many properties of the model manifolds hold for all aspherical manifolds, such as that their universal covers are Euclidean space or that their fundamental groups have solvable word problems. We will see in Chapter 2 that these are false.

It is not known whether their Euler characteristic has the same sign as the symmetric spaces of the same dimension have, i.e. whether $(-1)^n \chi(M^{2n}) \geq 0$, for closed aspherical manifolds. (This is sometimes called the Hopf conjecture, although Hopf only asked it for negatively curved manifolds.[16])

Finally, the Borel conjecture begets many others in the following indirect way: It implies that any method one would try to disprove it must fail. Thus any invariant of manifolds, defined by any method at all, no matter how clever or indirect, should be a homotopy invariant for aspherical manifolds. This means that the fundamental group must somehow catch lots of subtle geometry. Examples of this include the tangent bundle and various types of spectral invariants, but, in principle, one can consider any topological invariant at all.[17]

When studying this in detail, one is often led to problems that seemingly have nothing to do with aspherical manifolds. In Chapter 4 we will follow this road towards the Novikov conjecture, which in its analytic form has strong differential geometric implications – well beyond aspherical manifolds. In this form, the conjecture also develops analogues in quadratic form theory and in algebraic K-theory.

---

chain homotopy equivalence (with the usual dimensional shift) between the $\mathbb{Z}\pi$ chain complexes of singular chains on the universal cover and its dual.

[14] This can be compared with a theorem of Lawson and Yau (1972) for non-positively curved manifolds.

[15] Something that one would not ordinarily say of Borel.

[16] Recently, Avramidi (2014) gave some very striking evidence for the failure of this conjecture.

[17] An example of this includes simplicial volume *à la* Gromov (1982), which provides a homological explanation for the volume of certain locally symmetric manifolds. I mention this here because, unfortunately, it does not play a large role in what follows.

1.3 Notes

A good grounding in differential geometry is very helpful. For our purposes, Cheeger and Ebin (2008) is probably the best source. Milnor (1963), which is a rapid course in Riemannian geometry in *Morse Theory*, is adequate for most purposes in this book.

There are now a lot of approaches to Mostow rigidity and it has many extensions and generalizations. The original sources are Mostow (1968, 1973). I highly recommend the survey of Gromov and Lawson (1991). Probably the "easiest" proof (although one that is rather atypical) is Gromov's based on the ideas of "bounded cohomology." An excellent exposition of this can be found in Munkholm (1979). Zimmer (1984) gives a clear treatment of Margulis's superrigidity theorem.

The discussion here of the Borel conjecture is not the most direct or efficient. However, the equivalent statement that "the structure set of an aspherical manifold vanishes" reduces all of one's study to proving that some group is 0. This seems (to me) rather depressing. We prefer the point of view that the subject deals with actual *examples* and contains surprises. It makes it feel like one is actually studying *something* (Figure 1.1).

Figure 1.1 Six cartoon. © Sidney Harris, reproduced with permission. (http://sciencecartoonsplus.com/).

More seriously, the variants we consider shed light on some subtleties and possible approaches to the conjecture, and are, I think, natural questions that one would want to address for the same reason as one would want to know the truth of the Borel conjecture.[18]

And, finally, I hope that when the problem is ultimately solved, the spirit of the problem – as expanded on here – will continue to inspire future generations of mathematicians.

---

[18] So we prefer a *Comedy of Errors* to *Much Ado About Nothing*.

2

Examples of Aspherical Manifolds

This chapter discusses some of the basic examples of, mainly closed, aspherical manifolds that give content to our inquiry. After all, what good would the Borel conjecture be if there were no aspherical manifolds?

We give some constructions of ones that come from locally symmetric manifolds (i.e., Lie theory) including both arithmetic and non-arithmetic examples, and also of others that do not.

By contrast, the construction of noncompact aspherical manifolds is quite easy. There is an open aspherical manifold with fundamental group π iff π is countable and has finite cohomological dimension, as one can see by thickening a finite-dimensional $K(\pi, 1)$ complex (i.e. replacing all the cells in a CW-decomposition by handles). Remarkably, aside from finiteness conditions, this characterizes the groups that are retracts of (fundamental groups of) aspherical manifolds.

2.1 Low-Dimensional Examples

In low dimensions, almost all connected manifolds (even noncompact) are aspherical. The only connected nonaspherical surfaces are the sphere and the projective plane.

In dimension 3, among closed orientable 3-manifolds *all are aspherical* unless one of the following very good reasons holds:

(1) the fundamental group is finite (in which case, the universal cover is S^3 and the deck group is a subgroup of SO(3));

(2) the manifold is a nontrivial connected sum (and the separating 2-sphere is a nontrivial element of π_2); or

(3) the manifold is $S^1 \times S^2$.

All of this[1] is a consequence of the sphere theorem of Papakyriakopoulos (see e.g. Hempel 1976; Jaco 1980)

However, in understanding even closed 3-manifolds, it is essential that one consider manifolds with nonempty boundary as part of the story. Given an arbitrary 3-manifold, one first has a decomposition into irreducible pieces, under connected sum. This is unique up to the order of the decomposition. Then one breaks the manifold summands further into pieces, where the gluing is done along certain embedded incompressible[2] tori. This topological decomposition was discovered by Jaco and Shalen, and Johannson, and explained geometrically by Thurston and Perelman: After breaking the 3-manifold along this decomposition along a set of canonical tori (its torus decomposition), one is left with pieces, *all of which have geometric structure*,[3] i.e. a manifold with a complete metric, which is locally homogeneous.

Let's be more concrete. Suppose we start with a knot K in S^3, i.e. a smooth submanifold diffeomorphic to S^1. The complement is always aspherical (as before, by Papakyriakopoulos's sphere theorem). For the unknot, the complement is $S^1 \times \mathbb{R}^2$. It is often convenient to remove tubular neighborhoods of submanifolds, to obtain the "closed complement";[4] then we would obtain $S^1 \times \mathcal{D}^2$.

For all knots, we obtain an aspherical manifold with boundary as its complement X, whose boundary is a torus. The unknot is characterized by the property that $\pi_1(\partial X) \to \pi_1(X)$ is not injective: a nontrivial knot always has an incompressible torus embedded in its complement (i.e. an embedded $\mathbb{T}^2$ so that π_1 injects).

Sometimes there is another torus (i.e. not isotopic to the boundary) in the complement that is incompressible. When this happens, essentially what that means is that this knot can be thought of as being wrapped around another knot, i.e. that it has a companion (Figure 2.1). The process of finding companions must end – although not obvious, there is a geometric complexity that increases under companionship.

[1] At least for an infinite fundamental group. The description of what happens for a finite fundamental group depends on Perelman's solution of the geometricization conjecture.

[2] Recall that a surface in a 3-manifold is incompressible if its normal bundle is trivial, and its fundamental group injects into the fundamental group of the manifold.

[3] This is the celebrated geometricization conjecture. Actually, if an irreducible connected manifold contains any incompressible surfaces (and, in particular, if it has a nontrivial torus decomposition), then the geometricization of all of the pieces in its decomposition is a theorem of Thurston. For references, see the notes in §2.4.

[4] There are subtleties with doing this in the topological category. In the setting of locally flat manifolds, everything works the same (see Kirby and Siebenmann, 1977). When we discuss orbifolds, we will see that the analogous issue is not solvable in the topological setting, i.e. one cannot always find a "closed regular neighborhood" of the knot, and one needs a substitute for tubular neighborhood theory.

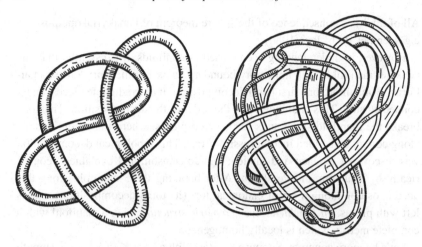

Figure 2.1 The knot on the left is a companion of the thinner knot on the right.[6]

So, now consider one of the deepest pieces, i.e. a knot with no companions. In that case, there are two cases: Torus knots are knots that lie on the surface of a torus that surrounds the unknot. These are parameterized by pairs of coprime integers (p, q) representing the homology class of the associated circles. *All of the remaining knots have hyperbolic complements*, i.e. have complete metrics of constant negative curvature and finite volume. (One can distinguish the two cases easily: the torus knots have fundamental group of their complement with nontrivial center – which precludes having a metric of negative curvature.[7])

In other words, the fundamental group of the complement Γ is naturally a discrete subgroup of $\mathrm{PSL}(2, \mathbb{C})$.

The same is true for the annular regions between the various embedded tori: they all have hyperbolic structures. Thus, a typical knot complement (and according to geometricization, this is typical) is a union of hyperbolic (or perhaps one of several other geometries (see Scott, 1983) manifolds glued together along their cusps. (See Chapter 3 for more of a discussion of the geometry at ∞ of noncompact locally symmetric spaces.)

This union itself does not have a locally homogeneous structure. Its fundamental group cannot be a lattice in any Lie group.

This is because, in any of the three-dimensional geometries (see Scott, 1983),

---

[6] Adapted from Thurston (1982).

[7] This is Preissman's theorem, which can be found in most introductory differential geometry textbooks. See Bridson and Haefliger (1999) for a proof not using *differential* geometry.

any $\mathbb{Z}^2$ is either *peripheral*, i.e. conjugate to the fundamental group of a boundary component,[8] or contains an element of the center of the fundamental group.[9] The $\mathbb{Z}^2$ coming from the torus of "companionship" is neither, and therefore this manifold does *not* have a locally symmetric structure.

The upshot is that it is very easy to obtain closed aspherical 3-manifolds whose fundamental groups are not lattices, e.g. the double of any knot complement other than torus knots. But they are obtained indirectly by gluing together lattices.

It is hard to make this precise, but till the early 1980s there was a general feeling that perhaps, somehow, lattices were the source of all closed aspherical manifolds. We will see that this is not the case as we go along, but let us start with the lattices themselves. Before we do, let us close this discussion by making one very useful observation about gluing aspherical objects:

Proposition 2.1 *Suppose that A, X, and Y are aspherical, $A = X \cap Y$, and that $\pi_1(A) \to \pi_1(X)$ and $\pi_1(A) \to \pi_1(Y)$ are injective. Then $X \cup Y$ is aspherical.*

Without the injectivity, the 2-sphere is a counterexample: it is a union of two disks along a circle, all aspherical, but not π_1 injective.

To see why the proposition is true, we shall construct the universal cover of $X \cup Y$ and *observe* that it is contractible. We begin by taking the cover of X. Over A (by injectivity) we get many copies of the universal cover of A (according to the cosets of $\pi_1(A) \to \pi_1(X)$). Each of these is glued to a copy of the universal cover of Y (which also contains many copies of the universal cover of A). We then proceed by gluing back copies of the universal cover of X, and so on. This is a union of contractible spaces glued together along (disjoint) contractible spaces, so this is contractible.

Remark 2.2 If one shrinks each copy of the universal cover of X to a point, and each copy of the universal cover of Y to a point while stretching and shrinking the copies of the universal cover of A to intervals, we get the Bass–Serre tree associated to this amalgamated free product description of $\pi_1 X \cup Y$.

This proposition is of critical importance. It enables us to construct interesting examples by gluing. We will either explicitly or tacitly apply it many times. A consequence of this is that we can take geometric models for given groups (i.e. $K(\pi, 1)$s for groups) and glue them together to construct models for various amalgamated free products and Higman–Neumann–Neumann (HNN)

---

[8] Quotients by lattices have a natural compactification (the Borel–Serre compactification) which makes them into the interiors of manifolds with boundary. It is this virtual boundary that I am referring to when I describe a subgroup of the fundamental group as being peripheral.

[9] Like in the situation of a circle bundle over a surface.

extensions: the quality of the union will depend on the quality of the complexes we begin with and of the inclusion of the subgroup. But, for example, it shows that the category of finite $K(\pi, 1)$, i.e. πs that are realized by finite aspherical complexes, is closed under amalgamated free products and HNN extensions.

This, in particular, allows the construction of finite aspherical complexes whose fundamental groups have unsolvable word problems, or other logical complications. The Davis construction, discussed below, will incorporate these features into fundamental groups of aspherical manifolds.

2.2 Constructions of Lattices

... Arithmetic and Non-arithmetic

Given a Lie group, even a quite explicit one like $O(n, 1)$ (the automorphisms of the quadratic form $x_1^2 + x_2^2 + \cdots + x_n^2 - x_{n+1}^2$, i.e. the isometry group of hyperbolic n-space) or $SL_n(\mathbb{R})$, it is not trivial to find uniform lattices; that is discrete subgroups of G such that G/Γ is compact.[10] Indeed, this is not always possible, e.g. for solvable Lie groups.[11]

However, if G is semisimple, Borel gave a general construction of uniform lattices (and Raghanuthan gave non-uniform lattices;[12] see Raghunathan (1972) for both). For $SL_n(\mathbb{R})$ there is an obvious lattice, namely $SL_n(\mathbb{Z})$, but it is not uniform, i.e. cocompact. If we think of $SL_n(\mathbb{Z})\backslash SL_n(\mathbb{R})/SO(n)$ as the space of flat tori (as in §1.1), then tori that are more and more eccentric (i.e. the result of identifying opposite sides in a rectangle with sides t and $1/t$) leave any compact subset of this space (the shortest geodesic is approaching 0 length).

Let's make this a bit more precise (or more general). To talk about the "integer points" in a Lie group, we should define it over the field $\mathbb{Q}$ (there are many distinct ways of doing this). Then, for simplicity, let's assume that the group is linear – there will be a maximal subgroup isomorphic to $\mathbb{Q}^{*k}$ in $G(\mathbb{Q})$; here the diagonal matrices and $k = n - 1$. If $k > 0$, then $G/G(\mathbb{Z})$ is not compact and one can take powers of a matrix in this $\mathbb{Q}$-split torus to leave any compact.

The converse holds, i.e. the nonexistence of such a $\mathbb{Q}$-split torus implies compactness (and this is a theorem of Borel and Harish-Chandra). We defer

---

[10] Recall that Γ is a lattice in G if, giving G its natural (Haar) measure, the quotient G/Γ has finite volume.

[11] The two-dimensional Lie group of affine isomorphisms of $\mathbb{R} \to \mathbb{R}$ (the "$ax + b$" group) contains no lattices.

[12] Note that $\mathbb{R}^n$ has uniform lattices, but no non-uniform lattices. The same is true for all nilpotent real Lie groups.

further discussion of this to Chapter 3, where the size of the torus will be seen to govern the "size" of G/Γ.

Another way to tell if a lattice is non-uniform is to see if it contains any nontrivial unipotent elements. (Consider the Lie group G as a matrix group, and then g is unipotent if its characteristic polynomial is $(t-1)^n$ for some n, i.e. if g differs from the identity by a nilpotent matrix.) No uniform lattice contains unipotent elements: the length of a geodesic represented by g in $\Gamma\backslash G/K$ is proportional to the supremum of $|\log(\lambda)|$ over eigenvalues of g representing the g. The converse had been a conjecture of Selberg, proved by Kazhdan and Margulis (and we refer to Margulis (1991) for the proof), and this property is often easy to check.

Finding appropriate $\mathbb{Q}$ structures for the case of SL_n is rather nontrivial and requires some development of the theory of division algebras. We shall leave this to the references, but for those who know some algebra, the group of units in an order in a division algebra of dimension n^2 does the trick.

Let us now return to the problem of constructing uniform lattices.

For $O(n, 1)$, looking at $O(n, 1)(\mathbb{Z})$ does not do the trick: one obtains a lattice, but not a uniform one.[13] However if we replace the quadratic form $x_1{}^2 + x_2{}^2 + \cdots + x_n{}^2 - x_{n+1}{}^2$ by $\mathbf{Q} = x_1{}^2 + x_2{}^2 + \cdots + x_n{}^2 - \sqrt{p}\, x_{n+1}{}^2$, then the real Lie group is the same: the quadratic forms are isomorphic over $\mathbb{R}$. However, $O(\mathbb{Q})(\mathbb{Q}[\sqrt{p}])$ has two different embeddings into real orthogonal groups, associated to the two embeddings of $\mathbb{Q}[\sqrt{p}]$ into $\mathbb{R}$, according to whether $\sqrt{p}$ is positive or negative.

The (real) orthogonal group associated to making $\sqrt{p}$ negative is the usual compact orthogonal group. Note that the orthogonal group has no nontrivial unipotent elements. This means that $O(\mathbb{Q})(\mathbb{Z}[\sqrt{p}])$ is a uniform lattice in $O(n, 1) \times O(n + 1)$. However, we can safely project to the first factor, as the second factor is compact, with at most a finite subgroup as kernel. In other words, the space $O(\mathbb{Q})(\mathbb{Z}[\sqrt{p}])\backslash O(n, 1)/O(n + 1)$ is a compact hyperbolic orbifold. Replacing $O(\mathbb{Q})(\mathbb{Z}[\sqrt{p}])$ by a torsion free subgroup of finite index gives a compact hyperbolic manifold.

This method produces many lattices. Lattices produced in this way are called *arithmetic*. Note that when written in coordinates, automorphisms defined using larger fields than $\mathbb{Q}$ give rise to Lie groups over $\mathbb{Q}$ – this is formally called "restriction of scalars." Using suitable quadratic forms over arbitrary totally real fields, we can get uniform lattices in any $O(p, q)$.

The general case follows, as Borel says, from the statement that "any real

---

[13] Considering the automorphisms of the slight variant $a_1 x_1{}^2 + \cdots + a_n x_n{}^2 - a_{n+1} x_{n+1}{}^2$, one obtains a uniform lattice iff this indefinite quadratic form does not represent 0 (i.e. does not vanish on any integral vector). However, the Hasse–Minkowski theorem says that this does not happen when $n > 4$.

semisimple Lie algebra has a form defined over a totally real field $E \neq \mathbb{Q}$ all of whose conjugates are compact." Borel proves this Lie-theoretic statement via tricky (for me) Lie algebra calculations in his paper (and the book of Raghunathan (1972) explains how to guarantee $\mathbb{Q}$ forms that produce the non-uniform lattices, as well).

For simple Lie groups of rank ≥ 2 (or even irreducible[14] lattices in semisimple groups) Margulis shows that these are all the examples, i.e. that all lattices are arithmetic. The reader should pause to reflect on how amazing this result is: one is given a structure with only local information defined over $\mathbb{R}$ (say a group of real matrices, or a finite-volume Riemannian manifold modeled on some $K \backslash G$) and one needs to find an algebraic number field and a form of the Lie group from this and then an isomorphism of one's given object with the arithmetic construction.

In the cases not excluded by Margulis (and the subsequent work of Corlette, Gromov, and Schoen that proves arithmeticity in some rank-1 situations by more analytic methods: see Gromov and Schoen, 1992), it is an important question of whether there are non-arithmetic lattices.

We mention here three such constructions, all of which are in $O(n, 1)$. (Some examples are also known in $U(n, 1)$ for small values of n (see, e.g., Deligne and Mostow, 1993), but these are isolated.)

2.2.1 Method One: Reflection Groups

The first is classical, and is based on constructing polyhedra in hyperbolic space so that reflections across its walls generate a reflection group on hyperbolic space. In the hyperbolic plane, the easiest example is a triangle with angles π/p, π/q, and π/r, so that $1/q + 1/r < 1$. (Below is an example with $p, q, r = 2, 3, 9$.) Even in dimension 2, Takeuchi (1977) showed only finitely many of these are arithmetic (and indeed gave a list of them).

It is known that such examples exist in small dimension, and do not exist in very high dimensions. Nevertheless, they perhaps motivate the Davis construction to be discussed in §2.3 below. Figure 2.2 shows a nice hyperbolic planar group generated by reflections.

---

[14] A lattice is reducible if, after passing to a subgroup of finite index, it is a product of two other lattices. An irreducible lattice in a product of real groups will project to a dense subgroup of each of the factors. So, for example, among lattices acting on a product of two hyperbolic planes, reducible ones will have deformations, but irreducible ones will be arithmetic (and have no deformations).

Figure 2.2 A hyperbolic triangle group.

2.2.2 Method Two: Closing Cusps

This method is due to Thurston, and is his famous Dehn surgery theorem
(see Thurston, 2002). Consider a hyperbolic manifold with cusps (e.g. a knot
complement, or most[15] link complements for "nonsplittable" links, i.e. links in
which components cannot be isotoped to lie in disjoint balls). Thurston shows
that for all "sufficiently large" surgeries, one obtains a compact hyperbolic
manifold.

What does this mean? Given a manifold whose boundary is a torus, we
can "close it up" by gluing in a solid torus $S^1 \times \mathcal{D}^2$. Although there is an
$\mathrm{SL}_2(\mathbb{Z}) = \pi_0 \mathrm{Diff}(\mathbb{T}^2)$ set of possible gluing diffeomorphisms, the diffeomorphism type of the manifold is determined by the image of the circle $\partial \mathcal{D}^2$. (One
can imagine the gluing as being done in stages: first glue in a thickened $\mathcal{D}^2$ to
get a boundary component that is an S^2 and then glue in a final ball, which has
no indeterminacy.) These are parameterized by the primitive (i.e. indivisible)
elements of $H_1(\mathbb{T}^2) \approx \mathbb{Z}^2$.

Thurston's theorem now asserts that, if one excludes finitely many possibilities at each cusp, then all the remaining possibilities of filling produce
hyperbolic manifolds. Moreover, as the boundary curves get longer and longer,
the hyperbolic manifold that is constructed gets closer and closer to the original cusped hyperbolic manifold in a very reasonable geometric sense: The
"surgery" can, up to very small perturbation, be imagined as taking place

---

[15] One needs to exclude phenomena analogous to companionship (which prevent any geometric
structure) or torus knots (which correspond to structures that are not hyperbolic).

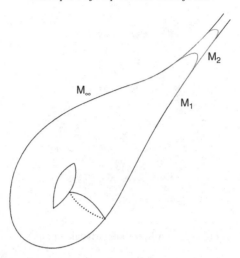

Figure 2.3 The limit of the volumes of the filled manifolds is the volume of the original cusped manifold. The cusped manifold is a pointed Gromov–Hausdorff limit of the filled manifolds.

further and further from the "core" of the original manifold.[16] (We will discuss the shape of noncompact locally symmetric manifolds at infinity more in Chapter 3.)

As a result, these manifolds have different volumes that converge to the volume of the original cusped manifold. This is a very crude reason for non-arithmeticity (although it does not do a single example!): for any G, the volumes of the arithmetic lattices $K\backslash G/\Gamma$ form a discrete subset of the positive reals.[17] Figure 2.3 gives a schematic of how the different "cusp closings" converge to the original cusped manifold.

It is very interesting to ponder this example from the representation theoretic viewpoint. One starts with a representation:

$$\rho : \Gamma \to \mathrm{PSL}_2(\mathbb{C})$$

that describes the original hyperbolic manifold with cusps. The filling gives nearby representations $\rho_n : \Gamma/\langle \gamma_n \rangle \to \mathrm{PSL}_2(\mathbb{C})$, where $\langle \gamma_n \rangle$ is the subgroup

---

[16] These examples provide a good set of examples for thinking about thick–thin decompositions, and the Cheeger–(Fukaya)–Gromov collapse theory.

[17] In some cases, they are even quantized (i.e. multiples of a given smallest one) using the Gauss–Bonnet theorem. I suspect that the converse holds, i.e. that only for Gs with $\chi(K\backslash G/\Gamma) \neq 0$ (this can be re-expressed in various ways – but, in particular for the case of hyperbolic manifolds, this is exactly that the dimension be even) are the volumes of torsion-free lattices quantized.

normally generated by the *n*th filling curves. These provide a family of nearby but inequivalent representations to Γ.

Of course, Mostow's rigidity theorem asserts the uniqueness (up to conjugacy) of the discrete faithful representation. These representations are perturbations that are not faithful but are discrete. (For a closed manifold, all nearby representations to the discrete faithful one are in fact equivalent to it.) This phenomenon is highly special to this Lie group. Superrigidity is a significant strengthening of the representation-theoretic aspect of Mostow rigidity in high rank, and would preclude anything like this in higher rank.[18]

This method also has had a number of applications to constructing aspherical manifolds (and groups) that are not lattices. We will mention some in §2.3, and the method will recur when we discuss the groups of Gromov that disprove a version of the Baum–Connes conjecture[19] in Chapter 8.

2.2.3 Method Three: Gromov–Piatetski-Shapiro (G-PS) Grafting

This is the only method[20] that is known to produce examples in all dimensions. We describe the idea, but none of the technicalities, for which we refer to the original paper (Gromov and Piatetski-Shapiro, 1988).

Suppose that you have two compact arithmetic manifolds, and that they have a common codimension-1 submanifold. In other words, we have M and M' that are not (virtually) isometric, but both contain a separating totally geodesic submanifold V. Then we can cut both M and M' along V, and glue one side of M to the other side of M'. This is clearly a hyperbolic manifold.[21]

This manifold cannot be arithmetic, essentially because it has a big enough piece of M that it would have to be M if it were, but it would similarly have to be M', but it can't be both!

How do we get such pairs?

We get uniform lattices from orthogonal groups, but it is possible for different quadratic forms to give the same lattice. The condition is that the forms be similar (i.e. equivalent to rescaled versions of one another). Now it is pretty easy: If one takes the orthogonal groups of the quadratic forms $x_1{}^2 + x_2{}^2 + \cdots + x_n{}^2 - \sqrt{2}\,x_{n+1}{}^2$ and $3x_1{}^2 + x_2{}^2 + \cdots + x_n{}^2 - \sqrt{2}\,x_{n+1}{}^2$ over $\mathbb{Q}[\sqrt{2}]$, one gets noncommensurable lattices for n even. (The case of n odd is another trick away.) Now these each have an involution associated to $x_1 \to -x_1$ whose fixed

---

[18] Which is a good thing, because superrigidity gives rise to Margulis's arithmeticity (and therefore to discreteness of the set of volumes).

[19] This is a C^*-algebra version of the Borel conjecture.

[20] G-PS call it "interbreeding."

[21] More precisely, it is clearly a compact manifold with constant curvature equal to -1, but such are, of course, hyperbolic.

set is a codimension-1 submanifold: essentially the orthogonal group of the lattice $x_2{}^2 + \cdots + x_n{}^2 - \sqrt{2}\, x_{n+1}{}^2$.

We will have use for the natural topological variant of this method for constructing interesting examples (such as counterexamples to certain orbifold variants of the Borel conjecture) in Chapter 7. One tries to find interesting aspherical objects with boundary and then obtains monsters by grafting[22] them together.

2.3 Some More Exotic Aspherical Manifolds

2.3.1 Method One: Davis's Reflection Group Method

This method was introduced by M. Davis (1983), whose self-proclaimed aim was to describe aspherical manifolds whose universal cover is not Euclidean space. There is a simple criterion (thanks to the Poincaré conjecture) for determining whether a contractible manifold is $\mathbb{R}^n$ or not; it is whether the manifold is *simply connected at infinity*.[23]

Recall that a manifold (or even locally compact space) is connected at infinity, if the complement of any compact subset has exactly one "noncompact" component (more precisely, one component with noncompact closure). Assume this is the case, then one can glob on all the compact components, to obtain a somewhat larger compact, whose complement has exactly one component.

Now let us consider a sequence of compact subsets that exhaust the space: $A_i \subset A_{i+1}$ and $M = \bigcup A_i$. The latter is simply connected at infinity if the inverse limit sequence

$$\pi_1(M - A_1) \leftarrow \pi_1(M - A_2) \leftarrow \cdots \leftarrow \pi_1(M - A_i) \leftarrow \cdots$$

is pro-equivalent to the trivial system, i.e. for each i, there is a $j > i$ so that $\pi_1(M - A_i) \leftarrow \pi_1(M - A_j)$ is trivial.

Note that this is *not* equivalent to there being "no loop that can be moved all the way to ∞." The system of $\mathbb{Z} \leftarrow \mathbb{Z} \leftarrow \cdots$, where all arrows are multiplication by 2, has that property, but is not pro-trivial. The inverse limit is indeed trivial, but the multiples of 2^n come from n stages ahead – and this image does not stabilize.

At least for high-dimensional manifolds, this pro-triviality (i.e. simple connectivity at infinity) is equivalent to there being an exhaustion by compact sets, all of whose complements are simply connected. In general, the inverse limit

---

[22] Or interbreeding them.

[23] This criterion, in dimension > 4, is due to Stallings (1962) (and extended to dimension 4 by Freedman; in dimension 3, it follows from the Poincaré conjecture).

of this sequence is independent of the defining compact sets, but this "fundamental group at ∞" only sometimes[24] plays the same role as the fundamental group for compact manifolds. In any case, it is a good approximation to "$\pi_1(\partial)$ if it only were the interior of a manifold with boundary ∂."

A good example of a contractible manifold (of dimension greater than 2) which is not Euclidean space is the interior of a contractible manifold whose boundary is non-simply connected. The boundary of a compact contractible manifold is automatically a homology sphere (i.e. has the homology of a sphere) – which is a sphere (according to the Poincaré conjecture) iff it is simply connected.

However, every homology sphere bounds a contractible topological manifold.[25] Some three-dimensional examples of homology spheres can be obtained by gluing together two nontrivial knot complements along their boundaries, interchanging longitudinal and meridional directions. Higher-dimensional examples can be obtained by spinning low-dimensional ones: puncture a homology sphere and cross it with a disk, and then take the boundary of this manifold.

Without relying on any theory, a simple example of a contractible 4-manifold whose boundary is non-simply connected is a Mazur manifold, constructed as follows: attach a $\mathcal{D}^2 \times \mathcal{D}^2$ to $\mathcal{S}^1 \times \mathcal{D}^3$ along a (neighborhood of a) nontrivial knot in $\partial(\mathcal{S}^1 \times \mathcal{D}^3) = \mathcal{S}^1 \times \mathcal{S}^2$ that represents a generator of $\pi_1 = \mathbb{Z}$.[26] (Mazur observed, see the crystal clear exposition in Zeeman (1962), that the product of this manifold with the interval [0, 1] is a ball.)

Davis's idea was to generalize the obvious construction of $\mathbb{R}^2$ from a square by repeated reflection and gluing (producing the checkerboard with an action of the product of two infinite dihedral groups, $D_\infty \times D_\infty$) to a construction of some contractible manifold by reflecting across the top simplices of a triangulation of the boundary of any contractible manifold with boundary, with an action of a Coxeter group (that is, a group generated by reflections, whose only relations are commutation of the reflections along incident faces; see next page), whose quotient is precisely this "seed" contractible manifold.

Davis also calculated that, if the seed has non-simply connected boundary,

---

[24] Essentially when the manifold is tame at ∞.

[25] This is classical and due to Kervaire if the homology sphere is of dimension 4 and higher. It is strictly speaking correct in the PL and topological categories – in the smooth category it might be necessary to take the connected sum with an exotic sphere (a differentiable manifold homeomorphic to the sphere). In dimension 3 this is true in the topological category by the work of Freedman, but it is *not* true in the PL and smooth categories, by Rochlin's theorem, that the signature of a closed spin (smooth) 4-manifold is divisible by 16. The most straightforward proof of this important theorem is probably the one given in Lawson and Michelsohn (1989).

[26] Of course, to get an example, one should specify a knot and calculate that one gets a nontrivial homology sphere; but Gabai's theorem on "Property R" guarantees this.

then the manifold he so constructed is also non-simply connected at ∞. In particular, this happens if one starts with a Mazur manifold.

It is a general fact that Coxeter groups are linear, and therefore virtually torsion free, so a finite index subgroup acts on this contractible manifold freely, giving the relevant compact aspherical manifold with exotic universal cover.

Now for a few more details and a generalization with some indication of applications.

A right-angled[27] Coxeter group is given as a pair (Γ, V), where Γ is a group and V is a generating set, by elements of order 2. All the relations of Γ are consequences of relations of the form $(vw)^2 = 1$. We shall take the barycentric subdivision of a triangulation of our seed X. Define an abstract group Γ, generated by involutions v, one for each top simplex. We impose the relation $(vw)^2 = 1$ (and hence v and w commute) if the two simplices share a face. Note that if a k-tuple of simplices have pairwise commuting associated generators, then the intersection of these simplices is nonempty (and conversely). Consider $Z = \Gamma \times X/\sim$ where we identify points $(\gamma, x) = (\gamma', x')$ iff $\gamma^{-1}\gamma'$ lies in the group generated by all the generators of all the simplices that x lies in. So, in the interior of the seed, there is no identification. On the simplex corresponding to a generator v, v acts trivially. Davis proved by an induction on the length of the words in a Coxeter group that one obtains in this way a contractible manifold by showing that it is an ascending union of contractible spaces glued along contractible subspaces.[28]

The Davis construction is most usefully put into the context of CAT(0) geometry,[29] both in its own right in understanding the geometry that such a group has, and also because of the role that negative and non-positively curved geometry plays throughout our story. Nevertheless, we defer this discussion for now, and will say a bit more about it in describing the next construction.

Another variant that has extremely important applications is using strange seeds to construct aspherical manifolds with other strange properties. Start with any aspherical seed that is a manifold with boundary. Triangulating the boundary, and constructing the reflection group, one obtains here an aspherical manifold with a cocompact Coxeter action, and therefore, by passing to the universal cover, a contractible manifold with a cocompact group action, so on taking the quotient, a compact aspherical manifold which inherits properties from the seed. For example, this is a good way (following Davis and Haus-

---

[27] We assume right angles for simplicity. Otherwise, the exponent in the power for the nontrivial relations would be different.

[28] Perhaps reminiscent slightly of the argument in §2.1.

[29] CAT(0) is a synthetic notion of non-positive curvature, named by Gromov in honor of Cartan, Alexandrov, and Toponogov (see Gromov, 1987). Wait a page!

mann, 1989) to produce an aspherical manifold with no smooth structure, or even no triangulation (using a seed that is a topological manifold that is non-triangulable, but whose boundary is triangulable, so that the construction can be done).

If K is any finite aspherical complex, one can take its regular neighborhood in Euclidean space[30] to obtain a manifold with boundary to use as a seed. This is a good way to produce aspherical manifolds whose fundamental groups are not residually finite or don't have a solvable word problem. An excellent book on Coxeter groups, their properties, and diversity is Davis (2008).

2.3.2 Method Two: Branched Covers (Gromov–Thurston Examples)

Gromov and Thurston (1987) gave some very interesting examples of compact manifolds with pinched negative curvature, i.e. curvature between -1 and $-1 - \varepsilon$ by a variant of the philosophy of Dehn filling. This elaboration of that philosophy paves the way to other interesting constructions of groups by "adding large relations."

The basic idea is that negative curvature is a condition of large links. After all, negative curvature means that geodesics spread faster than in Euclidean space. So, if one takes a triangulated two-dimensional polyhedron, and then metrize it so that every triangle is an equilateral triangle with side length 1, then assuming that each vertex is incident to at least seven triangles should give a type of negative curvature. (As an exercise with Euler characteristic, neither the 2-sphere nor the torus has such a triangulation.)

A suitable version of curvature is given by the notion of CAT(k) geometry. A metric space X is called geodesic if its metric is generated by the length of paths connecting pairs of points. Riemannian manifolds are a good example, but one can make others by using a metric and then taking lengths of paths. Now, suppose that we have a triangle in X; then we can construct a triangle in one of the model geometries with curvature k (i.e. rescaled hyperbolic space, Euclidean space, or a sphere). We say X is CAT(k) if the triangles in X are thinner than the corresponding model triangle, meaning that each leg is closer to the union of the other two in X than they are in the model. Figure 2.4 shows a δ-thin triangle.

This is equivalent to curvature less than k for Riemannian manifolds and is

---

[30] Any (finite) polyhedron can be simplicially embedded by general position in a much larger-dimensional Euclidean space. Subdividing, and taking the union of all of the simplices that touch this complex, one obtains a (compact) manifold with boundary that (simplicially collapses onto and therefore) deformation retracts to the polyhedron. This is called a *regular neighborhood*.

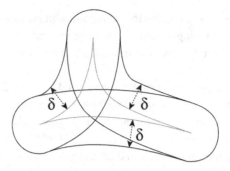

Figure 2.4 A δ-thin triangle.

a useful synthetic substitute for other metric spaces. If X is locally CAT(0), then its universal cover is contractible. (Points will be connected by unique geodesics, and the contraction will be radial.) A great example is a tree.

Back to the case of triangulated surfaces: six incident triangles for each point implies CAT(0), seven gives a negative[31] CAT curvature.

In the Dehn surgery theorem, we can think of the process of filling as gluing on (a family of) $\mathcal{D}^2$s along the translates of a geodesic on the boundary torus. Thurston's theorem tells us that we can have negative curvature (indeed, he gives constant, but that's too much in general) if the length of the geodesic is long enough.

Gromov and Thurston do something similar. They consider a hyperbolic manifold M with a totally geodesic submanifold V of codimension 2. They show that k-fold branched covers[32] (can be proved to exist, at least sometimes, and then) can be given metrics with curvature between -1 and $-1 - c/\log(k)$. The volume in this construction grows linearly: the metric is constructed quite explicitly and deviates from the hyperbolic metric only in a small neighborhood of the submanifold (as the heuristic suggests).

Philosophically, when k gets large, the curvature should be getting more negative. They essentially have to stretch the neighborhood to make it more pinched (i.e. so that the divergence of the geodesics has more time to occur).

The reason that these manifolds can't be made constant negative curvature is a nice application of Mostow rigidity. They all have $\mathbb{Z}_k$-actions, which would be isometric if they were constant curvature. Varying k and modding out by

---

[31] Depending on the length of the triangles.

[32] Recall that a branched cover of a manifold M along a codimension-2 submanifold V is a cyclic covering space of the complement $M - V$ that restricts to the usual cyclic cover of the circle to itself in the direction normal to V. This allows one to fill in V in the covering space, and obtain a manifold (with $\mathbb{Z}_k$-action – whose fixed set is V, and whose quotient is M).

the actions would produce infinitely many different hyperbolic orbifolds with bounded volume. However, above dimension 3, there are only finitely many hyperbolic orbifolds with any given volume bound (Wang's theorem: see Wang, 1972).

As the final technical point to mention, one can take branched cover along a codimension-2 submanifold iff it is trivial as a homology class. We can construct examples of this by the arithmetic construction we discussed earlier. If one uses the quadratic forms that arose in "grafting," then there is a $\mathbb{Z}_2 \times \mathbb{Z}_2$ action generated by two reflections. The fixed set of the action of the whole group is null-homologous in the fixed set of either of the involutions – which gives us the relevant M and V to start this construction.

In general, this method is about adding long relations and keeping negative curvature. This method is related to the ideas of small cancellation theory (as is CAT(-1) geometry in general) and in both its manifold and nonmanifold versions has led to many very interesting groups, some of which we will discuss below.

2.3.3 Method Three: Hyperbolization

The basic idea of hyperbolization is very simple, and there are many hyperbolization methods, i.e. implementations of this idea. We will be brief and leave the reader to study the (rather beautiful) literature (see the notes in §2.4). On the other hand, it is impossible to resist mentioning at least a few of the surprising examples.

The Kan–Thurston theorem asserts that any simplicial complex X has the homology type of a group π, i.e. there is a map $B\pi \to X$ which is an isomorphism (for all local coefficient systems on X). Baumslag, Dyer, and Heller gave a very nice approach to this theorem that gives a finite complex $B\pi$ if X is finite (Baumslag *et al.*, 1980).

The idea is to find a "simplex of acyclic groups" and glue these together. One simple version can be done as follows, using cubes instead of simplices. This doesn't make a difference since one can replace every simplicial complex by a "cubulated" complex. So we will instead look for cubes of acyclic groups.

Acyclic groups are easy to come by. A simple example is any free product with amalgamation $\pi = F *^{F'} F$, where F and F' are free groups of rank k and $2k$, respectively, and the first inclusion of F' to F induces a split surjection on first homology $\mathbb{Z}^{2k} \to \mathbb{Z}^k$ projecting onto the first k dimensions, and the second inclusion interchanging the first k and second k basis elements. In this case, gluing tells us that $B\pi$ is the 2-complex obtained by taking the double mapping cylinder of a wedge of $2k$ circles mapping to two wedges of k circles. A straightforward Mayer–Vietoris calculation then gives that Bπ is acyclic.

An "interval of acyclic groups" is simply given by the diagram of groups $\pi \to \pi \times \pi \leftarrow \pi$, where the first (respectively second) inclusion is given by the inclusion of the first (second) factor. From intervals of groups, we can obtain squares and the cubes of groups by taking products.

Note that if we have a cubical complex, we can then (ordering the vertices!) glue together the associated cube of acyclic groups. This will produce a complex (in this case, finite if X is, and of twice the dimension) which has all the desired properties.

Notice that this construction is aspherical in the category of simplicial complexes (or cubical complexes) and simplicial inclusions.

Hyperbolizations do exactly the same thing, but using aspherical *manifolds*[33] instead of complexes. In both of these constructions, it is critically important that fundamental groups inject for gluing purposes.

It is not possible to arrange for the map to be a homology equivalence (for then the 2-sphere would be homology-equivalent to an aspherical surface – which we know by classification is not the case).[34] However, other geometric properties can be achieved by suitable constructions of simplices or cubes of aspherical manifolds.

The seed is often chosen to be non-positively curved (or negatively curved), orientable, or even with stably trivial[35] tangent bundle. Points are hyperbolized as points, and the geometry is rigid enough that the links of these points are the same in X and its hyperbolized version. If X is a manifold, so will be the hyperbolized space, and the map $\mathbb{H}(X) \to X$ will be degree 1 and preserve characteristic classes. This implies that $\mathbb{H}(X)$ is cobordant to X,[36] so, for example, every cobordism class contains an aspherical manifold.

If M is a manifold with boundary, Gromov suggested hyperbolizing $M \cup c \partial M$ (where $c \partial M$ denotes the cone of the boundary of M). This will produce an aspherical complex with a single singular point, whose link is ∂M. One can show that if ∂M is aspherical, then one can remove this singular point to get a "relative hyperbolization" that ∂M bounds (mapping to M). Thus, not only is every manifold cobordant to an aspherical manifold, but also cobordant aspherical manifolds are cobordant through aspherical manifolds.

---

[33] We give up on acyclicity, however.

[34] In higher dimensions, I do not know how to eliminate any closed manifold from being homology-equivalent to an aspherical manifold. However, the Hopf conjecture would clearly preclude this.

[35] That is, trivial after adding on a trivial bundle.

[36] By Thom's classical work that shows that bordism is governed by tangential information (Thom, 1954).

Among the applications of this technique (besides ones we will see later) are aspherical manifolds that cannot be triangulated or smooth manifolds (whose universal covers are topologically $\mathbb{R}^n$) with CAT(0) metrics, but no Riemannian metric of non-positive curvature.

For example, a non-triangulable aspherical manifold comes from the following. The Poincaré homology 3-sphere[37] Σ bounds a 4-manifold W whose intersection form is E_8 (the unique eight-dimensional positive definite unimodular quadratic form over $\mathbb{Z}$ with $\langle x, x \rangle \equiv 0 \mod 2$ for all x).[38] Hyperbolize $W \cup c\partial$. Then remove the cone point, and glue on the contractible 4-manifold, constructed by Freedman, that Σ bounds. This gives a topological manifold X that, being homotopy equivalent to the hyperbolization, is aspherical. On the other hand, this manifold is "spin" in the sense that its first two Stiefel–Whitney classes must vanish (since they do for W, and hyperbolization is tangential), which then prevents smoothness – by the cobordism property X has signature 8, but Rochlin's theorem asserts that any smooth spin 4-manifold has signature a multiple of 16.

The complex X cannot be triangulated as a simplicial complex, as can be seen using either the Casson invariant[39] (or, even easier now, the three-dimensional Poincaré conjecture).

Ontaneda (2011) refined the construction of hyperbolization to produce arbitrarily well-pinched negatively curved hyperbolizations, so one can, for instance, construct manifolds with curvature $-1 - \varepsilon < k < -1$ in any cobordism class.

2.4 Notes

That surfaces tend to be aspherical is classical. For 3-manifolds, there were some early results by combinatorial methods. For example, Aumann (1956)[40] proved the result asserted in its title with its main topological tool being the gluing lemma. That 3-manifolds in general tend to be aspherical (and, for example, the complements of all knots, and all nonsplittable links) is due to Papakyriakopoulos (1957).

The tools introduced in that paper (the Dehn lemma, loop, and sphere the-

---

[37] See Kirby and Scharleman's (1979) for a beautiful description of this 3-manifold and many descriptions and properties of it.

[38] See Serre (1973) for more information.

[39] Casson showed how to count the conjugacy classes of SU(2) representations of the fundamental group of the homology 3-sphere, and that, when done properly, these reduce mod 2 to 1/8 of the signature of any smooth cobounding spin 4-manifold.

[40] Whose author later won a Nobel Prize (for work in game theory).

orems) were the core of 3-manifold topology (their power being most evident for the class of "Haken" 3-manifolds) until the Thurston revolution brought in a wealth of more (differential) geometric techniques. This development can be found in any standard book on 3-manifolds. (Good books for the torus decomposition and some of its pre-Thurston understanding are Hempel, 1976, and Jaco, 1980.)

The geometricization conjecture of Thurston is a picture of *all* closed 3-manifolds in terms of locally symmetric ones. The possible geometries are well described in Scott (1983). Very useful explanations of Thurston's theorem proving this picture correct in the situation where there is an incompressible surface are Morgan (1984) and Kapovich (2001) (from a different point of view than Thurston's original approach). A detailed explanation of Perelman's result can be found in Morgan and Tian (2014).

The study of locally symmetric manifolds started in the nineteenth century. These manifolds are now studied by mathematicians of many different stripes. Besides being interesting examples to geometers, the geometry and topology of many of these manifolds are the essence of such classical results of algebraic number theory as the Dirichlet unit theorem (which calculates the group of units in the integers of an algebraic extension of $\mathbb{Q}$, and which is the compactness of a certain torus) and the finiteness of the class number (which, for instance in the situation of a totally real field, follows from the existence of a compactification for Hilbert modular varieties – the cusps corresponding to elements of the class group). We will discuss arithmetic manifolds and hints of arithmeticity in Chapter 3. As mentioned earlier, Borel (1963) gave the first general construction of uniform lattices for all $K \backslash G$. It is much simpler to give non-uniform lattices. The books by Eberlein (1997) and Witte-Morris (2015) are extremely useful.

Non-arithmetic lattices, as we have seen, are ubiquitous (if not so easy to construct) in low dimensions. The question of exactly which semisimple Lie groups admit them is still open. As we mentioned, for rank greater than 1, Margulis's arithmeticity theorem assures us that there are no (irreducible) examples (see Zimmer, 1984; Margulis, 1991).

The only known construction that works in infinitely many dimensions is the Gromov–Piatetski-Shapiro (G-PS) grafting method we explained. Deligne and Mostow (1993) gave some examples in $U(n, 1)$ for small n. On the other hand, in $Sp(n, 1)$ and F_4, Gromov and Schoen (following on earlier work of Corlette) showed that arithmeticity does hold using analytic methods related to harmonic maps (Gromov and Schoen, 1992).

The G-PS manifolds play a role in counting the number of hyperbolic manifolds with volume less than V, in dimensions greater than 3 (when it is finite)

(Burger *et al.*, 2002) and with diameter less than D in all dimensions, including dimension 3 (Young, 2005).[41]

As emphasized in the main text, the examples of non-arithmetic lattices are suggestive of tools for constructing interesting aspherical manifolds that have nothing to do with lattices. Davis (1983, 2000) was motivated, as he explains therein, by Andreev's theorem about reflection groups in hyperbolic space.

Closing cusps has been applied both to manifolds and to nonmanifolds. See Hummel and Schroeder (1996) for the situation of closing cusps for, e.g., complex hyperbolic manifolds (and its impossibility in the quaternionic case). CAT(0) geometry was broadcast to the world by Gromov (1987) in his paper on hyperbolic groups. The main theme of that paper is developing a large-scale (or coarse) notion of negative curvature for groups, as a property of their Cayley graphs, and showing how this notion deepens and generalizes our understanding of hyperbolic manifolds. The most obvious examples of such groups are fundamental groups of negatively curved manifolds, and also free groups. But there are many more!

For instance, Gromov pointed out that one can cone very long words at will[42] (as a generalization of the idea of Thurston's Dehn surgery theorem) and maintain negative curvature, giving "easy" finitely generated torsion groups (just kill large powers of the elements of the group, one at a time).[43]

That paper also introduces hyperbolization (with some glitches regarding the procedure fixed in Davis and Januszkiewicz, 1991, Charney and Davis, 1995, and Davis *et al.*, 2001), which also give some new applications. The paper, all told, launched a major area of geometric group theory and numerous other investigations. See, for example, Ghys and de la Harpe (1990) for an exposition of much of the content of that paper. Bridson and Haefliger (1999) is an excellent source on non-positively curved spaces that are not necessarily manifolds.

Regarding more basic facts about discrete groups that arose in this chapter, see C. Miller (1971) for constructions of groups with unsolvable word problem and related matters. Baumslag *et al.* (1980) is the paper that gives the finite form

---

[41] Note that when a hyperbolic Dehn surgery is done, the filling takes place further and further down the cusp, and the diameter of the manifold increases with the length of the curve filled.

[42] What I mean is that one can represent a long word in $\pi_1 X$ by a long closed geodesic in X, and then we can attach a disk along this word, and maintain negative curvature. If the geodesic is long, then the geometry is that of locally having an n-gon with $n > 6$ at the new vertex.

[43] If one starts with a lattice in $Sp(n, 1)$ and does this, one gets an infinite torsion group with Property (T). This example also shows that while Property (T) implies finite generation, it does not imply finite presentation. (See Chapter 3 for the basics of Property (T).) On the other hand, this method does not solve the Burnside problem of giving finitely generated exponent p groups. However, even this can be achieved in the hyperbolic group setting, as was shown by Ivanov and Ol'shanskii (1996).

of the Kan–Thurston theorem along the lines described here. It is subsequently applied in Baumslag *et al.* (1983) to give remarkable information about the possible sequences of homology groups of a finitely presented group (it's obviously not arbitrary: there are countably many finitely presented groups and uncountably many sequences of even finite abelian groups!).[44]

Rochlin's theorem, mentioned in explaining the construction of a non-triangulable four-dimensional aspherical manifold, asserts that the signature (see Chapter 4) of a smooth spin 4-manifold is a multiple of 16. This was immediately understood to be an anomaly, and led to various examples of phenomena where dimension four behaves differently from the smooth perspective than higher dimensions. This turned out to be the tip of the iceberg with the advent of Donaldson's thesis (see Donaldson, 1983) – and the work that has followed it – which has yielded much more profound information about smooth 4-manifolds.

---

[44] It also contains the construction of an acyclic universal group, i.e. an acyclic finitely presented group containing every finitely presented group as a subgroup. (Note that there's no finitely generated group containing all finitely generated groups.) This group has been surprisingly helpful for various constructions. As one example relevant to this chapter, it was applied in an early version of Davis *et al.* (2001) for the construction of relative hyperbolization – although this was not necessary in the final version, which followed Gromov's original ideas more closely.

3

First Contact: the Proper Category

3.1 Overview

Having given some idea of the kinds of manifolds to which the Borel conjecture applies directly in Chapter 2, we consider now the effect of modifying Borel's heuristic. Taking light of Prasad's (1973) extension of Mostow rigidity to the case of nonuniform lattices, we ask whether topological rigidity holds in this context?

It was already noticed in the early 1980s that this is not the case. Making use of Borel's calculations of the stable cohomology of $SL_n(\mathbb{Z})$, Farrell and Hsiang observed that for $n > 200$ and Γ a torsion-free subgroup of finite index in $SL_n(\mathbb{Z})$, the quotient $SO_n \backslash SL_n(\mathbb{R})/SL_n(\mathbb{Z})$ is a not "properly rigid;" i.e. there are infinitely many manifolds M not homeomorphic to $SO_n \backslash SL_n(\mathbb{R})/SL_n(\mathbb{Z})$, but proper homotopy-equivalent to it.

Actually this happens iff $n \geq 4$ (and, moreover, the same is true for any number rings in place of $\mathbb{Z}$) as we will §3.7.[1]

The goal of this chapter is to explain this in its natural setting, using it as an excuse to explain some aspects[2] of the structure of $K \backslash G/\Gamma$,[3] Property (T),[4] L^2 cohomology[5], and some surgery theory that we will need in later chapters. Not as critical on utilitarian grounds, but nevertheless important, are discussions of

---

[1] Actually, we will only explain the failure of proper rigidity if $n > 3$; its affirmative solution depends on the "Borel conjecture with coefficients" and will have to wait till later.

[2] The next several footnotes are intended for the more expert reader.

[3] The discussion of which is also relevant to the proof of the Novikov conjecture for linear groups explained in Chapter 8.

[4] Which we will use, as is traditional, in the construction of expanders, which are relevant to the failure of forms of the Baum–Connes conjecture.

[5] Which is used in the proof of the flexibility theorem later that affirms a consequence of the Farrell–Jones conjecture and of the Baum–Connes conjecture unconditionally.

the cohomology of arithmetic groups (ultimately these discussions go to the very meaning of the conjecture),[6] and superrigidity.

The outline of the chapter is as follows: we will first explain the overall shape of $K \backslash G / \Gamma$ (which is a far-reaching generalization of the classical nineteenth-century reduction theory of binary quadratic forms) and give some information about the Borel–Serre compactification of this manifold (Borel and Serre, 1973). Then we will discuss some generalities about the cohomology of arithmetic groups and describe Borel's results on these groups.

Assembling all of this with some surgery theory, we will see a critical role played by the $\mathbb{Q}$-rank. The case of $\mathbb{Q}$-rank $= 0$ corresponds to the compact manifolds, i.e. the Borel conjecture in its usual sense, and if $\mathbb{Q}$-rank < 3, it turns out that these noncompact manifolds behave (for the purposes of topological rigidity) just like the compact case, and results explained later in the book will give their proper rigidity. Nonrigidity will immediately follow from the combination of surgery theory with Borel's calculations for very large n (as mentioned above, $n > 176$).

Both for the purpose of lowering n and for allowing a wider range of Lie groups (and for the purposes of later developments) we digress and explain several important properties of lattices in higher-rank groups, and of certain linear groups.

The first of these topics is strong approximation. This property of linear groups will give us control on certain finite quotients of linear groups. We will need this only in this chapter, so our discussion will be brief.

We then turn to Kazhdan's Property (T). Our focus will merely be on definitions, and we leave to other sources serious discussions of the scope of this property and its remarkable applications. These ingredients are then assembled and combined with superrigidity[7] to show that any lattice that has $\mathbb{Q}$-rank ≥ 3 has a finite sheeted cover that is not properly rigid.

This proper rigidity we thus obtain is somewhat weaker than one would hope: it asserts the existence of a proper homotopy equivalence $f: M \to K \backslash G / \Gamma$ that is not properly homotopic to a homeomorphism. We will need to work harder to ensure that M is not homeomorphic to $K \backslash G / \Gamma$ (by some other map), and that M is smoothable, and to get the set of such Ms to be infinite. For these we will use a mix of tools from comparison to the Lie algebra mod p, to the Baily–Borel compactification in the Hermitian case, to the use of "generalized modular symbols" of Ash and Borel (1990), in order to give a definitive solution for all $SL_n(O)$ (with O a number ring) and for all Γ of $\mathbb{Q}$-rank > 3. (Alas, at

---

[6] As the cohomology of groups gives rise to geometric consequences via the Novikov conjecture.

[7] The extension of linear representations from lattices to the semisimple Lie groups that contain them.

the time of this writing, for example, the proper rigidity properties of certain lattices in E_7 are still not well understood.)

We close the chapter by considering the morals of this story, a reexamination of the forest having focused on particular trees. Despite the failure of proper rigidity, we consider noncompact variations of rigidity that actually are true for these locally symmetric spaces. We also discover a role for functoriality in this problem – an aspect which could seem surprising given that the initial problem is purely about certain very specific and beautiful objects.

3.2 $K\backslash G\Gamma$ and its Large-Scale Geometry

... in which we encounter the Tits building and the Borel–Serre compactification[8]

If G is a connected Lie group, then it has a maximal compact subgroup K, which is unique up to conjugacy. Topologically, $K\backslash G$ is contractible. Give G a right invariant and K bi-invariant metric. If G is semisimple (i.e. has no normal solvable subgroups), then $K\backslash G$ gets a complete metric of non-positive curvature.

As discussed in Chapter 2, G often contains lattices. We shall assume (for simplicity) that G is given the structure of linear algebraic group defined over $\mathbb{Q}$. The first lattices one thinks of are $G(\mathbb{Z})$ and its congruence subgroups, i.e. matrices lying in $G(\mathbb{Z})$ that are $\equiv I \bmod n$. (We have to do this if we want to restrict attention to torsion-free lattices so that $K\backslash G/\Gamma$ is a manifold – the quotient space being a manifold means that the action of Γ on $K\backslash G$ is free: the isotropy of the action of Γ on the right has to be a compact subgroup of the discrete group Γ, and hence finite, and will be trivial when Γ is torsion-free. Conversely, when Γ has torsion, each element of finite order has a fixed point in $K\backslash G$, making the quotient an orbifold.)

The possibility of other algebraic number fields is not essentially eliminated by this condition, because of the method of restriction of scalars: the group $SL_n(\mathbb{Z}\sqrt{2})$ is a lattice in $SL_n(\mathbb{R}) \times SL_n(\mathbb{R})$. For uniform lattices, as we saw in §2.2, there are other arithmetic lattices that come from G having compact forms that are Galois conjugate to the given form – because a lattice in $G \times G'$ gives us one in G by projecting if G' is compact (or alternatively, G and $G \times G'$ are isomorphic after modding out by their maximal compact subgroups). For the noncompact case, these more subtle lattices don't play a role – since all the forms must be noncompact (because Γ contains unipotents and compact groups do not), so the definition of arithmeticity is somewhat less subtle in this case.

---

[8] With apologies to A.A. Milne

While our focus in Chapter 2 was on the compact case, here we are interested in what occurs in the noncompact case. An important theorem of Borel and Harish-Chandra[9] "blames" noncompactness on a "$\mathbb{Q}$-split torus" for G.

Let us follow this subgroup around in the simplest situation $SL_n(\mathbb{Z})$. We will see an even more precise picture than mere noncompactness.

In $SO(n)\backslash SL_n(\mathbb{R})$ we can consider the torus of diagonal matrices (such that the product of their entries is 1). As a space of tori, these are the "rectangular" tori. Taking the logs of these eigenvalues, we get a map to $\mathbb{R}^{n-1}$ (the elements of $\mathbb{R}^n$ that have the sum of their components equal to 0). The symmetric group Σ_n acts on this by permutation – without loss of generality, we can assume that the eigenvalues are listed in increasing order. This gives us a polyhedral cone in $\mathbb{R}^{n-1}$ and a subset of $SO(n)\backslash SL_n(\mathbb{R})/SL_n(\mathbb{Z})$. This subset gives us a very good large-scale picture of this quotient manifold: for example, this embedding is essentially undistorted, and every point in the quotient space is of uniformly bounded distance to a point of this sector. Moreover, this statement is true if $\mathbb{Z}$ is replaced by integers in a totally real field. Although the real Lie group this embeds in a product of $SL_n(\mathbb{R})$s, the effect of taking the quotient by the action of $SL_n(O)$ is to cuts it down to the size of the polyhedral cone that is the quotient of the maximal flat.[10] The proofs of these kinds of statements are the subject of "reduction theory," developed by C.L. Siegel (1988), A. Borel, and their successors (see Borel and Ji (2005) for a modern account).

For other lattices we will have to glue together copies of this sector according to some combinatorial description governed by the theory of Tits buildings – which records the combinatorics of the parabolic subgroups. All of this is first most easily observed in yet another, even simpler, example, the product of hyperbolic manifolds $\prod M_i$. After discussing this toy example, we will return to $SO(n)\backslash SL_n(\mathbb{R})/SL_n(\mathbb{Z})$ and the general case.

Each noncompact hyperbolic manifold M has a core, with cusps coming off. Pick a base point, and a sequence of points going towards infinity in each of the cusps. The geodesics connecting this base point to those points converge to a finite union of geodesic rays, each of which is isometrically embedded in the manifold (see Figure 3.1).

This union of geodesics looks like an asterisk with one "prong" for each cusp; we denote this by A. (This is the direct analogue of the polyhedral cone from the $SL_n(\mathbb{Z})$ case.)

One can imagine a map from M to A, roughly mapping each point to the

---

[9] See Borel and Harish-Chandra (1961).

[10] This is very much like the phenomenon that occurs in the Dirichlet unit theorem, where all of the directions in logarithm space for the various embeddings of the units just curl it up into a torus.

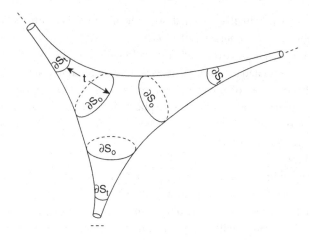

Figure 3.1 Adapted from Thurston's notes

point on the asterisk closest to it, (and then modifying it slightly on a compact set, arrange the map so that the inverse image of the base point is the core of M, and the inverse image of any point in one of the rays is a "flat manifold horospherical section" of the cusp.

Let me elaborate on the terminology.

The isometry group of hyperbolic space $\mathbb{H}^n$ is $O(n, 1)$ – which we will imagine via the ball model. The isometries form three classes: elliptic, hyperbolic, and parabolic. Each elliptic element has fixed points in the interior, and lies in a maximal compact. (The action of the isometry group is transitive, so what fixes one point is conjugate to what fixes any other point: hence, the maximal compact subgroup is unique up to conjugacy.)

Hyperbolic elements act via translation along a geodesic (with some rotation in the normal direction.[11]) A parabolic element has a unique fixed point on the boundary sphere at ∞.

Given such a fixed point, the *horosphere* going through that point can be defined as follows. Choose a unit speed geodesic γ going from p to a specific point at ∞. Now consider the sphere of radius R centered at $\gamma(R)$. The limit set of these spheres is an orbit $O(n, 1)_p / O(n, 1)_\infty$. The isotropy group is a *parabolic subgroup*, which is isomorphic to the semidirect product $O(n - 1) \ltimes \mathbb{R}^{n-1}$ which is the isometry group of $\mathbb{R}^{n-1}$.

(In general, parabolic subgroups are those subgroups that contain a *Borel*

---

[11] Following Thurston (2002), we do not distinguish between hyperbolic isometries and "loxodromic" ones.

subgroup, i.e. a maximal connected solvable group. They are the isotropy groups of points on the boundary of $K\backslash G$.)

Now let us return to our hyperbolic manifold with a number of cusps. Lifting the geodesics associated to the cusps gives a finite set of points on the boundary, which are fixed points of nontrivial parabolics. The subgroup of Γ fixing a (lifted) cusp acts as a lattice on the horosphere. The quotient is a flat manifold (which is a cross section of the cusp – choosing another point p on γ would give a parallel cross section).

The product of a number of hyperbolic manifolds both contains and maps to the corresponding product of asterisks, which is a polyhedral cone whose dimension in the $\mathbb{Q}$-rank of this product lattice.[12]

Note that the inverse image of a point in this cone depends strongly on which face that point lies on. It will be a product of some number of cores and some number of flat manifolds. (Note that by taking finite covers of this product, we can mangle the product structure, but will still get a similar union of flat pieces approximating the manifold.)

For SL_n the picture is similar. We've seen the cone, and the inverse of a point in the interior of the top simplex is a nilmanifold: isomorphic to $\mathrm{UT}(n,\mathbb{R})/\mathrm{UT}(n,\mathbb{Z})$, where $\mathrm{UT}(n,?)$ denotes the group of upper triangular matrices with (1s on the diagonal and) entries in '?'.

Recall that a point in the top simplex corresponds to a diagonal matrix, whose eigenvalues are distinct. This unitary group is the unipotent subgroup of the matrices that preserve the flag given by these subspaces. A point in a different simplex corresponds to some coincidences among eigenvalues. At these points, one has an incomplete flag and normal to it one has a "genuine" lattice part (corresponding to a product of SLs associated to the various combined eigenspaces) with a nilpotent bundle over that associated to the unipotents that are the identity module the flag.

As one moves towards infinity, the unipotent pieces have volume that decays rapidly to 0,[13] and that is what accounts for the finiteness of the volume of these nonuniform lattices. The lattice part stays bounded in size (but does not shrink[14]).

---

[12] Here by $\mathbb{Q}$-rank we merely mean the number of noncompact hyperbolic factors, whether or not they are arithmetic. As a consequence of Margulis's arithmeticity theorem, all, even non-arithmetic lattices, can be approximated by finite polyhedral cones, defining for us $\mathbb{Q}$-rank even when there is no $\mathbb{Q}$-structure! The reason is that there is such a structure for negatively curved manifolds, and everything is virtually a product of negatively curved homogeneous spaces and arithmetic ones.

[13] A nilmanifold is essentially "an iterated fiber bundle of torus on top of torus and so on". The layers shrink at different rates. Gromov (1978) has shown that manifolds with metrics of bounded curvature but diameter going to 0 are finitely covered by nilmanifolds.

[14] This is also similar to what occurs in the case of a product of hyperbolic manifolds – the

Another concrete case for which the calculations are not difficult is the case of Hilbert modular groups,[15] $\Gamma = \mathrm{SL}_2(O_F)$ where F is a totally real field of degree d. In that case, there are finitely many cusps (equal to $h(O_F)$, the class number of the ring[16]). This group acts on a product of d hyperbolic planes (where $d = [F; Q]$). The cusps are actually solvable manifolds.[17] The bounded part is a torus corresponding to O_F^*. The fiber is the torus $\mathbb{R}^d/O_F$ and the monodromy of this bundle is the action of O_F^* on O_F. The base torus stays of bounded size as one goes down the cusp (it takes some distance to work up the twist corresponding to a nontrivial unit), while the fiber torus decays exponentially by homothety as one goes down the cusp.

Now let us work in general, guided by these special case. If G is a linear algebraic group defined over $\mathbb{Q}$, we shall define a simplicial complex, the Tits building of G using the parabolic subgroups of G. The minimal parabolic is **B**, by definition, the Borel subgroup, and G itself is the maximal parabolic.

To a parabolic P we associate a simplex σ_P so that $\sigma_P \subset \sigma_Q$ iff $Q \subset P$. The group G corresponds to the empty simplex. The maximal simplices correspond to (conjugates of a) Borel[18] subgroup.

It is a very nice theorem of Solomon and Tits (proved rather geometrically: see, e.g., Abramenko and Brown, 2008) that this complex has the homotopy type of a wedge of spheres of dimension $q - 1$ (where $q = \mathbb{Q}$-rank).

The Borel–Serre compactification (Borel and Serre, 1973) of $K\backslash G/\Gamma$ is a compact manifold[19] with boundary so that $K\backslash G/\Gamma$ is its interior. Actually, it has a more refined structure: it has the structure of a manifold with corners – and this structure carries a great deal of geometry in it, but we will not need this.

The compactification takes place on $K\backslash G$, and is $G(\mathbb{Q})$- (but not $G(\mathbb{R})$-) equivariant. Associated to P we have a Euclidean space e_P so that $\dim e_P + \dim \sigma_P = q - 1$. These open cells are disjoint, but $e_P \subset \mathrm{cl}(e_Q)$ iff $P \subset Q$.

The corner structure comes like this. The unipotent subgroup of P acts on $K\backslash G$ as a free $(\mathbb{R}_+^*)^{\dim(\sigma_P)+1}$-proper action. Include each orbit into the

---

inverse images of points that are not in a top simplex have bounded diameter, which does *not* go to 0 as the point moves to infinity. Of course, the volumes of these point inverses go to 0 very rapidly, or the locally symmetric manifold could not be finite volume.

[15] See Freitag (1990) for a crystal clear explanation.

[16] For congruence subgroups, the number of cusps is the order of a *ray class group*.

[17] That non-nilmanifolds arise is because here G has rank greater than 1, and we are dealing with nonpositive curvature rather than strict negative curvature.

[18] It is not instantly obvious that this is a simplicial complex. A hint is that for simple algebraic groups, the conjugacy classes of parabolic subgroups are in a 1–1 correspondence with subsets of the nodes of the Dynkin diagram.

[19] Actually, when Γ has torsion, it is an orbifold.

$(\mathbb{R}_+{}^*)^{\dim(\sigma_P)+1}$-space $([0,\infty))^{\dim(\sigma_P)+1}$. One can thus compactify each orbit.[20] The relations among the parabolic subgroups enable one to glue these together to include $K\backslash G$ as the interior in a manifold with corners on which the $G(\mathbb{Q})$-action extends. Borel and Serre topologize this union as a manifold so that the action of Γ on it is continuous and proper discontinuous. In particular, they see that down in the quotient, they obtain a compactification.

They also observe that the boundary of $K\backslash G$ so obtained has the Tits complex as its nerve and therefore the Γ cover of the ∂ has the homotopy type of a wedge of spheres $\bigvee S^{q-1}$.

In the case of a lattice of $\mathbb{Q}$-rank 1, the picture is the one of isolated cusps, and the compactification glues onto the end a copy of the slice of the horosphere. For a product of these manifolds, one obtains the product of these compactifications (and, of course, the corner structure is evident in this case).

Moreover, using the fact that the universal cover of these closures are contractible, it is quite easy to see that the boundaries look like joins of the boundaries of the universal covers of the original compactified factors – and hence an infinite wedge of spheres, $\bigvee S^{q-1}$ (where q is the $\mathbb{Q}$-rank).

Note then the underlying homotopy type:

- If $\mathbb{Q}$-rank $= 0$, then we must be compact (and the homotopy type is that of $\varnothing$).
- If $\mathbb{Q}$-rank $= 1$, then the cover of the boundary is a union of copies of the universal cover of the boundary. Thus the Borel–Serre boundary is a (union of) aspherical manifold(s) whose fundamental group is a subgroup of Γ (of course, it's a lattice in the parabolic associated to that cusp).
- If $\mathbb{Q}$-rank $= 2$, then we get a pleasant surprise, the boundary is connected – which means that every compact subset of $K\backslash G/\Gamma$ has a unique component with compact closure (i.e., it has one end).

Moreover, the boundary is a closed aspherical manifold, since it has an aspherical cover, namely the regular cover associated to Γ, which is homotopy equivalent to a wedge of circles.[21]

This is actually a very interesting aspherical manifold that is not a lattice in any Lie group! However it is not really a surprise to us – the Tits building in this situation is a graph, and we have lattices associate to the nodes, glued together according to "boundaries" along the edges[22]. Like 3-manifolds, these

---

[20] Formally, one should take an associated bundle to viewing $K\backslash G$ as a $(\mathbb{R}_+{}^*)^{\dim(\sigma_P)+1}$-principal bundle using this action on the octant $([0,\infty))^{\dim(\sigma_P)+1}$.

[21] Note that aspherical is equivalent to all higher homotopy groups vanishing, but higher homotopy groups are unchanged in covering spaces.

[22] For example for $SL_3(\mathbb{Z})$ one gets two copies of $SL_2(\mathbb{Z}) \ltimes \mathbb{Z}^2$ thought of as block 3×3 matrices

boundaries have decompositions into geometric pieces, and it is not hard to generalize this construction to more complicated kinds of "graph manifolds".

The connectedness of this cover means that the map from fundamental group at ∞ to Γ is surjective. In other words, any loop in $K \backslash G / \Gamma$ can be pulled to ∞ (i.e. outside of any compact). However, to do this, one typically must increase the diameter of loops.[23]

If $\mathbb{Q}$-rank > 2, then we discover that the boundary is not aspherical (π_{r-1} is nonzero) – our first hint that all is not well with a proper Borel conjecture.[24] As we will see in the coming sections, because of this, when $\mathbb{Q}$-rank > 2, proper rigidity typically fails. At the end of the chapter we will try to learn some lessons from this failure.

3.3 Surgery

Surgery theory is a framework for studying the classification of high-dimensional manifolds. In this section we will describe some of the features of surgery theory, and in particular, a situation where there are "no obstructions". In particular, we will explain the observation of Farrell and Hsiang (1982) that for very large lattices the proper analogue of the Borel conjecture fails. Later sections will show that failure is actually ubiquitous and more dramatic than these examples show.[25]

Our presentation in this section is quick and dirty. Later on we will need and give more precise, and more conceptual, discussions: the need for better calculations requires alternative descriptions, from whose vantage point the very nature of our central problem changes.

Atiyah (1961) observed that:

Theorem 3.1 *If one has a homotopy equivalence between closed manifolds $h: M' \to M$, then there is a kind of equivalence between their stabilized tangent bundles, namely* stable isomorphism of spherical fibrations.

Let me explain. Assume first that $M \ni m$ and $M' \ni m'$ are smooth so that they have tangent bundles, TM and TM' respectively, in the usual sense.

(with a 2×2 block either on the top left or bottom right). These intersect along the Heisenberg group $U(3, \mathbb{Z})$ in $SL_3(\mathbb{Z})$. The fundamental group of the boundary is this amalgamated free product. The kernel of the map of this group to $SL_3(\mathbb{Z})$ is an infinite-rank free group.

[23] This will be (part of) the reason why we will ultimately succeed in proving a "bounded" topological rigidity for higher-rank locally symmetric manifolds – see the discussion in the morals, §3.8.

[24] Of course, the resolution *could have been* that there are some special non-aspherical manifolds that are rigid. There are some, but Borel–Serre boundaries turn out not to be among these.

[25] But as we said, there are also versions of rigidity that do apply to nonuniform lattices.

An equivalence between tangent vector bundles in the usual sense would be a continuous family of linear isomorphisms (not necessarily the differential, Dh, of the map) $TM'_{m'} \to TM_{h(m)}$. A stable isomorphism of such vector bundles would be such a family $TM'_{m'} \times \mathbb{R}^d \to TM_{h(m)} \times \mathbb{R}^d$ for some d. A *stable isomorphism of spherical fibrations* is such a family of maps, not necessarily linear, but which is a degree-1 proper homotopy equivalence on each fiber. (This means that the map induces a homotopy equivalence between the fiberwise one-point compactifications, i.e. the stable spherical fibrations. Note that the one point compactification can be thought of as being the unit sphere of one stabilization further.)

This implies that some invariants of the tangent bundle are homotopy invariant, such as Stiefel–Whitney classes.[26] However, this equivalence relation on bundles is very weak: over a space X of finite type,[27] there are only finitely many such equivalence classes.[28] However, characteristic classes, such as the Pontrjagin classes, allow for an infinite number of conceivable tangent bundles for manifolds within that homotopy type.

Just as (oriented) bundles can be thought of as maps into Grassmanians,[29] BSO, there is a classifying space for (oriented) spherical fibrations BSF, i.e. maps $E \to X$ whose homotopy fiber is a sphere are classified by maps $X \to$ BSF, so that we can interpret Atiyah's theorem as saying that the composite map

$$M \to \text{BSO} \to \text{BSF}$$

is a homotopy invariant of compact manifolds M. The proper analogue of Atiyah's theorem holds as well.

So, given $h: M' \to M$, taking into account the automatic equivalence of their stable tangent bundles in BSF, gives us a refined tangential data for a homotopy equivalence:

$$\nu(h): M \to F/O,$$

where F/O is the fiber of the map BSF $\to$ BSO. This invariant of h is called the *normal invariant of h* (since it is a stable invariant, and the stable normal

---

[26] This fact also follows from the Wu formula that gives a homotopy-theoretic description of the Stiefel–Whitney classes in terms of the action of the Steenrod operations on the cohomology of a manifold.

[27] That is, with the homotopy type of a finite CW-complex.

[28] This follows immediately from an obstruction theory – induction over the skeleta of a triangulation – making use of Serre's result that the stable homotopy groups of spheres are finite.

[29] That is, there is a universal bundle, and every bundle is the pullback of this bundle under a map that is well-defined up to homotopy.

bundle is adequate for its definition, rather than the more subtle, unstable tangent bundle).

Another way to say this is that the two tangent bundles combine to give a map from M to the homotopy pullback of

$$\begin{array}{ccc} & & \text{BSO} \\ & & \downarrow \\ \text{BSO} & \to & \text{BSF}, \end{array}$$

which, of course, is homotopy equivalent to $\text{BSO} \times F/O$, as we leave to the reader.

Now, I should say that there is a similar discussion possible in the category of nonsmooth, triangulable, or even topological, manifolds, which gives rise to classifying spaces – so in the topological case, we have $\nu(h): M \to F/\text{Top}$. *A first view of surgery theory* is that it is about the difficulty in realizing maps into F/O or F/Top from homotopy equivalences.

However, there is one situation where there is no obstruction at all:

Theorem 3.2 (π–π theorem)   *Suppose that M is a connected manifold with nonempty connected boundary, $\dim M \geq 6$, and $\pi_1(\partial M) \to \pi_1(M)$ is an isomorphism. Then every homotopy class of maps $M \to F/\text{Cat}$ (for $\text{Cat} =$ Diff, PL, Top) is realized by a homotopy equivalence of pairs $(M', \partial M') \to (M, \partial M)$.*

A relative version of this theorem actually implies a uniqueness result for the pair $(M', \partial M')$.[30] This theorem is immediately relevant to our situation, since the Borel–Serre compactification, when $\mathbb{Q}$-rank$(\Gamma) > 2$, satisfies the hypothesis of this theorem.

We shall now review some results about the nature of these classifying spaces.

First of all, the homotopy groups of BSF are finite, so the map $G/O \to \text{BSO}$ is a rational homotopy equivalence.

The reason for this is not difficult: the homotopy groups of BSF corresponded to spherical fibrations over the sphere. A spherical fibration over S^n can be thought of (just like a bundle) as the result of gluing together two trivial bundles over the two hemispheres $\mathcal{D}^n_\pm$. The gluing is a map $S^{n-1} \to$ self-homotopy equivalences of the fiber sphere S^i, which is the iterated loop-space $\Omega^i S^i$ of a sphere. A little thought then shows that the homotopy groups of BSF are therefore the same as the stable homotopy groups of spheres, and these are finite thanks to a theorem of Serre (see Serre, 1951).

---

[30] It will be unique up to h-cobordism, or, if we work with simple homotopy equivalences, then it will be unique up to Cat-isomorphism.

Characteristic class theory also tells us that Pontrjagin classes give us a rational homotopy equivalence BSO → $\prod K(\mathbb{Z}, 4i)$.

The theorem of Kervaire and Milnor (1963) on the finiteness of the number of smooth structures on a sphere can be translated into the statement that the homotopy of Top /O is finite, or that $F/O \to F/\text{Top}$ is a rational equivalence.[31] Thus:

Theorem 3.3 *There is a rational homotopy equivalence*

$$F/\text{Cat} \to \prod K(\mathbb{Q}, 4i).$$

Remarkably, Sullivan gave a complete and precise analysis of F/Top,[32] which we will explain in Chapter 4. See, for example, Rourke and Sullivan (1971) – in itself a historically interesting paper – for part of the proof of the following, and Madsen and Milgram (1979)) for a complete explanation.

Theorem 3.4 *At the prime 2, there is an equivalence:*

$$F/\text{Top}_{(2)} \to \prod K(\mathbb{Z}_{(2)}, 4i) \times K(\mathbb{Z}/2, 4i - 2).$$

Away from 2, there is an equivalence:

$$F/\text{Top}[1/2] \to \text{BSO}[1/2].$$

Remark 3.5 In writing things this way, we are using localization theory for simply connected spaces (or of H-spaces) which enables one to assign to such a space X, the localization of X as a set P of primes. This space $X_{(P)}$ is functorially associated to X, and its homotopy (and homology) groups are those of X, but tensored with $\mathbb{Z}[1/q]$, where q runs over the primes *not* in P. So $X_{(2)}$ has as homotopy groups those of X, tensored with the group of rational numbers with odd denominators.

Localizing at a set of primes has the effect of ignoring contributions of the other primes. Part of the theory explains how to combine the information at the various primes with rational information to give information about ordinary homotopy classes of maps [; X]. We refer the reader to Hilton *et al.* (1975) for

---

[31] This is an outright lie of the worst kind: it is a misleading truth. To set up such an equivalence, one needs to be able to do enough topological topology (i.e. topology in the topological category) to be able to mimic many smooth constructions. In particular one requires topological transversality – which is indeed a theorem from Kirby and Siebenmann (1977). With transversality however, it is a simple matter to prove that rational Pontrjagin classes are topological invariants (a transparent consequence of the statements thrown about in the main text) – as we explain in §4.5. That was a major result of Novikov, for which he earned a Fields medal. In the next section we will return to this train of thought. In any case, for now, please bear with the inaccuracies above.

[32] Actually, Sullivan did the PL case, but once the work of Kirby and Siebenmann mentioned in the previous footnote became available, the result for Top immediately follows.

an exposition of this theory and Bousfield and Kan (1972) for a more modern approach.

Warning Sullivan's map to BSO[1/2] is not transparently related to the tangent bundle of the underlying smooth manifolds (when one has a homotopy equivalence between closed manifolds) – and then forgetting their smooth structure – however, rationally it contains the same information as should be reasonable given our discussion above.[33]

Let us now combine our discussion into a proposition:

Proposition 3.6 *If $M = K \backslash G / \Gamma$ is a locally symmetric manifold of dimension greater than 5 and $\mathbb{Q}$-rank$(\Gamma) \geq 3$, then there are infinitely many smooth manifolds proper homotopy-equivalent to M that are not homeomorphic to M (detected by their rational Pontrjagin classes) if, for some i, $H^{4i}(M; \mathbb{Q}) \neq 0$.*

(The reader who is familiar with Siebenmann's thesis can also reverse the argument we have given to prove the converse to this proposition.)

We can assume M is replaced by the Borel–Serre compactified version. If the $\mathbb{Q}$-rank$(\Gamma) \geq 3$, this is a π–π manifold, so Wall's theorem reduces it to a classifying space question – and the cohomological condition is exactly equivalent to the set of homotopy classes of maps $M \to F/\text{Top}$ to be infinite (and infinitely many of these classes will automatically be smoothable).

Following Farrell and Hsiang, we presently observe that for $n \geq 176$, Borel's work gives on cohomology of arithmetic groups gives us this conclusion for $SO(n) \backslash SL_n(\mathbb{R}) / SL_n(\mathbb{Z})$ (or more precisely a lattice in $SL_n(\mathbb{Z})$ that is of finite index and torsion free). (We remark that for $\mathbb{Z}[i]$, Borel's results would have allowed the choice of $n > 32$.)

The proper setting for this work is the relation between cohomology of arithmetic groups and representation theory, but we will avoid a general discussion focusing on just the contribution of the trivial representation – which Borel (1974) showed was the whole story in a "stable range".

The result is that:

Theorem 3.7 *For $K < \mathbb{Q}$-rank$(\Gamma)/4$, $H^k(K \backslash G / \Gamma; \mathbb{R})$ is represented by differential forms on $K \backslash G$ that are right G-invariant.*

In particular, the lattice itself is irrelevant! (We will see that however, above

---

[33] It turns out that BO $\to$ BTop is an isomorphism on homotopy groups rationally (the injectivity of this map being Novikov's theorem on topological invariance of rational Pontrjagin classes, and the rational surjectivity following from the finiteness of the number of differential structures on the sphere).

this value of k, the cohomology group can indeed change with the choice of lattice Γ.)

Here's a way to think about this. Suppose L is a compact Lie group containing K; then, by the Hodge theorem, we can compute $H^*(K\backslash L)$ by means of harmonic forms, but by integrating with respect to L, and using the uniqueness of harmonic representatives, we can essentially identify the cohomology with the forms on $K\backslash L$ that are invariant under the action of L.

Now if G is a real semisimple group, with K its maximal compact, we denote by $G_{\mathbb{C}}$ its complexification, and by G' the maximal compact of $G_{\mathbb{C}}$. The Cartan decomposition for G' and $G_{\mathbb{C}}$ only differ by a multiplication by i. This implies that the G-invariant forms on $K\backslash G$ are essentially the same as the G'-invariant forms on $K\backslash G'$. We call $K\backslash G'$ the *compact dual* of $K\backslash G$.

For a uniform lattice, this copy of the cohomology of $K\backslash G'$ actually *embeds* in $H^k(K\backslash G/\Gamma; \mathbb{R})$.

For nonuniform lattices, this is not the case, and it is not easy to tell which of these cohomology are actually present in $H^*(K\backslash G/\Gamma)$ (e.g. the top class never survives). However, here Borel's theorem tells us that in the range mentioned above this *is* actually a complete description of the cohomology.

For $\mathrm{SL}_n(\mathbb{R})$, the complexification is $\mathrm{SL}_n(\mathbb{C})$, whose maximal compact is $\mathrm{SU}(n)$. Thus the compact dual is $\mathrm{SO}(n)\backslash \mathrm{SU}(n)$. Thus the cohomology is that of a product of spheres of dimensions $5, 9, 13, 17, \ldots$ The smallest dimension that is a sum of these and a multiple of 4 is 44, giving the result for $n > 176$.

For $\mathrm{SL}_n(\mathbb{C})$, thought of as a real Lie group, the complexification is $\mathrm{SL}_n(\mathbb{C}) \times \mathrm{SL}_n(\mathbb{C})$. Thus, the compact dual of $\mathrm{SU}(n)\backslash \mathrm{SL}_n(\mathbb{C})$ is $\mathrm{SU}(n)$ and therefore a product of spheres of dimension $3, 5, 7, 9, \ldots$ The first relevant cohomology is in dimension 8, so for $n > 32$ these produce examples.

This method shows failure of proper rigidity for $\mathrm{SL}_n(O_F)$ for $n > 32$ if F has a complex embedding, and $n > 176$ when F is totally real. These counterexamples are "stable" in at least two senses: (1) they do not go away if we stabilize the manifold by taking products with Euclidean space, $\mathbb{R}^k$; and (2) they survive on passing to any further finite cover.

However, this method is insensitive to the lattice in SL_n, and for example, this cannot lead to the idea that as the volume of the symmetric space goes up, so does the size of this set of manifolds, which actually seems to be the typical behavior.

More precisely, we will soon see that there is a finitely generated abelian group structure on this set of topological manifolds, and that (via a nonlinear map related to the Pontrjagin classes but distinct from it) it is $\cong \oplus H^{4i}(\Gamma; \mathbb{Q})$

after $\otimes \mathbb{Q}$.[34] We shall see that frequently the rank of this abelian group (even rationalized) grows with Γ.

However, the impatient reader who wants to move on to matters more directly concerned with the *validity* of rigidity can now skip to the end of this chapter or to the next (with occasional references to the skipped sections, especially about Property (T)).

3.4 Strong Approximation

Our first order of business is to give a fairly straightforward argument that, in the case of $SL_n(O_F)$, $n > 4$, there is always a finite sheeted cover with a substantial amount of cohomology. In §3.7, we will use this to give an essentially elementary replacement for the work of Borel used in the previous section to disprove the proper Borel conjecture for $n > 4$. (The argument for $n = 4$ will not be quite as elementary and will require material from §3.6.) We will write down the argument in the case of $\mathbb{Z}$, but the arguments are completely general. Following this we will discuss strong approximation, which gives a good understanding of the quotients of quite general linear groups. Ultimately, this will imply that all $\mathbb{Q}$-rank > 2 lattices have finite covers that are not properly rigid.[35]

We begin by noting that $SL_n(\mathbb{Z}) \to SL_n(\mathbb{Z}_p)$ is a surjection. The kernel $SL_n(Z; p)$ consists of matrices of the form $(I + pA)$, where $A \in M_n(\mathbb{Z})$ is such that $(I + pA)$ is invertible. The key thing as noted by Lee and Szczarba (1976) is that this congruence kernel has a homomorphism $\to M_n(\mathbb{Z}_p)$, assigning A to $I + pA$. Note that $\det(I + pA) = \pm p^n p_A(-1/p)$ and hence we need that A have trace $0 \bmod p$. (Of course, this is the Lie algebra of G in general.)

Now we can write down explicitly a 3-cycle in the congruence subgroup that is p-torsion and detected by projection to this abelian p-group. It is a $\mathbb{Z}^3$ in $SL_5(\mathbb{Z})$. There is a $\mathbb{Z}^2$ which consists of matrices that are 1s on the diagonal and the top row is $(1, 0, 0, pa, pb)$. This commutes with the Heisenberg group (Heis) of upper diagonal matrices in $SL_3(\mathbb{Z}) \subset SL_3(\mathbb{Z}) \times SL_2(\mathbb{Z}) \subset SL_5(\mathbb{Z})$. We obtain a $\mathbb{Z}^3$ by taking the product of the $\mathbb{Z}^2$ with the central $p\mathbb{Z}$ in the level-p congruence subgroup of the Heisenberg group.

This $\mathbb{Z}^3$ gives us a cycle in $H^3(SL_n(\mathbb{Z}; p); \mathbb{Z})$ which is nontrivial, because it

---

[34] The smooth version maps to the topological one so that the map is finite-to-one, and the image need not be a subgroup, but it contains a lattice in this cohomology group by an argument we will give in §3.7.

[35] But it will not imply stability in the second sense of the previous section. Indeed we will see a rank-3 reducible lattice where every proper homotopy equivalence to any finite sheeted cover becomes properly homotopic to a homeomorphism in a further cover.

is detected by mapping to $M_n(\mathbb{Z}_p)$ (by the Künneth formula), but is p-torsion, because the central $\mathbb{Z}$ is of order p in $H^1(\text{Heis}_3(\mathbb{Z};p);\mathbb{Z})$ – i.e. the homology of the level-p congruence subgroup of the Heisenberg group – since the 3×3 matrix

$$\begin{pmatrix} 1 & 0 & p^2 \\ 0 & 1 & 0 \\ 0 & 0 & 1 \end{pmatrix}$$

is a commutator in this group. Consequently we have found *an element of order p in $H^4(\text{SL}_n(\mathbb{Z};p);\mathbb{Z})$* by the universal coefficient theorem.

We will see in §3.7 below that for p sufficiently large this element is the first Pontrjagin class of some manifold proper-homotopy equivalent to $\text{SO}(n)\backslash\text{SL}_n(\mathbb{R})/\text{SL}_n(\mathbb{Z};p)$. Actually, *these elementary calculations* with Lie algebras and playing with congruence subgroups suffice to show that for $\mathbb{Q}$-rank > 6 one can always find a congruence cover where there are arbitrarily large finite number of manifolds that can be distinguished by p_1 – the first Pontrjagin class.[36]

Reduction modulo primes for linear groups over fields of characteristic 0 is a very powerful method and produces many useful homomorphisms. This is, for instance, used to prove (see e.g. Wehrfritz, 1973) that such groups are residually finite (Malcev) and also virtually torsion-free (Selberg).

Let us describe some easy homomorphisms if $\Gamma \subset \text{GL}_n(F)$ is a finitely generated group over a field F of characteristic 0. Consider the generators of Γ as lying in a finitely generated ring over $\mathbb{Z}$. Its field of fractions is a finite (algebraic) extension of a field of finite transcendence degree. We can then "specialize" values for the transcendentals so that these matrices all lie in an algebraic extension (as the determinant will be a rational function that is not identically 0). Then the matrix entries really are algebraic numbers with finitely many primes in their denominators, and we can therefore reduce modulo large primes. However, for simplicity of exposition, we will imagine that our groups lie just over the integers, perhaps with finitely many denominators.

These congruence subgroups provide a natural sequence[37] of subgroups that converge to the trivial group. Amazingly, *the image of a linear group under such reductions is*, with finitely many exceptions, *governed by the Zariski closure of*

---

[36] As explained in §3.7, Novikov's theorem that rational Pontrjagin classes are topological invariants can be refined for p_1 to the statement that in $H^4(\text{BSTop}; \mathbb{Z}[1/2])$ it is definable for oriented topological bundles.

[37] Which corresponds to a tower of covering spaces if one chooses a sequence of moduli that divide one another. A different choice, which does not form a directed system but rather is just a sequence of covers, is the congruence kernels as one varies over different primes. Those still converge to the universal cover, for example, in the pointed Gromov–Hausdorff sense.

the group. (This is the content of the *strong approximation theorem*.) Thus, any Zariski-dense finitely generated subgroup of $SL_n(\mathbb{Q})$ surjects onto $PSL_n(\mathbb{Z}_p)$ for all but finitely many primes. Indeed, like in the Chinese remainder theorem, one can map onto almost any finite product $\times PSL_n(\mathbb{Z}_{p_i})$.

Slightly more precisely, let S be a finite set of primes. We consider $\mathbb{Z}[1/S]$ the ring of rational numbers whose denominators have all prime factors in S. Suppose that $\Gamma \subset GL_n(\mathbb{Z}[1/S])$ with Zariski closure G. Strong approximation asserts that the closure of Γ in $\prod G(\mathbb{Z}_p)$ is of finite index. Informally, strong approximation says that the closure of a linear group in the congruence topology is essentially determined by its closure in the Zariski topology.

A nice application of this is due to Lubotzky (1996). Recall that the start of the Gromov–Piatetski-Shapiro examples was the construction of a separating hypersurface in a hyperbolic manifold. Millson (1976) had noticed that on taking a finite cover, this hypersurface lifts to several components.

Actually this virtual disconnectedness is true in general, as the fundamental group of the hypersurface is not Zariski-dense in $O(n, 1)$ – it lies in a smaller $O(n - 1, 1)$ – and therefore not congruence dense. A suitable deep finite congruence cover will therefore have the hypersurface disconnected.

As each of the sides is Zariski-dense in the group, these both have full image, which means that the complement of the union of the lifts of the hypersurface have two components.

A corollary of Van Kampen's theorem and these observations directly gives:

Theorem 3.8 *Every hyperbolic manifold with a separating hyperbolic hypersurface has a finite index subgroup whose fundamental group surjects to a free group.*[38]

This then implies that such a lattice has *many* subgroups of finite index – indeed super-exponentially in the index (since nonabelian free groups do).

Another nice application of strong approximation, also due to Lubotzky (1987), is the following.

Theorem 3.9 *Any finitely generated group linear group in a field of characteristic* 0 *always has subgroups of index divisible by d (for any given d).*

We refer to Lubotzky and Segal (2003) for a more thorough discussion of strong approximation, its literature and applications.

---

[38] Explicitly, let M be a manifold containing two hypersurfaces A and B whose union does not separate M and $*$ be a base point of $A \cup B$. Then, making a curve transverse to $A \cup B$, one can write a product $aabba^{-1} \cdots \in F_2$ recording the order and directions of the intersections. This gives a (surjective) homomorphism $\pi_1(M) \to F_2$.

3.5 Property (T)

In this brief section we will discuss the notion of Property (T), discovered by Kazhdan during the 1966 Moscow ICM (during a game of ping-pong with Atiyah). While it seems at first like a technical property about unitary representations, it has had applications – surely not all foreseen at that point – to many areas of mathematics, and (via the notion of expander graph) theoretical computer science.

We shall also discuss the opposite notion, amenability, originally introduced by von Neumann in his analysis of the Banach–Tarski paradox. These are both fascinating subjects deserving (and having received) book-length treatments; here they are merely introduced in recognition of the role they will play several times in what follows.

We will begin on the amenable side of the universe, since it is more familiar. For finite groups G, averaging the values of a real-valued function on G is a general and straightforward algebraic procedure that involves no limiting procedures. If G is compact then, at least for continuous functions, this can be done by integration with respect to Haar measure.

Remarkably, using weak-$*$ limits it is possible to define averaging processes on some infinite groups. Even for $\mathbb{Z}$ this is a remarkable statement: we are asserting that there is a functional

$$A: L^\infty(\mathbb{Z}) \to \mathbb{R}$$

that assigns a number to any bounded sequence of real numbers, agrees with ordinary limit when it exists, and is positive, linear, and translation invariant. Positivity means that $A(f) \geq 0$ if $f \geq 0$. Linear is obvious and translation invariant means that A is invariant under the action of $\mathbb{Z}$ on $L^\infty(\mathbb{Z})$ by translation. Positivity and linearity can be achieved by extending any f (since $\mathbb{Z}$ is discrete, any function is continuous) to $\beta\mathbb{Z}$, the Stone–Čech compactification and evaluating this extension on any point in $\beta\mathbb{Z} - \mathbb{Z}$.

The invariance requires using a bit of the geometry of $\mathbb{Z}$, but this is the key! Replace the sequence by its averages (i.e. like Cesàro means). Let $g(n) = 1/(2|n + 1|) \sum f(m)$ (where the sum is over the interval $I_n = [-|n|, |n|]$.

Observation A, defined as the limit of the sequence $g(n)$, is translation invariant because the number of elements in the symmetric difference $I_n \triangle T I_n$ is $o(\#I_n)$.

Remark We made the construction using the Stone–Čech compactification. Sometimes (as hinted above) people construct A as a weak-$*$ limit of the averaging functionals that define the values of g; sometimes non-principal ultrafilters are used in making this construction. These are just cosmetic differences – although they have somewhat different feels (point-set topology versus functional analysis versus logic).

Note the averaging procedure (and the limiting procedure) is well defined when the sequence has a limit. However, in general, it is very dependent on our choices. For example, suppose we had replaced the intervals $I_n = [-|n|, |n|]$ by intervals $J_n = [n - |n|, n + |n|]$; we still would obtain an averaging function that satisfies all the above properties, yet would have a much less democratic[39] feel than the I_n seem to have – the values of f at most integers (e.g. those outside of union of the J_n) will then be completely irrelevant.

Democracy put aside, the above consideration suggests defining a *Folner sequence*[40] to be a sequence of subsets A_n of Γ, so that for any γ, $\#(\gamma A_n \triangle A_n)/\#A_n \to 0$. (This need only be checked for generators.) Under those conditions we can define a left-invariant positive linear functional by the procedure above. Folner (1956) proved the converse, that a group has a mean iff there is a sequence of such sets. Groups that have such a mean, or equivalently, an exhaustion[41] by subsets whose "boundaries" are asymptotically negligible, are called *amenable*.

(The boundary of a set in Γ is precisely the the union symmetric difference of the set with its translates under a generating set of Γ. If we consider the volume of a set the number of elements it contains, then the last sentence is just a restatement in words of the formula of the previous one.)

There is a close connection between amenability and unitary representation theory. Consider the unitary action of Γ on $L^2\Gamma$. It has a nontrivial fixed vector iff Γ is finite.

However, $v_n = (1/\sqrt{\#A_n})\sum \gamma$ where the sum is taken over A_n is a sequence of *almost-invariant vectors*. That is, $\|v_n\| = 1$ but for every γ, $\|\gamma v_n - v_n\| \to 0$. One can describe this as saying that the trivial representation is weakly contained in the regular representation – another equivalent of amenability.

Yet another interpretation of amenability can be given in terms of the Laplacian on functions $\nabla: L^2\Gamma \to L^2\Gamma$ defined as follows. We shall consider Γ as a graph, as usual, choosing a finite symmetric generating set S, and connecting two elements g and g' if there is an $s \in S$ such that $g = sg'$ (so that Γ acts on the right by isometries). Define the Laplacian by $\nabla f(x) = f(x) - (1/\#S \sum f(sx))$.

---

[39] And more fickle, in that J_n is disjoint from the later sets averaged over.

[40] These considerations do not explain why we would give *this* name to this class of subsets, only that we call attention to them. The last sentence in the paragraph is necessary for that point.

[41] It is a very elementary fact that if a discrete metric space X has a Folner sequence of subsets, then it has an exhaustion by Folner sets B_i; i.e., $B_i \subset B_{i+1}$ and $X = \bigcup B_i$.

It compares f to its average. Note that ∇ is a (bounded) self-adjoint and positive operator (by direct calculation of $\langle \nabla f, f \rangle$).

Theorem 3.10 (Kesten, 1959) $0 \in \mathrm{Spec}(\nabla)$ *iff* Γ *is amenable. This is equivalent to each of the following two statements:*

(1) *The symmetric random walk on Γ does not have exponentially decaying return probabilities, i.e. $p_{2n}(e,e) \neq O(c^n)$ for any $c < 1$, where e is the identity element of the group.*

(2) *The number of words (in the symmetric set of generators S) of length at most $2n$ representing the trivial element $W(n)$ satisfies $W(n)^{1/2n} \to \#S$.*

Note that the statement $0 \in \mathrm{Spec}(\nabla)$ does not mean that there are any eigenvectors with eigenvalue 0 (although that would be the simplest explanation), i.e. ker ∇ need not be nontrivial, because of the possibility of a nondiscrete spectrum. Indeed, 0 is an eigenvalue[42] iff Γ is finite.

However, the almost-invariant vectors are test functions of norm 1 with $|\nabla f_n| \leq \sum \#(\gamma A_n \nabla A_n)/\#A_n$ (summed over the elements of S) showing that it is not true that $\langle \nabla f, f \rangle > c\|f\|^2$ for any $c > 0$.

The connection between random walk, heat flow, and the Laplacian is important. Note that $\nabla = I - M$, where M is the Markov operator, defined by

$$\mathrm{M} f(x) = \mathrm{E}(f(\gamma x)),$$

where E means, as always, the expectation value of a random variable, and here it is f of a random neighbor of x (i.e. the translate by a random generator of Γ). Note $\|\,\mathrm{M}\,\| \leq 1$, and equality holds iff Γ is amenable. The probability of return is given by

$$p_n(e,e) = \langle \delta_e, \mathrm{M}^n \delta_e \rangle.$$

So if $0 \notin \mathrm{Spec}(\nabla)$, we get exponential decay of the return probabilities. (The converse is tricky.) The expression $W(n)/\#S^{2n}$ is simply another calculation of $p_{2n}(e,e)$ and hence statement (2) is equivalent to (1).

Property (T) is *opposite* to amenability (not its negation!) and it is quite nontrivial that there are any infinite groups at all that have this property.

Definition 3.11   A group Γ has Property (T) if *every* unitary representation that has almost-invariant vectors has a fixed vector. (In other words, given a generating set S, there is a Kazhdan constant ε – that typically depends on S – such that, for any nontrivial irreducible representation ρ (or, equivalently, any

---

[42] There is a natural generalization of ∇ to differential forms, and then as we will discuss in §3.6, ∇ frequently has nontrivial kernel acting on L^2-forms.

representation with no nontrivial fixed vectors ρ), the only v with $||\rho(s)v - v|| \leq \varepsilon||v||$ is $v = 0$. [43]

An amenable discrete group has Property (T) iff it is finite – one can construct almost-invariant vectors by averaging over a sequence of Folner sets.

Margulis showed that higher-rank lattices have only finite or finite index normal subgroups by the crazy strategy of showing that all quotients are amenable and have Property (T). Obviously, arbitrary quotients of Property (T) groups have Property (T).

Kazhdan observed, in his original 1967 paper, via consideration of induced representations, the following.

Proposition 3.12 *A locally compact group G has Property (T) iff any (and hence every*[44]*) lattice* $\Gamma \subset G$ *does.*

He also showed

Proposition 3.13 *A discrete group with Property (T) must be finitely generated.*

For suppose that $\Gamma = \bigcup \Gamma_n$ is an ascending union of proper subgroups. Then $\bigoplus L^2(\Gamma/\Gamma_n)$ is a unitary representation which has almost-invariant vectors (each γ ultimately acts trivially, so a sequence of vectors that are nontrivial only in the components indexed by a large n form an almost-invariant sequence of vectors), but it will have an invariant vector only if some $\Gamma_j = \Gamma$.

Theorem 3.14 (Kazhdan) *Products of real simple Lie groups of rank greater than 1 have Property (T).*

He deduced that lattices in these groups were finitely generated.

We already know enough to see that $O(n, 1)$ does not have Property (T), because we know lattices that have nontrivial $\mathbb{Z}$ quotients, and note that Property (T) is (obviously!) inherited by quotients. Less simple is that $U(n, 1)$ also does not have Property (T). This is shown in Kostant (1975), as is the following positive result.

Theorem 3.15 (Kostant) $Sp(n, 1)$ *has* Property (T)*, as does the real rank* 1-*form* $F_{4(-20)}$ *of the exceptional complex Lie group of type* F_4.

This gives us now negatively curved examples of Property (T) groups. We

---

[43] The notation is supposed to indicate that the trivial representation T is separated from all the other irreducible representations (by the parentheses).

[44] Assuming there is at least one!

can add large powers of all the elements one at a time,[45] and maintain negative curvature, giving (uncountably many![46]) Property (T) groups that are torsion.

The early history of Property (T) only had examples that came out of representation theory. Now there are completely different mechanisms for this of both algebraic and analytic geometric origin – so now there are many other Property (T) groups known. Before saying a little more about this, we digress to give another characterization of Property (T) (see Shalom, 2000; Bekka *et al.*, 2008).

Theorem 3.16 (Delorme–Guichardet, Shalom) *A group has* Property (T) *iff every action of* Γ *on a Hilbert space by affine isometries has a fixed point. If the group does not have* (T) *then there is an action where not only is there no fixed point, but the displacement* $\sum ||v - \gamma(v)||^2$ *has a realized minimum on the unit sphere (where* $\sum$ *is over the generating set).*

All amenable groups have affine isometric actions that are metrically proper, i.e. actions for which the orbits of vectors $\to \infty$ in norm (as $\gamma \to \infty$) as was shown by Bekka, Cherix, and Valette – yet another way in which Property (T) and amenable are at opposite poles.

A consequence of this theorem is that:

Corollary 3.17 *If a group* Γ *acts simplicially on a tree (without inversions) without fixing any vertex, then* Γ *cannot have* Property (T).

This excludes nontrivial amalgamated free products and HNN extensions, as well as giving another argument for the finite generation of Property (T) groups (see Serre, 2003). We shall prove the corollary by noting that, if Γ acts on a tree T, then it acts on $L^2(T)$.

Proposition 3.18 (Cartan) *If* Γ *acts on a tree* T *and it has a bounded orbit, then it has a fixed point.*

Cartan was actually working on other spaces of nonpositive curvature.[47] The proof goes like this. Given a bounded set in a tree, it lies in a *unique* ball of smallest radius. As this the bounded set is Γ-invariant, so is that ball, and therefore its center is fixed.

If the action of Γ on T has no bounded orbit, then $L^2(T)$ has no fixed vectors, which is incompatible with Property (T).

---

[45] This is an application of the "Dehn filling" idea as in the previous chapter.

[46] And hence the fact that Property (T) does not force finite presentability.

[47] I believe that Cartan's application was the uniqueness of the maximal compact in a semisimple group by considering the action of such a group on G/K, a complete manifold of nonpositive curvature. Incidentally, the analogous fact in the case of Lie groups over local fields makes use of the curvature properties of Tits buildings.

Appendix: Property (T) and Expanders

Expander graphs are graphs that are hard to disconnect, i.e. require the removal of many edges to separate a large number of vertices from the rest. It (now) seems obvious that such graphs should be valuable for the construction of things like communication networks. But, in fact, they have legion applications in theoretical computer science (Hoory *et al.*, 2006) and pure mathematics (Lubotzky, 1984, 2012).

We consider finite d-regular graphs Γ_i (for simplicity – a bound on valence is really all that's necessary). We consider the Cheeger constant of these graphs

$$h(\Gamma) = \inf(\#\partial A / \#A),$$

where A is a subset of Γ with fewer than half of the vertices, and ∂A is the set of vertices of A that share an edge with $\Gamma - A$. If we allowed A to be big then setting $A = \Gamma$ we'd always get 0 as our infimum.

This notion makes sense for infinite graphs, as well as finite ones, if we *impose* the condition that A is finite in the infinite case. Note that Γ is amenable as a group iff $h(\Gamma) = 0$ viewing Γ as a (Cayley) graph – and that this condition is equivalent to $0 \in \mathrm{Spec}(\nabla)$.

However, for expansion, we are interested in *finite* graphs, and we want the reverse, i.e. that $h(\Gamma_i) > \varepsilon > 0$.

To summarize:

Definition 3.19   An expander sequence of d-regular graphs is a sequence Γ_i (of d-regular graphs) such that $h(\Gamma_i) > \varepsilon > 0$.

These were first introduced and studied *explicitly* by M. Pinsker (in Bassalygo and Pinsker, 1973) – and in a paper presented at the 7th International Teletraffic Conference. He showed that they exist, by arguing that random graphs are expanders. They have been an important tool in theoretical computer science ever since, and you can find much interesting material and history in Hoory *et al.* (2006).

More recently, it was pointed out in Gromov and Guth (2012) that Pinsker was preceded by a paper of Kolmogorov and Barzdin that studied expanders as models for the brain (nodes on the surface and axons going through the bulk, without disjoint axons getting too close to one another), but then, alas, having an upper bound on size[48] to fit into our heads. Expanders were their examples of graphs that would be hard to fit in our heads.

Why this genericity of expansion should be true is clear if one considers a toy variant. Consider the graph Γ with n vertices determined by two permutations, using each permutation to connect $[i]$ to $[\pi i]$ (note that $[i]$ is also connected to $[\pi^{-1}i]$). Given a subset A, then the expected number of edges leaving A is $\#A(1 - \#A/\#\Gamma)^4$ suggesting a bound of at most 1/16 independently of $\#\Gamma$. Of course, there are many choices of A, and we have to compute the expected extremal. This means one should look at subsets A of size $n/2$ that contain significantly fewer edges leaving them, say $n/20$, and then estimating tail probabilities in a binomial distribution. The details are left to the reader.[49]

If one is interested in using this for building a network (or an error-correcting code or . . .), then random methods are not so useful – buildings surely must be built from blueprints.[50] The applications in mathematics often require knowing that certain graphs form expander sequences.[51]

Now, for finite graphs, 0 is always in the spectrum of ∇. Constant functions have $\nabla f = 0$. And 0 has multiplicity greater than 1 iff Γ is disconnected (different constant functions on the different components). Graphs that are connected but easily disconnected should therefore be characterized as having an eigenvalue near 0. This is the content of the following basic theorem.

Theorem 3.20 *A sequence of d-regular graphs is an expander sequence iff there is an $\varepsilon > 0$ so that the spectrum of ∇ restricted to functions with $\int f = 0$ (the orthogonal complement of the constants) is bounded $> \varepsilon > 0$.*

We will denote by $L^2(\Gamma)^\circ$ this subspace of $L^2(\Gamma)$.

This theorem is inspired by Cheeger's theorem in Riemannian geometry (see Cheeger, 1970) that bounds the isoperimetric constant of a Riemannian manifold in terms of the spectrum of the Laplacian. Note that for a subset A, the modified characteristic function, $f_A = 1_A - \#A/\#\Gamma$ has $\int = 0$, and ∇ related to $\#\partial A/\#A$. The isoperimetric constant is approximately realized by a level set of an eigenfunction for a small eigenvalue.

---

[48] There is a bound to how much of an expander can be fit without distortion, even in Hilbert space. This will be of critical importance later for purposes of the Novikov conjecture. For science fiction purposes, the cognitive capacities of aliens elsewhere in the multiverse can be expected to be greater than ours, in the Kolmogorov–Barzdin model, only if the number of spatial dimensions increases (or they have better programming of their neural nets).

[49] Actually, to the active reader. An inactive reader can find them written down in many places.

[50] I expect this to be my *bon mot* quoted years after I have otherwise been forgotten, showing how shortsighted people were back at the beginning of the third millennium. Indeed, I almost deleted this comment during revision.

[51] Many of these are closer to the Selberg example explained below than the Property (T) examples we begin with now. This is a good moment to mention that there are now many constructive methods of getting expanders that do *not* come out of Property (T).

A consequence of this theorem is that a random walk on an expander sequence is rapidly mixing.[52]

The following important result of Margulis (1988) is now perhaps anticlimactic, given our discussion.

Theorem 3.21 *Suppose that* Γ *is a group with* Property (T). *Then the Cayley graphs of* Γ/Γ_i, *for a sequence of normal subgroups of finite index in* Γ *(using a common generating set S coming from* Γ*), gives a sequence of expander graphs.*

To see why the isoperimetric inequality is true, consider $\oplus L^2(\Gamma/\Gamma_i)^\circ$ (where the superscript $^\circ$ means that we are considering the orthogonal complement of the constant vectors) and, since there are no fixed vectors, there can be no almost invariant vectors, which means that ∇f_{A_i} is large, which means that ∂A_i is also large.

Concretely we can set $\Gamma = SL_n(\mathbb{Z})$ for any $n > 2$ (and use the elementary matrices as a generating set) and obtain the expander sequence $SL_n(\mathbb{Z}/m)$ – where m is varying.

Note, by the way, that the representations arising in this proof are all (sums of) finite-dimensional representations of the group Γ, so we are nowhere near the full power of Property (T). Lubotzkyc and Zimmer have suggested the notion of Property τ, which is Property (T) for finite-dimensional representations, or even restricting further to a class of finite quotients (say ones factoring through some finite quotient or some congruence quotient).

A good example of this is $SL_n(\mathbb{Z})$ for $n = 2$. We shall work with a congruence subgroup of this group, which is a free group. Obviously, it does not have Property (T), as it has a $\mathbb{Z}$ quotient, and just as obviously covers corresponding to the subgroups $k\mathbb{Z}$, for a surjection of this group to $\mathbb{Z}$, have isoperimetric constant $\to 0$ (consider the inverse image of the interval $[0, k/2]$ in the cycle) and, again just as obviously, the bottom of the spectrum of these quotients of $SO(2)\backslash SL_2(\mathbb{R}) \to 0$ (by considering functions that are 1 on $[0, k/2]$ and -1 on $[k/2, k-1]$).

However, when we restrict our attention to the family of *congruence quotients*, then a theorem of Selberg asserts that, for all of these manifolds, $SO(2)\backslash SL_2(\mathbb{R})/SL_2(\mathbb{Z}; k)$ has $\lambda_1 > 3/16$. One can translate between graphs and manifolds, and actually this is a family of expander graphs whose girth[53] grows[54] (logarithmically) with k.

---

[52] Which perhaps suggests its application to de-randomization.

[53] The girth of a graph is the length of the shortest cycle in the graph; it is an analog of the length of the shortest geodesic (= twice the injectivity radius) of a compact manifold.

[54] Note that if we use the Property (T) expanders, relations in the fundamental group give bounded cycles everywhere in the graph. Random graphs will frequently have some short cycles, but relatively few of them.

Finally, we close our discussion by mentioning one of the more recent methods for proving Property (T), because it turns our discussion on its head and uses expander properties as a way of obtaining Property (T).

Theorem 3.22   *Let Γ be a group generated by a finite symmetric set S, with $e \notin S$. Let $L(S)$ be the graph with vertex set S and in which $\{s, s'\}$ is an edge if $s^{-1}s' \in S$. Suppose that $L(S)$ is connected and has spectral gap greater than $1/2$. Then Γ has Property (T).*

As a nontrivial consequence of this, in some models of random groups, having Property (T) is generically the case – a far cry from the essentially Lie-theoretic origin of the first examples. Moreover, this method produces groups with very strong fixed-point properties, often stronger than those true for lattices in high-rank groups. See the notes in §3.9 for some more discussion of this important direction.

3.6 Cohomology of Lattices

The cohomology of lattices is a topic of endless fascination that can be studied from many viewpoints, from the geometric[55] (construction of explicit cycles) to the analytic (e.g. Hodge theory and L^2-cohomology) to the number-theoretic (such as Langlands functoriality). In this section we will touch briefly on a few methods for producing cohomology classes motivated by purely utilitarian needs. For simplicity, we will divide our discussion into four parts:

(1) Property (T) and H^1;
(2) Matsushima formula and connection to representation theory;
(3) generalized modular symbols and geometric cycles;
(4) L^2-cohomology.

3.6.1 H^1 and Property (T)

We have already tacitly discussed $H^1(\Gamma; \mathbb{R})$ when discussing Property (T). Its vanishing is necessary if Γ satisfies (T), because otherwise $\mathbb{Z}$ is a quotient of Γ, and (T) is inherited by quotients.

Actually we had, less obviously, given a cohomological interpretation of

---

[55] Not to mention the heroic geometric group-theoretic work of Agol (2013), Haglund and Wise (2007, 2012), and Kahn and Markovic (2012) that gives positive first Betti number (and even more, homomorphisms to $\mathbb{Z}$ with finitely generated kernels) for finite covers of lattices in $O(3, 1)$. See the wonderful exposition by Bestvina (2014).

Property (T) in characterizing those groups by the fixed-point property: any action of Γ on a Hilbert space H by affine isometries has a fixed point.

This statement can be expressed cohomologically. Any affine action has a unitary part $\rho\colon \Gamma \to \mathrm{U}(H)$. (It can be obtained by letting $\rho(\gamma)(v) = \lim t\gamma(t^{-1}v)$ as $t \to 0$.) Affine actions are associated to cocycles, and cohomologically trivial ones are the ones with fixed points (i.e., are actually unitary after conjugating by a suitable translation).

Thus, the Delorme–Guichardet fixed-point theorem can be viewed as the cohomological statement that:

Theorem 3.23 Γ *is a group with* Property (T) *iff, for any unitary representation* ρ *of* Γ, $H^1(\Gamma; \rho) = 0$.[56]

The reason is this. The 1-cochains with values in the representation, $C^1(\Gamma; \rho)$ is made of H-valued functions on Γ, and an element $\alpha \in C^1(\Gamma; \rho)$ lying in $\ker d\colon C^1(\Gamma; \rho) \to C^2(\Gamma; \rho)$ means that $\alpha(\gamma\gamma') = \rho(\gamma)\alpha(\gamma') + \alpha(\gamma)$. Associated to a cocycle is the affine isometric action on H where γ acts by $\gamma v = \rho(\Gamma)v + \alpha(\gamma)$. This cocycle is a coboundary of a vector $v \in H = C^0(\gamma; \rho)$ if $\alpha(\gamma) = v - \rho(\gamma)v$. Then $\gamma v = v$ for every γ and the action has a fixed point (and one can conjugate the action by a translation to a unitary action).

Part of the interest in such statements is because of their connection to deformations. The infinitesimal version of rigidity asks about deformations of the defining representation $\rho\colon \Gamma \to G$. Reasoning about deforming the defining representations and working modulo the deformations given by inner automorphisms leads one to want to prove vanishing of such cohomology groups.

Kazhdan's approach to Property (T) gives a representation-theoretic method, but other cohomological vanishing theorems have been proved by Hodge-theoretic methods or Bochner arguments. These methods were employed by Calabi and Vesentini (1960), Calabi (1961), Weil (1960, 1962, 1964), and Selberg (1960, 1965) to prove early local rigidity theorems. They still are useful – as rigidity moves into new settings (such as for non-lattices, and fixed-point properties for actions on spaces other than Hilbert spaces).

Another consequence of rigidity of representations is that the defining representations of such a group cannot have "essential" matrix coefficients that are transcendentals, because transcendentals can always be deformed (or specialized). (By "essential" I am ignoring the possibility of conjugacy of an algebraic

---

[56] The Shalom improvement we had mentioned above replaces this cohomology by its reduced version, where one mods $\ker \partial$ by the *closure* of $\mathrm{Im}\,\partial$. Often reduced and unreduced groups are different, and the reduced ones are easier to study, but it sometimes happens that they vanish simultaneously (at least in low dimensions) – see also Block and Weinberger (1992), where a similar phenomenon occurs in a characterization of amenability.

matrix by a transcendental one.) To grossly simplify, this is why superrigidity (a vast generalization of Kazhdan's Property (T) for Γ) leads to arithmeticity theorems.[57]

It is worth noting that an immediate consequence of the theorem as stated is that all finite-dimensional irreducible representations are separated from unitary representations that don't contain them. In addition, although this is obvious in any case, as cohomology with coefficients in representations includes cohomology of covers, philosophically this study naturally leads us to consider the behavior in towers simultaneously with the cohomology of a given space, a theme we will return to in §3.6.4.

3.6.2 Matsushima Formula

The yoke binding representation-theoretic theory and cohomology is tightened by the Matsushima formula that extends the earlier observations connecting the cohomology of the compact dual to that of all locally symmetric manifolds with a given universal cover.

Unlike those previous observations, it has the virtue of being sensitive to the lattice. We will not directly make use of this material, but an awareness of it will make some discussions make more sense (or seem better motivated[58]).

The discussion is much simpler in the case of cocompact lattices, so we start by making this assumption.

The complex of differential forms on $K\backslash G/\Gamma$ can be identified with the cochain complex $C^*(\mathfrak{G}, \mathfrak{K}; C^\infty(G/\Gamma))$, where $\mathfrak{G}, \mathfrak{K}$ are the Lie algebras of G and K, respectively, and we use the Chevalley–Eilenberg complex for relative Lie algebra cohomology. Thus $H^*(K\backslash G/\Gamma) \cong H^*(\mathfrak{G}, \mathfrak{K}; C^\infty(G/\Gamma))$.

It turns out (this is a kind of elliptic regularity result) that we can break $C^\infty(G/\Gamma)$ into pieces according to the decomposition of $L^2(G/\Gamma)$. This is a sum of pieces that are G-invariant irreducible representations, with finite multiplicity. Ultimately one gets a formula of the form

$$H^*(K\backslash G/\Gamma) \cong \bigoplus m(\Pi, \Gamma) H_C^*(\mathfrak{G}, C^\infty \Pi),$$

where the right-hand side is a sum over irreducible representations of G, with multiplicities according to the number of times that they appear in $L^2(G/\Gamma)$ and the cohomological term (which only involves G and its representations, and not the lattice) being continuous cohomology with coefficients in the smooth vectors in the given representation. When G is simple, the cohomological term vanishes whenever $* < \mathrm{rank}_\mathbb{R}(G)$ and there are no terms other than the compact

---

[57] And, for example, Property (T) itself implies that all finite-dimensional unitary representations are equivalent to ones defined over an algebraic number fields.

[58] Or less unmotivated.

dual (which is the contribution of the trivial representation). (In the semisimple case, this vanishing holds below the lowest $\mathbb{R}$-rank of any of the factors.) These facts are responsible for the independence of rational cohomology in the stable range of the lattice – at least in the uniform case.

The place where the lattice enters is in the nontrivial representations because of the multiplicities $m(\Pi, \Gamma)$. These will frequently grow as Γ shrinks (note that if Γ' is a normal subgroup of Γ, the finite group Γ/Γ' acts on any of these Πs, and since these representations don't have a trivial part, the multiplicities must be nontrivial). A geometric approach to this is the following. If Γ is arithmetic, then it has non-normal subgroups that have a large number of symmetries (i.e., that do not cover the original manifold).[59] When one pulls a harmonic form up to such a cover, it can well be non-invariant under this action – causing the amount of cohomology to grow. If this would never happen, it would mean that the pullback to the universal cover would be invariant under $G(\mathbb{Q})$, which is exactly equivalent to it coming from the compact dual.

When Γ is nonuniform, then the above analysis of cohomology does not work directly, but Borel (1974) showed that nevertheless there is a range depending on the $\mathbb{Q}$-rank where it does hold. This is enough for the applications to SL_n when we let $n \to \infty$, (which is important for K-theory), but this is not enough for our immediate needs. Some highly unstable classes in the nonuniform case that are always beyond the range of this isomorphism are the topic of the next subsection.

3.6.3 Generalized Modular symbols

A different and transparent example of how cohomology grows in covers that is visible in hyperbolic geometry occurs for nonuniform lattices (in all dimensions).

If M is a noncompact finite volume hyperbolic n-manifold, then $cd(\pi_1 M) = n - 1$ (because M has the homotopy type of an $(n-1)$-dimensional complex, and it contains a $\mathbb{Z}^{n-1}$ in the fundamental group of the cusp).[60] It can certainly happen though that $H_{n-1}(M) = 0$ (e.g. this is true for all the hyperbolic knot

---

[59] This is related to the large *commensurator* of an arithmetic group. $G(\mathbb{Q})$ acts on the disjoint union of the $K \backslash G/\Gamma'$ where Γ' is commensurable with Γ, but each of these individual manifolds is only acted on by the normalizer of their own *fundamental group in* $G(\mathbb{Q})$. If Γ is arithmetic and Γ' is a $G(\mathbb{Q})$-conjugate of Γ (but not necessarily in the normalizer of Γ), we can take a subgroup Γ'' of finite index in $\Gamma \cap \Gamma'$ that is normal in Γ' but not in Γ. The group Γ'/Γ'' acts on the Γ'' cover, which is a cover of $K \backslash G/\Gamma$, but the action does not cover the projection to $K \backslash G/\Gamma$. These *hidden symmetries* are responsible for the algebra of Hecke operators that acts on cohomology groups of arithmetic manifolds.

[60] Of course, this is, in the arithmetic case, a special case of the result of Borel and Serre that the cohomological dimension differs from $\dim(G/K)$ by the $\mathbb{Q}$-rank.

complements in the 3-sphere). However, the fundamental group of the cusp is a proper "small" subgroup of the fundamental group, i.e. it is not Zariski-dense – it obviously lies in a proper parabolic, so by strong approximation we can find finite congruence quotients of $\pi_1 M$ onto which the cusp maps to a proper subgroup.

This means that these covers have multiple cusps (by covering space theory). Once you have more than one cusp, then $H_{n-1}(M) \neq 0$, because each cusp gives a cycle,[61] and the one relation among these is that the sum of all of these cycles vanishs. Associated to a pair of cusps there is a (number of) proper geodesic(s) lines going from one cusp to the other. These will have intersection number 1 and -1 on these two cusps (depending on ordering, and using a standard, say inward normal, convention for orientation of boundaries) and 0 with the other cusps. Each such proper geodesic gives a functional on homology which proves the nonvanishing of the individual cusps. (In fact, picking one cusp as a base, the lines connecting that cusp to all the others give #(cusps – 1) independent cycles.) As we go deeper in the group (or up a tower), the number of cusps increases and hence the size of the homology.

Of course, when the $\mathbb{Q}$-rank > 1, then this doesn't make sense as stated: the Borel–Serre boundary is connected in all covers, and $\pi_1^\infty \to \pi_1$ surjects. However, when we pay closer attention to the corners within the Borel–Serre boundary, which correspond to proper parabolic subgroups, none of these surjects, and the covers do indeed cause *these corners* to become multiple components, and then give rise to cycles.

Theorem 3.24 (See Ash and Borel, 1990; Schwermer, 2010) *Let G be an algebraic group defined over $\mathbb{Q}$, and let P be a $\mathbb{Q}$-parabolic subgroup of G. If $P(\mathbb{R}) = M(\mathbb{R})A(\mathbb{R})N(\mathbb{R})$ is the Langlands decomposition of this parabolic,[62] then there are nontrivial cycles in $K \backslash G / \Gamma$ of the form $N(\mathbb{R})/N(\mathbb{R}) \cap \Gamma$ in dimension $\dim(N(\mathbb{R}))$ if Γ is sufficiently deep. Passing to a congruence subgroup Γ' then there are at least #($\Gamma' \backslash \Gamma/(\Gamma' \cap P)$) (double cosets) linearly independent cycles obtained this way.*

Using congruence subgroups we then get a large (i.e. growing like a positive power of the volume, but definitely sublinear in it) rank of Betti number.

Remark 3.25 (In place of proof) Generalized modular symbols are examples of *geometric cycles*. Geometric cycles are associated to Lie subgroups of G, and give rise to some explicit cycles, when the lattice intersects them in a lattice.

---

[61] To get a well-defined cycle, one should adopt an orientation convention, i.e. making use of the normal direction pointing towards ∞.

[62] So that M is reductive, A is abelian and N is nilpotent.

To get an embedding, one often has to pass to a finite cover, and then when one passes to deep enough covers, they will (by strong approximation) typically produce a number of disjoint cycles.

The standard way to check that these cycles are nontrivial is to find another geometric cycle of the dual dimension that intersects it with nonzero intersection number. In the above theorem, Levi subgroups are the source of duals.

As in the case of modular symbols, pulling these up covers can give growth to the Betti numbers.

We did this with H^1 and the Millson example that uses codimension-1 geometric cycles in arithmetic hyperbolic manifolds associated to quadratic forms, and, following Lubotzky,localization observed that this even gave maps onto free groups. In this case this implies that there is then a further tower of covers (*not* converging to the trivial group) for which b_1 grows linearly with the index.[63]

3.6.4 L^2-cohomology

None of the methods discussed till this point has the potential of giving Betti numbers that grow linearly with volume (or, equivalently, with the index of the cover). However, the Euler characteristic tells us that this must happen *sometimes*. If $\chi(K \backslash G / \Gamma) \neq 0$, then, by multiplicativity of χ in finite covers, as one goes up any family of covers, some Betti number must increase linearly.[64]

In this section we will review the relevant facts about L^2-cohomology and especially a remarkable theorem of Lück that explains exactly when this rare situation occurs in towers of regular covers.[65]

This story begins without any particular interest in finite sheeted covers, but rather with the consideration of arbitrary regular covers.[66] For infinite complexes, there are alternatives to the usual simplicial chain complex: one can consider, for example, locally finite chains, which gives rise to Borel–Moore homology. This gives a non-homotopy-invariant homology theory: it is invariant under proper homotopy equivalences.

A more subtle choice is to consider the complex of L^2 simplicial chains[67] (or

---

[63] This family of covers shows very different geometry than that associated with congruence covers.

[64] Clearly, no Betti number for covers of a finite complex can grow faster than linearly, since these are bounded by the number of cells, which grows exactly linearly in the number of sheets of the cover.

[65] It actually also applies to sequences of regular covers that Gromov–Hausdorff converge to the universal cover. (See the notes in §4.11 for a recollection of Gromov–Hausdorff space.)

[66] Indeed, it can be developed in terms of arbitrary group actions.

[67] We are tacitly weighting all simplices equally in our discussion.

cochains). If the complex is locally finite (as it will be in all of our applications), then the ∂ map is a bounded map. Its homology is an invariant of X. It is functorial with respect to maps that are Lipschitz and "uniformly proper," i.e. if one has a bound on the size of the inverse image of simplices (or else the pushforward of an L^2 chain need not be L^2).

It is perhaps worthwhile to consider the case of $\mathbb{R}$. The chain complex is then identified with $0 \to L^2(\mathbb{Z}) \to L^2(\mathbb{Z}) \to 0$, where the boundary map sends $f \to (t-1)f$, where t is a generator of $\mathbb{Z}$. Obviously $H_1 = 0$, but H_0 is a large infinite-dimensional space (for example δ_0 is not in the image) but it doesn't seem to have much structure to say anything about.

There are two parts to the solution of this problem. The first is basic. We considered L^2 to enable the use of Hilbert-space methods, in which case we should insist that the constructed homology groups be Hilbert spaces. The way to achieve this is to insist that we never quotient out by non-closed subspaces, i.e. to take the closure of the image of ∂ when forming the homology groups. We will denote this version, i.e. where we take the quotient by closures, by $\mathcal{H}$.

(An equivalent alternative to using closures is to form a Laplacian from the chain complex in the usual formal way following Hodge, and define homology to be the kernel of the Laplacian. The "torsion" (closure $\operatorname{Im} \partial$)/$\operatorname{Im} \partial$ thrown away by this method corresponds to spectrum of ∇ near 0 that does not consist of harmonic forms.

The second part is to note that, following Atiyah (1974), when we are dealing with a universal cover,[68] the action of Γ on these Hilbert spaces is appropriate for defining a normalized dimension (using the theory of von Neumann algebras) that can be (in principle) an arbitrary nonnegative real number.[69] This will then define $b_i^{(2)}(X)$ (we suppress the Γ from our notation, unless needed) – the L^2-Betti numbers of X:

$$b_i^{(2)}(X) = \dim_\Gamma \underline{\mathcal{H}_i^{(2)}}(X).$$

We proceed informally. The idea shall be that we want to see what fraction of the regular representation some other unitary Γ representation is. We restrict attention to unitary representations that are closed subrepresentations of some multiple of the regular representation, as ours naturally are (viewing the quotient as the orthogonal complement to the image of the boundary).

We want $\dim_\Gamma L^2(\Gamma) = 1$. If P is a Γ-equivariant projection of $\bigoplus L^2(\Gamma) \to$

---

[68] or even a regular cover.

[69] In general the indices of $L^2(\mathbb{Z})$-modules can be any real number. However, not all of these arise as dimensions of kernels and cokernels of elliptic operators. In the special case of the de Rham operator on general finite complexes (or compact manifolds), this question is the very fruitful area of the Atiyah conjecture, which has deep positive and negative results. For other operators, such as the signature operator on manifolds with boundary, it is very easy to obtain transcendental numbers as such dimensions, even if the fundamental group in $\mathbb{Z}$.

V, then the dimension is a trace of P. To figure out what the trace should be, consider first the case when Γ is finite. In that case, V is finite-dimensional in the ordinary sense, and

$$\dim_\Gamma V = \dim(V)/\#\Gamma.$$

We can consider the matrix of the projection to have coefficients in $\mathbb{C}[\Gamma]$. This dimension is then the sum of the coefficients of the identity (element of Γ) along the diagonal, i.e. the coefficient of the identity in the trace. Note that when $\Gamma' \subset \Gamma$ is a finite index subgroup, we have:

$$\dim_{\Gamma'} V = [\Gamma : \Gamma'] \dim_\Gamma V;$$

here $L^2(\Gamma)$ is a sum of $[\Gamma : \Gamma']$ copies of $L^2(\Gamma')$ when thought of as a Γ' representation. It turns out that the dimension of any nontrivial representation is positive in this sense.

This has the property that $\dim_\Gamma V \oplus W = \dim_\Gamma V + \dim_\Gamma W$. Very useful is the property (almost obvious from the above heuristic)

$$\dim_{\Gamma'} V = \dim_\Gamma \mathrm{ind}_{\Gamma'}^\Gamma V.$$

The usual homological algebra shows that $K = X/\Gamma$ a finite complex, then one has

$$\chi_\Gamma(X) = \chi(K).$$

Atiyah went further and showed that if one takes any elliptic operator on a compact manifold, then the Γ-dimension of the kernel and cokernel on the universal cover make sense, and one has an equality of indices upstairs and down, but this is rather more delicate – it requires more geometry and analysis than the result on Euler characteristics, which is a result of pure algebra.

It is easy to see that for any infinite complex (and hence for any infinite group acting freely) $\mathcal{H}_{(2)}^0(X) = 0$; a constant map is L^2 iff it is 0. Applying the Euler characteristic relation, we see from setting K to be a finite graph that #generators of the free group F acting freely and cocompactly on it equals $1 - \dim_F \mathcal{H}_{(2)}^1$ (regular tree).

On the other hand, if Γ is amenable, then Cheeger and Gromov (1986) showed that for $X = E\Gamma$, the universal cover of $B\Gamma$, we have that $\mathcal{H}_{(2)}^i(X) = 0$. They deduced from that that the same is true for any Γ with an infinite amenable normal subgroup. And therefore $\chi(K) = 0$ if K is an aspherical complex whose fundamental group has an infinite normal amenable subgroup.

All of this connects to finite covers for residually finite groups by a beautiful theorem of Lück.

Theorem 3.26 (Lück, 1994)   *If K is a finite complex with residually finite fundamental group Γ and universal cover X, and letting Γ_i be a descending chain of normal subgroups (with K_i the associated covers) then*

$$\lim H_k(K_i)/[\Gamma : \Gamma_i] = b_k^{(2)}(X).$$

Thus for finite complexes one can ascertain linearity of the growth of Betti numbers in terms of $b_k^{(2)}(X)$ in terms of the universal cover, i.e. are there *any* L^2-harmonic k-forms. This is, interestingly enough, a statement that does not depend on the uniform lattice that is acting, or the sequence of normal finite index subgroups used in defining, the normalized Betti numbers.

It turns out that one can use harmonic analysis[70] on Lie groups to obtain that the only cohomology $\mathcal{H}_k^2(K\backslash G)$ that can be nonzero is when $k = \frac{1}{2}\dim(G/K)$: see Olbrich (2002). In this dimension it will be nonzero iff the Euler characteristic $\chi \neq 0$ (which can also be determined from the χ(compact dual) and which is iff $\mathrm{rank}_{\mathbb{C}}G = \mathrm{rank}_{\mathbb{C}}K$). So, for $\mathrm{SL}_n(\mathbb{R})$ this only happens for $n = 2$, but for $U(m,n)$ it's always true (and for $O(m,n)$ it depends on parity considerations of m and n).

This theorem is adequate for the purposes of understanding uniform lattices; however for nonuniform lattices, while there is a finite complex for $K(\Gamma, 1)$, – thanks to the Borel–Serre theory, it is *not* $K\backslash G/\Gamma$, which even has the wrong dimension. Thus the universal cover is not $K\backslash G$ and we cannot directly use the above calculation to learn about the growth of Betti numbers in towers. It is nevertheless true that the L^2-Betti numbers for nonuniform lattices are proportional (with the ratio of volumes being the proportionality constant) to those of the uniform lattices!

The most conceptual proof I know is due to Gaboriau (2002), who introduced notions of L^2-invariants for equivalence relations. Using this he showed that both Γ and ∇ act *to preserve measure* and that they also *commute* with each other on the same space X[71] with finite co-volume, then for every k,

$$b_k^{(2)}(\Gamma)/\mathrm{vol}\,(X/\Gamma) = b_k^{(2)}(\nabla)/\mathrm{vol}\,(X/\nabla).$$

We note that as a consequence of the theorems in this section, if $M = K\backslash G/\Gamma$, then $(-1)^{\frac{1}{2}\dim(G/K)}\chi(M) \geq 0$.

The Hopf conjecture asserts that this is true for all closed aspherical manifolds. It is not even known for (variable) negatively curved manifolds, although

---

[70] OK – one can if one is Borel.

[71] In this situation $X = G$, the ambient Lie group; the lattices can be viewed as acting in a commuting fashion by having one act on the left and the other on the right. The invariant measure exists, because the Lie group G is unimodular (whenever it has a lattice). This idea of Gromov is called *measure equivalence*.

Gromov (1991) did use L^2 ideas combined with Hodge theory to prove a Kähler version of this conjecture.[72]

In the next chapter we will discuss some other uses of L^2 to probe the Borel philosophy.

3.7 Mixing the Ingredients

We now wrap up our discussion and show the ubiquity of the failure of the naive proper analogue of the Borel conjecture. (Before jumping to conclusions, however, please go to §3.8 on morals!) All of the results and arguments in this section are joint work with Stanley Chang, and more details can be found in Chang and Weinberger (2003, 2007, 2015).

Our first result argument shows that we can use the completely elementary results about H^4 of congruence subgroups of $SL_n(O)$ to show proper-nonrigidity for all $n > 4$.

Theorem 3.27 *For every $n > 4$, for every lattice in $SL_n(O)$, the associated locally-homogeneous manifold has a finite sheeted cover that is not properly rigid. Moreover, we can arrange for this cover, there is a proper homotopy-equivalent manifold that is smooth, and is distinguished (topologically) from the locally symmetric manifold by having a different p_1.*

Proof We shall just use the groups $SL_n(O; p)$ studied above (every lattice contains these for large p, by the congruence subgroup theorem: Bass *et al.*, 1967). We turn to our classifying spaces armed with our knowledge about p-torsion in $H^4(SL_n(O; p); \mathbb{Z})$:

$$
\begin{array}{ccccccc}
F/O & \to & BSO & \to & \prod K(\mathbb{Z}, 4i) & \to & K(\mathbb{Z}, 4) \\
\downarrow & & \downarrow & & & & \downarrow \\
F/\text{Top} & \to & BS\text{Top} & & \to & & K(\mathbb{Z}[1/2], 4).
\end{array}
$$

The leftmost square consists entirely of rational homotopy equivalences because BF has finite homotopy groups according to Serre's theorem on the finiteness of stable homotopy groups of spheres. The map $BSO \to \prod K(\mathbb{Z}, 4i)$ is the total Pontrjagin class (interpreting cohomology classes as maps to Eilenberg–Mac Lane spaces).

The homotopy of BSO is known, thanks to Bott periodicity, and we have a $\mathbb{Z}$ in every fourth dimension. We shall ignore the prime 2. Bott periodicity,

---

[72] On the other hand, recent work of Avramidi (2018) calls this conjecture into question in general.

via its connection to the Chern character (see e.g. Hatcher, 2017), implies that $p_k : \pi_{4k}\mathrm{BSO} \cong \mathbb{Z} \to \mathbb{Z}$ is multiplication by $(2k - 1)!$.

Note that a Pontrjagin class p_k can be defined in a topologically invariant fashion in $\mathbb{Z}[1/N]$ if we invert all primes that arise in $\pi_i(\mathrm{Top}/\mathrm{O})$ for $i \leq 4k + 1$. So $\pi_3(\mathrm{Top}/\mathrm{O}) \cong \mathbb{Z}/2$ and then the groups vanish till $\pi_7(\mathrm{Top}/\mathrm{O}) \cong \mathbb{Z}/28$ and forever after, they are isomorphic to the group of differentiable structures on spheres studied by Kervaire and Milnor. Thus, the question of which primes need to be inverted becomes related to Bernoulli numbers. However, we will just use p_1 and be happy to invert the prime 2 to obtain topological invariance.

Now, to lift a map $K\backslash G/\Gamma \to K(\mathbb{Z}, 4)$ to F/O, note that in every dimension d there is an $N(d)$ so that there is a map from the d-skeleton $K(\mathbb{Z}, 4)^{[d]} \to F/\mathrm{O}$ (making use of the rational homotopy equivalence $\mathrm{BSO} \to \prod K(\mathbb{Z}, 4i)$) so that the composition $K(\mathbb{Z}, 4)^{[d]} \to F/\mathrm{O} \to K(\mathbb{Z}, 4)$ is multiplication by $N(d)$. Letting $d \geq \dim(G/K)$, and multiplying by $N(d)$, e.g. choosing $p > N(d)$, we obtain a normal invariant that we can do smooth surgery to and obtain a smooth proper homotopy equivalence $f : M \to K\backslash G/\Gamma$ distinguished by the fact that $p_1(M) - f^*p_1(K\backslash G/\Gamma)$ is of order p.

Notice that $f^*p_1(K\backslash G/\Gamma)$ depends only on the map that f induces on π_1, i.e. only on the homotopy class of the map, not the proper homotopy class. By Mostow rigidity, all automorphisms of Γ come from isometries of $K\backslash G/\Gamma$ to itself, and hence $p_1(K\backslash G/\Gamma) \in H^4(K\backslash G/\Gamma)^{\mathrm{Out}(\Gamma)}$. Consequently, this manifold M cannot be homeomorphic to $K\backslash G/\Gamma$ – it is not merely a proper homotopy equivalence that is not properly homotopic to a homeomorphism. □

Note that the above proof used the idea that smooth invariants are topological invariants if we ignore a few primes (whose number depends on dimension). It is an important fact that F/Top has an H-space structure, and $S^{\mathrm{Top}}(M)$ has an abelian group structure (for all manifolds) so that the map $S^{\mathrm{Top}}(M) \to [M : F/\mathrm{Top}]$ is a group homomorphism.[73] For π–π manifolds (of dimension greater than 5) this map is an isomorphism. The group structure on F/Top is exactly the one that makes the maps arising in Sullivan's description of F/Top into H-maps.

Proposition 3.28 *For all $d > 4$ there is an $M(d)$ such that if M is a smooth manifold, the image of the map $S^{\mathrm{Diff}}(M) \to S^{\mathrm{Top}}(M)$ contains a subgroup of index bounded by $M(d)^{\mathrm{rank}H^*(M;\mathbb{Z})}$.*

Proof This is a formal consequence of the statement that there is an $M(d)$ so

---

[73] Siebenmann proved this in the last essay of Kirby and Siebenmann (1977). It is a consequence of a periodicity theorem that is a cousin of Bott periodicity for BO. For a geometric explanation, see Cappell and Weinberger (1987) and Weinberger (1994).

that the composition $M(d)$: $F/\text{Top} \to F/\text{Top} \to \text{Top}/O$ is null-homotopic on the d-skeleton. The d-skeleton of F/Top is a finite complex, and Top/O is a cohomology theory with finite homotopy groups. Therefore $[F/\text{Top}_{(0)}: \text{Top}/O] = 0$ and hence[74] the inverse limit of N^* (over the integers)[75] on $[F/\text{Top}: \text{Top}/O]$ is trivial. Consequently, we can find the $M(d)$ that induces 0, as was our goal. $\qquad\square$

Remark 3.29 It is a consequence of the work of Kervaire and Milnor on differentiable structures on the sphere and smoothing theory that the map $S^{\text{Diff}}(M) \to S^{\text{Top}}(M)$ has finite kernel (whose order is also bounded by $M(d)^{\text{rank}H^*(M;\mathbb{Z})}$). The above proposition shows that, although the image is not a subgroup, the cokernel has a similar bound.

As a result, we have that, for π–π manifolds, $S^{\text{Diff}}(M) \to \bigoplus H^{4i}(M;\mathbb{Z})$ is finite-to-one and has image that contains a lattice in the target (with even some information on the torsion, if we are so inclined).

We now give a general converse to the rigidity that holds[76] in $\mathbb{Q}$-rank ≤ 2, but only for the topological category.

Theorem 3.30 *If $\mathbb{Q}$-rank$(\Gamma) > 2$, then there is a finite index subgroup Γ' of Γ for which $S^{p,\text{Top}}(K\backslash G/\Gamma')$ is nontrivial.*

Remark 3.31 Indeed, we can make this an elementary abelian 2-group of arbitrarily large size by pushing strong approximation slightly harder than we do in the discussion below.

Remark 3.32 If Γ is arithmetic then the $\mathbb{Q}$-rank(Γ) is defined as usual in terms of $\mathbb{Q}$-split tori. If it is reducible, then we add on to the arithmetic pieces the number of non-compact manifold factors it has. These non-arithmetic factors are all negatively curved, and they have the same general shape as $\mathbb{R}$-rank $- 1$ non-compact symmetric spaces: they have cusps that can be compactified, and these boundaries are aspherical, with cusp subgroups that are of infinite index.

Proof We shall use Sullivan's decomposition of F/Top at the prime 2: F/Top has a $K(\mathbb{Z}/2,2)$ factor, so we need to produce Γ' with large $H^2(;\mathbb{Z}/2)$. Let us assume that we are in the arithmetic case, leaving the modifications for the reducible case to the reader.

Recall that, according to Lubotzky's theorem, Γ has a subgroup of even index

---

[74] As the homotopy groups are all finite, there is no issue of $\lim^1$; it is also true in our case for the reason that we can work with a fixed finite skeleton.

[75] Note that $F/\text{Top}_{(0)}$ can be thought of as an infinite mapping telescope of self-maps $F/\text{Top} \to F/\text{Top}$ induced by multiplications by the integers (no matter what H-space structure is used).

[76] We will explain this in a later chapter.

– hence a normal subgroup of even index. Hence there's a finite group of even order H that is a quotient of Γ. Let Γ' be the inverse image of some involution in H.

If the Lie group G has no rank-1 factors, then it has Property (T), and $H_1(\Gamma')$ is necessarily finite. If there are rank-1 factors, but Γ is irreducible, we can deduce the same thing from superrigidity. In any case, we then see that $H_1(\Gamma')$ has an even-order cyclic summand. Consequently, we have $\mathrm{Ext}(H_1(\Gamma'), \mathbb{Z}/2) \neq 0$; by the universal coefficient theorem, there is an injection $0 \to \mathrm{Ext}(H_1(\Gamma'), \mathbb{Z}/2) \to H^2(\Gamma'; \mathbb{Z}/2)$, giving us a nontrivial element in the structure set as desired. □

The first remark is proved by producing quotients making use of many primes, and then having a large elementary abelian subgroup of which to take the inverse image.

Problem 3.33 The above reasoning shows that we can make rank $H_1(\Gamma', \mathbb{Z}/2)$ large by taking a deep lattice. This rank is necessarily $O(\mathrm{vol}(K\backslash B/\Gamma'))$ and in the rank-1 case it can actually grow linearly (although this doesn't produce any exotic structures). However, if one takes a descending chain[77] or assumes that we are irreducible in a semisimple group of rank ≥ 2, is it the case that this rank is $o(\mathrm{vol}(K\backslash B/\Gamma'))$?

As a converse to the low-rank proper rigidity, this theorem has a couple of weaknesses: these elements (at least in the irreducible case) die on passing to further covers. Also, it would be nice to know that some particular structure sets (groups, in fact) are infinite – and we would be interested in knowing whether we can say anything about how these groups grow in size as we move up a tower. We now address these questions.

Remark 3.34 $H^i(X; \mathbb{Q}) \to H^i(Y; \mathbb{Q})$ is one-to-one for any finite cover Y of any space X. So, if we detect a structure set using a *rational* Pontrjagin class, then these survive forever.

First of all, we note a case where we can prove that the proper structure set is infinite for a simple reason of this sort.

Proposition 3.35 *If $K\backslash G/\Gamma$ is a Hermitian symmetric space with $\mathbb{Q}$-rank$(\Gamma) >$ 2, then $S^{p,\mathrm{Top}}(K\backslash G/\Gamma) \otimes \mathbb{Q} \neq 0$.*

Proof We argue as above, but we will need to see that $H^4(K\backslash G/\Gamma; \mathbb{Q}) \neq 0$; the proposition will be proved by p_1. The obvious cohomology class to use is the square of the Kähler class. However, one needs to check that this class is nontrivial.

---

[77] And we are not in the case of a surface!

Let's now be a bit more explicit. Given a projective embedding, one can pullback the generator of $H^2(\mathbb{CP}^N)$: this is the Kähler class. The way it evaluates on homology is by intersecting with any linear hyperplane $\mathbb{CP}^{N-1}$. Using a projective embedding of the Baily–Borel compactification[78] of $K\backslash G/\Gamma$, it would then suffice to find a codimension-2 linear subprojective space that does not intersect the singularity set of the Baily–Borel compactification, since its intersection with $K\backslash G/\Gamma$ will be a subsurface on which the square of the Kähler class is nontrivial. This merely requires that the codimension of the singularities of the Baily–Borel compactification to be larger than 2. This can be seen by inspection, as noted in Jost and Yau (1987). □

At the cost of weakening the hypothesis on $\mathbb{Q}$-rank to one that is *not* necessary for rigidity, one can prove a much stronger theorem.

Theorem 3.36 *Suppose* $M = K\backslash G/\Gamma$ *is a locally symmetric manifold with* $\mathbb{Q}$-*rank*$(\Gamma) > 3$, *then* $\lim S^{p,\mathrm{Top}}(K\backslash G/\Gamma') \otimes \mathbb{Q}$ *is of infinite rank (where the limit is taken with respect to arbitrary finite covers* $K\backslash G/\Gamma'$ *of* $K\backslash G/\Gamma$).[79]

We shall denote the limit, $\lim S^{p,\mathrm{Top}}(K\backslash G/\Gamma')$ by $S^{\mathrm{virtual}}(K\backslash G/\Gamma)$.

We shall, for simplicity, only deal with the case of $\Gamma = \mathrm{SL}_n(O_F)$ with $n > 3$ (note that the $\mathbb{Q}$-rank of such is $n - 1$) – which includes a very interesting $\mathbb{Q}$-rank $= 3$ example, and tells the complete story for this important class of lattices.

In light of our previous remarks and the theory of generalized modular symbols, all that we need to do is find proper $\mathbb{Q}$-parabolic subgroups for $\mathrm{SL}_n(O_F)$ whose unipotent radicals have dimension $\equiv 0 \bmod 4$.

Parabolics are associated to, perhaps incomplete, flags in F^n. If we use the flag $F^k \subset F^n$, then the dimension of the associated unipotent subgroup (of automorphism inducing the identity on the associated graded to this flag) is $dk(n - k)$, where $d = [F : \mathbb{Q}]$.[80] If n is even, we can use $k = 2$, and if $n = 1 \bmod 4$, we can use $k = 1$. If $n = 3$, then the $\mathbb{Q}$-rank is 2, so we can assume that $n > 4$, so we can set $k = 4$.

The general case in the theorem is a similar case-by-case analysis.[81]

---

[78] See Baily and Borel (1966) for the completion of Hermitian locally symmetric spaces as projective varieties.

[79] One can form this limit with respect to various families of covers, and the limits can change. For example, one can show that if $\mathbb{Q}$-rank$(\Gamma) > 5$, then if one takes the sequence of squarefree congruence covers, the limit has an infinitely generated torsion subgroup that frequently dies when included in the limit over all finite index subgroups.

[80] The presence of this factor d implies that if the lattice in the theorem were obtained by restriction of scalars from a number field of even degree, we would obtain the same growth of S^{virtual} in the problematic $\mathbb{Q}$-rank $= 3$ case not covered in the theorem.

[81] I am deeply appreciative of the help that Dave Witte-Morris gave us with these calculations that we had done incorrectly at first.

Finally, given the infinite rank of $S^{\text{virtual}}(K\backslash G/\Gamma) \otimes \mathbb{Q}$, it becomes reasonable to ask what is the growth rate of rank $S^{p,\text{Top}}(K\backslash G/\Gamma')$ as one moves up a tower. In general, it seems like the rank grows like some power of the volume $[\Gamma : \Gamma']^\alpha$ for some $\alpha \leq 1$.

The question of (approximately linear) growth follows easily from Lück's theorem combined with the results of Cheeger and Gromov and of Gaboriau explained in §3.6.

Theorem 3.37 *Assuming that* $\dim(G/K) > 4$, *and* $\mathbb{Q}$-rank$(\Gamma) > 2$, *the ranks* $S^{p,\text{Top}}(K\backslash G/\Gamma') = o([\Gamma : \Gamma'])$ *iff* rank$_{\mathbb{C}}(G) >$ rank$_{\mathbb{C}}(K)$ *or* $\dim(G/K)$ *is not divisible by* 8.

Indeed for G semisimple with no rank-1 factors, one can prove that

$$\text{rank } S^{p,\text{Top}}(K\backslash G/\Gamma') \otimes \mathbb{Q} = o(\text{vol } K\backslash G/\Gamma')$$

(i.e. we do not have to assume that they are part of a tower). Presumably, this also holds in that case as well for irreducible lattices. And it is interesting to speculate on the nature of the torsion for these lattices (both in a tower and those that are not).

3.8 Morals

What do we learn from this discussion? Certainly that in large $\mathbb{Q}$-rank, the proper Borel conjecture fails.

But that's a summary, not a moral.

The reason that the proper Borel conjecture fails is interesting. It turns out that only in $\mathbb{Q}$-rank ≤ 2 are the symmetric spaces "aspherical" in the "relevant sense," i.e. the sense relevant to *proper* rigidity. We observed that these low $\mathbb{Q}$-rank lattices are the only ones where the space at ∞, i.e. the Borel–Serre boundary, is aspherical.

What is a good way of thinking about "aspherical in the relevant sense"? We need to lose some geometry and move towards a categorical answer.

For proper maps, we are working in the proper category, and it makes sense to look for a properly aspherical space.

What should proper "aspherical" mean? This space should be defined to be a terminal object in the category of spaces and maps that are "1-equivalences,"[82]

---

[82] That this is the right thing to look at is suggested by the functoriality properties that we will see that S^{Top} (i.e. surgery theory) is blessed with. More primitively, the π–π theorem – which our discussion crucially depended on – suggests the very special role that the fundamental group (which is equivalent to 1-equivalence classes of connected spaces) will play.

i.e. where one can solve all one-dimensional lifting problems in a way that is unique up to homotopy (in the category). If $\mathbb{Q}$-rank ≥ 3, the terminal object should be "the core of $K\backslash G/\Gamma'' \times [0, \infty)$ – which is *not* $K\backslash G/\Gamma$. (There's a pretty straightforward proper map from latter to the former, but no proper section to this map.)

If we give up on doing *any* new geometry at infinity, might we be able to survive this lack of proper asphericity, and get some rigidity theorem for all $K\backslash G/\Gamma$?

One way we can do this is by insisting that our maps are homeomorphisms outside some compact set (and that we allow homotopies to be relative to the complement of a somewhat larger compact set). In this case, the symmetric space is a terminal object and we will see in Chapter 4 that the ordinary Borel conjecture for a closed aspherical manifold constructed by a Davis construction applied to the Borel–Serre compactification implies this is relative to infinity rigidity (i.e. relative to the complement of some large, unspecified compact set), so it is a consequence of the Borel conjecture.

And, indeed this case is, essentially, a theorem of Bartels *et al.* (2014b).[83]

In any case, this discussion enables (and forces) us to expand our attention to all aspherical manifolds with boundary, with the boundary aspherical or not, provided we work relative to the boundary. (Or rel∞ in the noncompact case.)

There's another sense in which $K\backslash G/\Gamma$ is aspherical, if we make a category where maps are Lipschitz and not allowed to move any point too far. In that case, the large-scale geometry discussed in the first section comes to bear, and one can indeed prove that the $K\backslash G/\Gamma'$ are "boundedly rigid."[84] This bounded category (and other "controlled analogues") will play a large role when we discuss the Novikov conjecture in the upcoming chapter.

3.9 Notes

This chapter covered a lot of ground, and all of the topics discussed need more systematic treatments. Happily, many exist for them. A very good general reference for arithmetic manifolds is Witte-Morris (2015).

The subject of compactifications of $K\backslash G/\Gamma$ is an important one, and two of these played a role in our discussions, the Borel–Serre and the Baily–Borel (see

---

When we study groups with torsion, it will turn out that the terminal object is not rigid, and we will be led back to geometry and orbifolds, i.e. to enlarging the category.

[83] Their paper covers the case of arithmetic lattices. Non-arithmetic lattices can be handled by the same idea of reduction to the arithmetic case used above in §3.2 when we defined the $\mathbb{Q}$-rank for the non-arithmetic case.

[84] See Chang and Weinberger (2007).

Borel and Serre, 1973 and Baily and Borel, 1966). Both of these are extraordinarily important. The Borel–Serre compactification gives finite generation of group homology, the calculation of their cohomological dimension, and that these groups are duality groups in the sense of Bieri and Eckmann[85] (see Bieri and Eckmann, 1973). The Baily–Borel compactification shows finite generation of the spaces of modular forms via projective embedding. The literature on these and many others is surveyed and explained in Borel and Ji (2005).

We shall sometimes have need for Tits buildings defined for Lie groups over other fields. For example, if one wants to study $SL_n(Z[1/p])$, it acts ergodically on $SL_n(\mathbb{R})$ and we need to supplement $SL_n(\mathbb{R})$ with $SL_n(\mathbb{Q}_p)$ to get discreteness. Tits buildings give a structure that replaces the symmetric space $K\backslash G$. The group $SL_n(Z[1/p])$ acts properly on the product of the real symmetric space and the building. An immediate consequence of this theory is that the virtual cohomological dimension of such groups is finite. There are many references for the theory of buildings, each with a different emphasis; for our purposes, Tits (1974) and Abramenko and Brown (2008) are especially recommended. I also highly recommend the paper Alperin and Shalen (1982) which is a model of this type of application.

Atiyah's theorem (about how much of the tangent bundle is homotopy invariant) is better phrased in terms of stable normal bundles (for an embedding in a very high-dimensional Euclidean space), rather than tangent bundles. In that case, the conceptual explanation, due to Spivak (1967), is that *as a spherical fibration*, this stable normal bundle is definable for any Poincaré complex; that is, for any finite complex that satisfies Poincaré duality.[86] The idea of this fibration is quite simple: Poincaré complexes can be characterized as those complexes X for which a regular neighborhood of X, when polyhedrally embedded in Euclidean space looks like, in a homotopy-theoretic sense, the situation that arises for tubular neighgborhoods of smooth manifolds. This means specifically that the homotopy fiber[87] of the inclusion of the boundary of this neighborhood into the neighborhood is a homotopy sphere (like the epsilon-sphere bundle mapping to the smooth manifold, to which a tubular neighborhood deform retracts). Spivak also gives a homotopy-theoretic characterization of this fibration.[88]

---

[85] Actually they motivated the definition of Bieri–Eckmann duality by being a first nontrivial class of examples of this.

[86] See Spivak (1967) and Wall (1968) for what this notion means in detail: it generalizes the fact that the homology and cohomology groups must be isomorphic, but it also demands that it be implemented via a fundamental class, and also hold with arbitrary local coefficient systems – in particular any finite cover of a Poincaré complex is a Poincaré complex.

[87] Recall that any map can be replaced (at the cost of replacing the spaces involved by homotopy equivalent ones) by a fibration, as observed by Serre in his thesis.

[88] It is the unique stable spherical fibration whose top homology class is spherical (i.e. lies in the image of the Hurewicz homomorphism).

The surgery classification of manifolds was begun by Kervaire and Milnor (1963) in the case of smooth manifolds that are homotopy-equivalent (and therefore homeomorphic) to the sphere, and then extended to the simply connected case by Browder and Novikov and reformulated using classifying spaces by Sullivan. (References for surgery theory include Wall (1968), Browder (1972), Ranicki (1992, 2002), Weinberger (1994), Lück (2002a), and Chang and Weinberger (2020).) Although we have not yet dealt with the classification of closed manifolds (see Chapter 4!) the simply connected case can be deduced from the discussion given here: if $h: M' \to M$ is a homotopy equivalence, one can always deform it so that it will be transverse to a point p, and with $h^{-1}(p)$ a single point. In that case, there is a neighborhood isomorphic to a ball, whose inverse image is a ball. Deleting the interiors of these balls, we get a structure on the complement. On the other hand, any structure on the complement restricts to a homotopy sphere on the boundary, and, thanks to the Poincaré conjecture (in the PL and Topological categories), it can be completed to be a structure on the closed manifold. Thus $S^{\mathrm{cat}}(M) \cong S^{p,\mathrm{cat}}(M - p)$ for M simply connected and where cat is equal to Top or PL.

The classifying space F/Top is its own fourth loop space (more correctly, it is $\mathbb{Z} \times F/\mathrm{Top}$ that is its own loopspace[89]) as can be seen from the description given in the text and using Bott periodicity at the odd primes. It turns out that this is the first step towards a functorial view of surgery theory, which cannot at all be explained in "without obstructions" terms, as our first pass went: the structure space[90] S measures the difference between completely analogous local and global obstructions, i.e. $\mathbb{Z} \times F/\mathrm{Top}$ is a cohomology theory associated to a spectrum whose homotopy groups are surgery obstruction groups.

Surgery theory is nicest in the topological category. The canonical reference for the foundational theorems in this setting is Kirby and Siebenmann (1977) – which is the original source for them. Unlike the smooth category where the foundations are built on Sard's theorem, Morse's lemma, and the fundamental existence theorem for ordinary differential equations (with smooth coefficients), the topological category is distinctly more difficult to get off the ground. The proofs of the basic theorems, essentially theorems about the topology of $\mathbb{R}^n$, require deep global results in the smooth or PL categories either about non-simply connected manifolds (*à la* Novikov and Kirby) or about manifolds

---

[89] This extra $\mathbb{Z}$ has significant geometric implications, hinting to an amazing world of non-resolvable homology manifolds.

[90] Indeed, the structures that we considered are promoted to being homotopy groups of a space rather than merely a set. This idea first arose in work of Casson (1967) and was developed and advocated by Quinn in his thesis.

"controlled over a metric space" – introduced by Quinn (1979, 1982b, 1982c, 1986) – a major theme in the coming chapters.

However, the theory ends up having an *even nicer* formulation than the topological category when one includes homology manifolds, but here the local issues currently seem even more difficult and the global theory is in much better shape than the local (see Bryant *et al.*, 1996). I will discuss this a bit in Chapter 4 discussing the functoriality of surgery and also in our discussion of the Wall conjecture (the "existence Borel conjecture").

That there is a lot of homology in congruence subgroups is something that I learnt from Ruth Charney (1984). Torsion in homology of arithmetic groups is quite mysterious. For SL($\mathbb{Z}$), in the limit, this is determined by the solution to Quillen–Lichtenbaum conjecture by Rost and Voevodsky[91] (by the work of Dwyer and Friedlander, 1986), but for congruence groups and other arithmetic groups the picture is still obscure.[92] Bergeron and Venkatesh (2010) and Calegari and Venkatesh (2019) have suggested that the analogue of the L^2-Betti story holds – something hard to tell in general because of issues involving regulators. (Test question:[93] in the stable range of Borel's theorem, how do the images of the cohomology lattices corresponding to different lattices in the same group relate to one another?) The stabilization by going up the congruence tower has been studied by Calegari and Emerton (2012) introducing a notion of completed cohomology making a connection to p-adic Lie groups).

A problem whose solution would seem to be illuminating in this direction is the following: Can one estimate the ratio of $b_i(X;\mathbb{Z}/p)/\mathrm{vol}\,(X)$ (where vol (X) is *some* simplicial notion of volume, say the number of simplices) for a simplicial complex by random sampling. That this is possible for rational Betti number is the idea of the Lück approximation theorem; see also Farber (1998), (where this point is clearer – his condition for Lück's theorem to hold for non-normal covers is precisely that the relative volume of the set of points where the covers do not look "universal" goes to 0), and Abert *et al.* (2017) and Elek (2010), where it is explicit.[94] This would then suggest that in the situation where rank$_\mathbb{C} G$ − rank$_\mathbb{C} K$ = 1 there would be growing torsion in covers, but not because of large elementary abelian subgroups.

We refer to Lubotzky and Segal (2003) for a survey of the strong approximation theorem. We note that, although the original proof (by Weisfeiler (1984) – see also Matthews *et al.* (1984) and Vasserstein) used the classification of

---

[91] See the survey Weibel (2005).

[92] But see Calegari (2015).

[93] It could be someone knows the answer to this and will email me!

[94] In this context, the paper of Clair (2003) where Lück's theorem is made more quantitative expressly in terms of injectivity radius of the covers.

finite simple groups, this is no longer necessary (as pointed out there) thanks to work of Nori (1987), Larsen and Pink (2011), and Hrushovski and Pillay (1995) (using algebraic geometric and/or model-theoretic ideas.)

I remember learning about amenability from Bob Brooks. When I was a student, he told me about Kesten's work and explained that, although there are no L^2 harmonic functions on a universal cover, amenability controls whether 0 is in the spectrum. This material appeared in Brooks (1981) and is a manifold version of the statement asserted for the discrete group. Other papers by Brooks, Sunada, and others compared the spectral geometry of the manifolds to the spectral geometry of the associated finite graphs. This can all be viewed part of the L^2-cohomology story (including an appropriate de Rham theorem for comparison of smooth and simplicial models) when one jazzes up the story to include foliated spaces rather than just covers (see Bergeron and Gaboriau, 2004). There are a number of excellent sources on amenability, the Banach–Tarski paradox, and its connection to random walks and to operator algebras (see e.g. Lubotzky, 1984; Paterson, 1988; Wagon, 1993).

The geometric group theory of amenability and non-amenability has led to the consideration of some remarkable groups. Non-amenable groups that don't contain free groups (the von Neumann conjecture) were first constructed by Ol'shanskii and Ju (1980) – the torsion groups satisfying Property (T) produced by the method of adding large relations are also examples. On the other hand, Whyte's thesis (see Whyte, 1999) gives a "true" analogue of the von Neumann conjecture that can be used, for instance, to extrapolate between the characterization of nonamenability in terms of random walks (i.e. vanishing of 0th L^2-homology) and that in terms of the existence of "Ponzi schemes"[95] (Block and Weinberger, 1992; 1997)) to all other L^p-homology with $p > 1$.

Grigorchuk's (1984) group of intermediate growth (i.e., so that the number of group elements that can be expressed as the product of n generators grows more than a polynomial, but less than exponentially) was the first non-solvable amenable group. Bartholdi and Virag (2005) actually proved the amenability of some related groups by consideration of their random walk. Both of these are examples of automata groups – see the survey by Zuk (2003). Recently, Juschenko and Monod (2013) have given (uncountably many) *simple* amenable groups.

Property (T) is the subject of a very useful book (Bekka *et al.*, 2008) and is at the center of numerous problems.

It is now liberated from its original representation-theoretic roots by the

---

[95] The Cayley graph of a non-amenable group always supports a scheme wherein each vertex exchanges a uniformly bounded amount of money with its neighbors, so that each vertex ends up net positive. This is impossible on Cayley graphs of amenable groups.

method of Garland (1973), Ballman and Swiatkowski (1977), and Zuk (2003), and also by Shalom's work using bounded generation and algebraic methods (related to K-theory) to show that a number of interesting groups (like linear groups over Laurent series rings) have Property (T): see his ICM talk (Shalom, 2006) for information. The analytic method is useful in studying strengthenings of Property (T), e.g. to include group actions on other Banach spaces, or on other general spaces with curvature conditions. Stronger forms of Property (T) can be invaluable to extensions of the rigidity program in other directions, such as the Zimmer program, which, broadly defined, tries to study nonlinear actions of large groups (e.g. lattices) on manifolds – perhaps, but not necessarily, preserving some geometric structure (such as a volume form).[96]

I will leave the reader to consult Kowalski (2008) and Lubotzky (2012) for recent applications of expanders to discrete groups and to number theory, as well as to references on new proofs of Selberg's 3/16 theorem (at least > 0 theorem!) and the connections to additive combinatorics that enable all this. There have also been other constructions of explicit families of expanders, such as the zig-zag product of Alon and Wigderson (explained very nicely in Hoory et al. (2006)) and the new Ramanujan graphs[97] constructed by Marcus et al. (2015).

That both amenability and Property (T) have characterizations in terms of L^2-homology/cohomology should have made it possible to make a segue between this section and the one on L^2 and growth of Betti numbers, but this seemed forced, so I chose not to push this.

Lück's (2002b) book gives a good overview of how L^2 interacts with groups and compact manifolds. Entirely missing (and not relevant to our concerns in this essay) are relations of L^2-cohomology to intersection cohomology of compactifications and other stratified applications. There has been much work since that book was written, both internal to the subject (such as interesting examples of transcendental L^2-Betti numbers (see Grabowski, 2014, and references therein)[98] and of connections to other parts of topology.

Atiyah (1974) introduced L^2-Betti numbers and that L^2 indices to deal with the kernels of elliptic operators on universal covers and to get finite quantities measuring the sizes of these typically infinite-dimensional spaces. Connes

---

[96] In the notes to Chapter 8, we will mention a bit more about this. Here we content ourself with a citation of the Bourbaki talk (Cantat, 2017 on the theorem of Brown–Fisher–Hurtado showing that certain lattices don't have any effective C^2-actions on low dimensional manifolds. $SL_n(\mathbb{Z})$ does not act effectively and C^2 on any manifold of dimension less than $n - 1$.

[97] Ramanujan graphs are graphs that have optimal spectral gaps for their Laplacians. The first examples were contructed by Lubotzky et al. (1988), using Deligne's solution of the Ramanujan conjecture – the circumstance that led to their name.

[98] On the other hand, even now, there is no known example of a torsion-free group where L^2-Betti numbers are not integers.

later proved an index theorem for foliations – see Connes (1982) and Moore and Schochet (2006) for a version closely related to the one most relevant here – that gives something like an average version of the indices one sees over the leaves (see Connes, 1994). If one views both theorems in the situation of limits of coverings (the Benjamini–Schramm limit of a sequence of finite covers being a transversely measured foliated space), then the cohomology related to Connes's theorem is exactly the one occurring in the Bergeron–Gaboriau theorem mentioned above.

Another interesting convergence arises when one thinks about the information given in the L^2-theory of symmetric spaces. Results about the limits of normalized Betti numbers (via thinking about the Matsushima formula and multiplicities) were first derived by DeGeorge and Wallach (1978, 1979) using the Selberg trace formula.

That symmetric spaces tend to be concentrated around the middle from the L^2 perspective has been well known for a while. I cannot track it down. Clearly Singer was aware of this when he conjectured in 1977 that the same might be true for the universal covers of arbitrary aspherical manifolds as an approach to the Hopf conjecture (discussed in the text). A very useful exposition (which does not make excessive demands on the reader's knowledge for harmonic analysis and which goes further and explains what occurs for additional invariants related to the spectrum near 0 for forms as well as functions) is Olbrich (2002). Atiyah and Schmid (1977) connect the use of the L^2 index theorem to representation theory and use this connection to re-prove some of the main results of Harish-Chandra.

Cheeger and Gromov were led to L^2 methods for an opposite reason than they arose for us: they wanted a substitute tool to use when there are not enough finite covers (see e.g. Cheeger and Gromov, 1985b) – for example, if one is studying the geometry of a manifold whose fundamental group is not residually finite. However, in some sense these are two sides of the same coin: the individual manifold (or lattice) might be hard to understand, but this limiting object is more transparent and brings order to the finite world. Their paper (Cheeger and Gromov, 1986) builds foundations for the theory and proves the generalization of Rosset's theorem on vanishing of Euler characteristic. Cheeger and Gromov (1985a) gives a direct but very delicate proof that the proportionality of Betti number to volume from locally symmetric manifolds remains true when one moves from the uniform to the nonuniform case. I personally prefer the method (Gaboriau, 2002) mentioned in the text.

The remarks about the rate of growth of $S^{p,\mathrm{Top}}(K\backslash G/\Gamma)$ are clearly essentially the same as questions about the rate of growth of Betti numbers. Besides the linear case, more remains at the level of conjecture. It seems reasonable to believe that the elements that don't come from the compact dual grow at a rate that's a power of volume, and that there is power upper bound – see Sarnak and Xue (1991), Xue (1992), Abert *et al.* (2017) – (and of course note that this is exactly the situation for the generalized modular forms, or, indeed, every use of geometric cycles I am aware of). The torsion story is harder to be sure of. In any case, if one climbs up the congruence tower of squarefree numbers, then the method based on comparison to the Lie algebra will give (at least if $\mathbb{Q}$-rank > 5) an infinitely generated torsion group in the limit. I suspect that quite generally $S^{virtual}(K\backslash G/\Gamma)$ will have infinitely generated torsion and infinitely divisible elements, but I do not have anything to show in justification of this suspicion.[99]

Finally, regarding the morals of the story told in this chapter: most stories are not improved by having their morals stated explicitly, and, further, most morals seem fairly obvious when just said outright,[100] and the tales told to illustrate them often seem more interesting than they are. And, perhaps this is true in our special case as well. In any case, as we proceed, we will now feel the need to be more and more functorial (and the geometric ideas that gave life to the problem, like good parents, will still be there, but in the background, giving guidance and perhaps providing inspirations, but never overwhelming independent development).

And the bounded rigidity of $K\backslash G/\Gamma$ mentioned there is proved in Chang and Weinberger (2007).

---

[99] But I do hope to hear more about this in coming years.

[100] Yet, I note, that they are frequently deep truths in the sense of Fermi, truths whose negations are also true.

4

How Can It Be True?

4.1 Introduction

After Chapter 3 showed the ubiquity of the failure of *proper* rigidity for non-compact locally symmetric manifolds, in this chapter we begin to examine the problem for closed manifolds. Rather than seriously engaging the question of how to prove the Borel conjecture, we focus on how it can possibly be true.

After all, every method for distinguishing manifolds is a potential obstacle that needs to be overcome. For example, in Chapter 3 we saw the import of characteristic classes, so a major focus of this chapter must be about why it is that (e.g. what are some mechanisms for) the characteristic classes of a manifold homotopy-equivalent to $K \backslash G / \Gamma$ must be the same as those of $K \backslash G / \Gamma$ (if we are in the compact case). This is essentially the topic of the Novikov conjecture[1] and it will be the main focus of this chapter and the next.

But there are invariants not at all related to the characteristic classes that can be used to distinguish manifolds.[2] The classical example is the theory of lens spaces: lens spaces are quotients of the sphere $S^{2n-1}/\mathbb{Z}_k$ where S^{2n-1} is thought of as the unit sphere of $\mathbb{C}^n$, and $\mathbb{Z}_k$ acts via a unitary representation on $\mathbb{C}^n$. Each representation is a sum of irreducible one-dimensional representations, and to obtain a quotient *manifold* we assume that each of the irreducible pieces has $\mathbb{Z}_k$ acting *freely*, i.e. can be described as rotation by a *primitive* root of unity. These are called the rotation numbers in the definition of the lens space. A lens space

---

[1] It is important to note that Novikov was not at all thinking about the Borel conjecture when formulating his problem. It arose very naturally in the course of his work on the topological invariance of rational Pontrjagin classes as we will see in §4.5 below.

[2] Recall that in the situation of simply connected manifolds, the Browder–Novikov theorem (see the notes from Chapter 3) tells us that, beyond homotopy type, the G/Top characteristic class is a complete invariant (so that ordinary characteristic classes only lose a finite amount of information).

might be denoted by $\text{Lens}_k(a_1, \ldots, a_n)$ or some such similar notation where the a_i are integers prime to k, and denote the rotation numbers.

Changing the order of the a_i is costless. Changing a_i to $-a_i$ is an equivalence of the underlying real representation, but not the complex one, and changes the natural orientation on the manifold. But one can do this an even number of times and keep the orientation.

If we just care about the underlying manifold, we can change the group action by multiplying all of the rotation numbers by the same s prime to k. Usually we will assume that we preserve an identification of the fundamental group with $\mathbb{Z}_k$, i.e. that we have a fixed homotopy class of maps to $K(\mathbb{Z}_k, 1)$ that we are preserving, equivalently that we are interested in conjugacy of the group actions.

The number of π_1 and orientation preserving homotopy types among lens spaces in a fixed dimension is $\varphi(k)$ (the Euler φ function) – with the homotopy type being determined by the product of the rotation numbers.[3] However,diffeomorphism (or homeomorphism, in this case) is exactly equivalent to orientation-preserving real linear equivalence, i.e. the changes we described above – according to a beautiful and deep theorem of de Rham (see Cohen, 1973).

In dimension 3, *all* orientable manifolds have trivial tangent bundles, and there is no more tangential information to be had.[4] (In higher dimensions, the Pontrjagin classes do distinguish some lens spaces from each other – they are essentially symmetric functions in the squares of the rotation numbers, but there aren't enough of these to determine these numbers themselves (see Milnor, 1966).

There are two essentially different proofs of this theorem; that is, proofs based on different principles: de Rham's original argument that it is now natural to view from the point of view of algebraic K-theory; and another argument, due to Atiyah, Bott, and Milnor (see (Atiyah and Bott, 1967), 1968) that involves (equivalence classes of) quadratic forms associated to the lens spaces defined either in terms of spaces that they bound (cobordism theory) or via some measure of how lopsided (around 0) the spectrum of some self-adjoint operators are ("spectral asymmetry") (Atiyah *et al.*, 1975a,b).

Both of these proofs pose challenges to the Borel conjecture; we will discuss

---

[3] A pair of lens spaces can be compared by a map that preserves their fundamental groups. Orienting them both, we can ask the degree of this map. A form of the Borsuk–Ulam theorem tells us that this degree is prime to k – the congruence class is independent of the map, and is the ratio of the product of the rotation numbers defining the two lens spaces.

[4] Actually, it is possible for the normal invariant of map between oriented 3-manifolds to be nontrivial, because of the extra information that goes beyond the tangent bundle itself – but this is information only at the prime 2, and it always vanishes for homotopy equivalences.

the Atiyah–Bott–Milnor argument later in the chapter, and de Rham's challenge in the next.

Let's start with the Novikov conjecture, the response to the challenge of characteristic classes.

4.2 The Hirzebruch Signature Theorem

Before discussing how the Borel conjecture can be true, it is worth asking, along the same lines, how the Poincaré conjecture can be true? After all, over S^{4k} there are an infinite number of vector bundles – distinguished by the Pontrjagin class p_k. Why aren't these the Pontrjagin classes of homotopy spheres?

The answer is given by the Hirzebruch signature theorem that gives a homotopy-theoretic interpretation of a certain combination of characteristic classes.

Definition 4.1 If M^{4k} is a closed oriented manifold, then the *signature* of M is defined as the *signature* of the quadratic (i.e. symmetric bilinear) form $H^{2k}(M; \mathbb{Q}) \otimes H^{2k}(M; \mathbb{Q}) \to H^{4k}(M; \mathbb{Q}) \to \mathbb{Q}$; that is, it is the difference in dimensions between a maximal positive-definite subspace of $H^{2k}(M; \mathbb{Q})$ and a negative-definite subspace.

Note that Poincaré duality tells us that this quadratic form is non-singular. Such a form (over a field) can be diagonalized – and the signature is the number of positive eigenvalues – the number of negative ones.

By its definition it just depends on the oriented homotopy type of M.

Theorem 4.2 ((Hirzebruch, 1995)) *There are homogeneous graded polynomials $L_k(p_1, \ldots, p_k)$ in the Pontrjagin classes, so that*

$$L_k(p_1, \ldots, p_k) = 2^{2k}(2^{2k} - 1)B_k/(2k)!p_k + \text{terms involving the lower classes,}$$

(where B_k is the kth Bernoulli number) so that $L = 1 + L_1 + \cdots + L_k + \cdots$ is multiplicative for sums of bundles, and

$$\text{sign}(M^{4k}) = \langle L_k(p_1, \ldots, p_k), [M] \rangle,$$

where we have denoted by p_i the Pontrjagin classes of the tangent bundle of M.

The last statement is what gives the theorem its name, the Hirzebruch signature theorem. Hirzebruch actually gives a formula for the L from which the first statement follows.

Note that signature depends on orientation, exactly as the right-hand side does. An immediate consequence of the formula is that a manifold that is stably parallelizable (i.e. with trivial normal bundle) has signature equal to 0.

Another significant consequence of the theorem is that if $N \to M$ is an r-sheeted cover (not necessarily connected or regular), then $\mathrm{sign}\,(N) = r\mathrm{sign}\,(M)$, as the tangential information is the same for a manifold and its cover, just the fundamental classes are multiplied.

The signature can be defined for spaces more general than topological manifolds.[5] For example, for manifolds with boundary, the relevant quadratic form can be defined, but it has a torsion subspace that should be removed. In that case, for example, signature is not multiplicative in finite coverings.

Wherever one has Poincaré duality, there are signatures. And in all cases, one can ask the question about multiplicativity in finite covers.

For L^2-cohomology, the $*$ operator often gives rise to a form of Poincaré duality. For instance, for infinite regular covers, this gives rise to a generalization of multiplicativity: the L^2-signature (which is a kind of normalized signature of covers[6]) equals the signature of the base.[7] On the other hand, there are other complete manifolds where L^2-cohomology is self-dual not coming from covering spaces, and then one can define signature-type invariants which need not be multiplicative.

Intersection homology provides another example: for interesting classes of spaces such as compact complex algebraic varieties, it gives a form of Poincaré duality.[8]

Finally, whenever one has a representation $\rho \colon \pi_1(X) \to \mathrm{U}(n)$, there is an associated flat bundle on x^{4k}, and a Hermitian form $H^{2k}(X;\rho) \otimes H^{2k}(X;\rho) \to \mathbb{C}$, and hence a signature[9] $\in \mathbb{Z}$. A rather surprising consequence of the Atiyah–Singer index theorem is that for X a manifold, $\mathrm{sign}_\rho(X) = n\mathrm{sign}\,(X)$.

We shall see later (as a consequence of controlled topological ideas) that this is indeed a consequence of the fact that the Poincaré duality is a local state-

---

[5] Aside from dimension 4, topological manifolds with trivial tangent bundle can be smoothed. And signature is multiplicative in finite covers of closed topological manifolds, as the reader should be able to prove by the end of the chapter.

[6] And in the residually finite case, is a limit of normalized signatures of finite covers, *à la* Lück.

[7] This is due to Atiyah, and resembles the statement we discussed in Chapter 3 about Euler characteristic, but it is somewhat deeper than it. The result about Euler characteristic is a statement about finite complexes, but this is one about manifolds. Atiyah's proof was based on the ideas of the Atiyah–Singer index theorem.

[8] Which can sometimes, e.g. in the work of Cheeger on the Hodge theory of Riemannian pseudomanifolds and the work of many on the Zucker conjecture, be interpreted in L^2 terms.

[9] In the Hermitian setting, there is not much of a difference between a Hermitian form and a skew-Hermitian form: you can go from one to the other by multiplying by i. As a result one can get get signature-type invariants in dimension $2\bmod 4$. We will not see this playing a direct role for closed manifolds of dimension $4k + 2$ – but this does play a role in the Atiyah–Bott–Milnor story for lens spaces of dimension $1\bmod 4$.

ment[10] and thus is true in the intersection homology and topological manifold settings as well.

Remark 4.3 (On the proofs) There are essentially two different proofs of the signature theorem. The original proof (Hirzebruch's) deduces the theorem axiomatically from three properties of signature:

(1) Signature is cobordism invariant, i.e. if $M = \partial W$ (where M and W are compact), then sign $(M) = 0$.
(2) Multiplicativity sign $(M \times N) = $ sign (M)sign (N).
(3) sign $(\mathbb{CP}^{2k}) = 1$.

Then the result follows essentially from the work of Thom on cobordism theory. (As Hirzebruch (1971) writes: "How to prove it? After conjecturing it I went to the library of the Institute for Advanced Study (June 2, 1953). Thom's *Comptes Rendus* note had just arrived. This completed the proof.") This method[11] remains important in purely topological settings. Often it is important to make use of quantities over the form $\Omega_*(?) \otimes_{\Omega_*(*)} \mathbb{Z}$, where $\Omega_*(?)$ is the homology theory whose chains are maps of oriented manifolds in "?", and which is viewed as a module over $\Omega(*)$ by multiplication and $\mathbb{Z}$ is viewed as a module over $\Omega(*)$ by means of the signature (or some other invariant of manifolds, on some occasions).

These considerations, systematically employed[12] by Sullivan (and the key to his analysis of the structure of F/Top, for example) exist embryonically already in this work of Hirzebruch.

Moreover, the π–π theorem, at the core of the flexibility results in Chapter 3, gives a starring role to cobordism and "?" of the form $K(\pi, 1)$. (Please pause and think this through.)

All that being said, this method is hard to apply to the flat bundle result mentioned above. For example, naively, one runs into the fact that no multiple of S^1 with a nontrivial flat complex line bundle with non-root of unity monodromy bounds in a way that extends (flatly) over the surface: the bordism group of manifolds equipped with flat bundles is huge.

The other main method for proving the Hirzebruch signature theorem was motivated by it – it goes via the Atiyah–Singer index theorem. Here sign (M) or a twisted cousin of it is viewed as the index of an elliptic operator on M, and such an index can be calculated cohomologically.

Originally, this wasn't a completely disjoint proof in that the first proof of

---

[10] And is thus true in the intersection homology setting for varieties (or Witt spaces).
[11] Of spending time (at IAS or) in libraries.
[12] Or exploited?

the index theorem went via cobordism theory (see Palais, 1965). However, subsequently two different proofs of the signature theorem have been found – one K-theoretic (Atiyah and Singer, 1968a, 1971) and one via study of the heat equation (see Atiyah *et al.*, 1975a,b; Gilkey, 1984). The K-theoretic perspective shall play a large role starting in Chapter 5 and will be discussed further there, but consequences of the index theorem shall already play an interesting role in this chapter, and the heat equation approach is also relevant to our story (e.g. in §4.9).

For now, so we can return to our story, let us be content with observing that the Hirzebruch theorem answers the question with which we started this section. If Σ is a homotopy sphere, then its signature is 0 (it has no middle cohomology at all), and so, since the coefficient of p_k in L_k is nonzero, we conclude that p_k must be 0 and none of those bundles we feared actually arise as tangent bundles of homotopy spheres.

4.3 The Novikov Conjecture

In the 1960s, Novikov (1966) suggested a generalization of one of the key consequences of the Hirzebruch theorem.

Conjecture 4.4 (Novikov conjecture; most primitive form) *Suppose that $\alpha \in H^i(\mathrm{B}\Gamma; \mathbb{Q})$ and let M be a closed oriented manifold of dimension $4k + i, f :$ $M \to K(\Gamma, 1)$ a map, then the quantity*

$$\mathrm{sign}\,_\alpha(M, [M]) = \langle f^*(\alpha) \cup L_k(M), [M] \rangle \in \mathbb{Q}$$

is an oriented homotopy-invariant.

To see its implication for the Borel conjecture, suppose that M and M' are closed aspherical manifolds with the same fundamental group Γ. Then if $h: M' \to M$ is a homotopy equivalence, and $h^*(p_i(M)) \neq p_i(M')$, then $h^*(L_i(M)) \neq L_i(M')$ and we can find a cohomology class in the cohomology of M' which equals that of M which equals that of $K(\Gamma, 1)$ that pairs nontrivially on this difference (using Poincaré duality) and get a contradiction.

For many purposes the following equivalent dual formulation is useful:

Conjecture 4.5 (Novikov conjecture; dual formulation) *Suppose M is a closed oriented n-manifold, and a map $f : M \to K(\Gamma, 1)$ is given. Then*

$$f_*(L(M) \cap [M]) \in \oplus H_{n-4i}(\mathrm{B}\Gamma; \mathbb{Q})$$

is an oriented homotopy-invariant (for manifolds with reference maps to $\mathrm{B}\Gamma$).

Let's be concrete and consider the case of $\Gamma = \mathbb{Z}$ (imagining that Γ is $\pi_1 M$). In that case, there is a natural homotopy class of map $f : M \rightarrow K(\mathbb{Z}, 1) = S^1$ to use. The conjecture calls attention to the invariant of M^{4k+1} given by $\langle f^*([s^1]) \cup L_k(M), [M] \rangle$. Playing with this a little and using the Hirzebruch formula, we see that the higher signature is given by $\operatorname{sign} f^{-1}(*)$ for any regular value $*$.

Orientations are easily obtained from orientations on M and the circle. That this quantity is independent of the regular value is because of the cobordism invariance of the signature. The puzzle Novikov places before us is why this quantity is a homotopy invariant. After all, the property of being a homotopy equivalence is a global property, and does not descend to submanifolds – a homotopy equivalence $h : M' \rightarrow M$ does not need to induce a homotopy equivalence $h|_{h^{-1}f^{-1}(*)} : h^{-1}f^{-1}(*) \rightarrow f^{-1}(*)$.

Remark 4.6 This is a key difference with the problem of homeomorphisms, which are indeed hereditary homotopy equivalences. If $h : M' \rightarrow M$ is a homeomorphism, then restricted to any $U \subset M$, $h|_{h^{-1}(U)} : h^{-1}(U) \rightarrow U$ is a proper homotopy equivalence. For hereditary homotopy equivalences, Novikov's theorem on topological invariance of rational Pontrjagin classes holds (Siebenmann, 1972).

Remark 4.7 *This* case of the Novikov conjecture does have a straightforward algebraic topological explanation (as was first observed,[13] I believe, by Rochlin). The cohomology of the infinite cyclic cover of M has the structure of a module over $\mathbb{Q}[\mathbb{Z}]$, a p.i.d. The linking pairing on the torsion submodule on H^{2k} satisfies Poincaré duality over $\mathbb{Q}$, and its signature can be identified with the invariant under discussion. However, in §4.4, we will discuss other methods of much wider scope.

4.4 First Positive Results

We shall discuss two and a half methods that give some useful and interesting positive information about the problem. The first is "codimension-1 splitting" and is a high-dimensional variant of the powerful tools used in the pre-Thurston period of three-dimensional topology. It gives very good information about "Haken manifolds," even in high dimensions.

It starts by trying to answer the question we asked at the end of the last section, given that homotopy equivalence is not hereditary, why should the

---

[13] These ideas about infinite cyclic covers are the bread and butter of knot theory.

signature of certain submanifolds be unchanged? Splitting theorems show that the homotopy equivalence can (often) be homotoped to one which is hereditary on codimension-1 submanifolds. This will also be a first occasion to consider algebraic K-theory that arises as an obstruction.[14]

After this, we will turn to Lusztig's thesis, which introduced a nice family of flat line bundles on manifolds with free abelian fundamental group, and brought the Atiyah–Singer index theorem for families to bear on the problem. Finally, we will give a variant of the last method that gives a proof for high-genus surfaces, not based on their Haken nature, or any family of line bundles, but rather based on a beautiful surface fibration discovered by Atiyah and Kodaira (and also explain why we call it half a method).

4.4.1 The Splitting Problem

The splitting problem in its simplest form supposes we have $h: M' \to M$ a homotopy equivalence and V a locally separating codimension-1 submanifold of M. The problem is to homotop h to a new map that restricts to the inverse image of V as a homotopy equivalence.

Unfortunately the answer to this is sometimes negative (in high dimensions), and we will build up to it by explaining first the problem where the main obstruction to this first arose, fibering.[15] During our first run we will not be comprehensive, but will only introduce some of the *dramatis personae*.

It is reasonable to extend the range of the discussion of splitting to allow the target to be a non-manifold as follows: Let X^n be a Poincaré complex; that is, a space that satisfies Poincaré duality in a suitably strong form with respect to all local systems.[16] We suppose that Y^{n-1} is a locally two-sided $(n-1)$-dimensional Poincaré complex, Poincaré-embedded in X. This means that there is a (perhaps disconnected) complex Z, such that $Y \cup Y$ is a boundary for Z, i.e. $(Z, Y \cup Y)$ is a Poincaré pair, which simply means that Z with those two copies of Y satisfies the Poincaré duality appropriate to manifolds with boundary. We insist that X is the result of gluing together the two copies of Y.

The splitting problem can now be phrased as: given a homotopy equivalence $h: M \to X$ and $Y \subset X$ Poincaré-embedded, can we homotop h so that h is transverse to Y,[17] and restricts to a homotopy equivalence between Y and its inverse.

---

[14] But we will discuss this more seriously in the Chapter 5.

[15] Very closely related to the invariants introduced by de Rham in *his* proof of the classification of lens spaces that we will discuss in Chapter 5.

[16] Details can be found in Chapter 2 of Wall (1968).

[17] Note that for transversality one does not need manifolds: a "normal structure" to the sub-object that is a vector bundle (or perhaps somewhat weaker than that) is enough.

One example where this is relevant is the following fibering problem:

Problem 4.8 Suppose M is a manifold with a surjection $\pi_1 M \to \mathbb{Z}$ (i.e. a map $f \colon M \to S^1$ with connected homotopy fiber). When is there a fibration of M over the circle, i.e. a map to S^1 realizing this data (e.g. homotopic to this map)?

In the case when $\pi_1 M \to \mathbb{Z}$ is an isomorphism, a beautiful necessary and sufficient condition was given by Browder and Levine (1966): fibering is possible iff the associated infinite cyclic cover has finitely generated homology. (After all, if the manifold fibers, the fiber would be homotopy equivalent to this cover.)

The first hypothesis for the general problem should be an analogue of this kind of finiteness. It is convenient to ask that the infinite cyclic cover, written F, should have as cellular chain complex $C_*(F) \sim C_*$ a finitely generated projective complex.[18] A space with finitely presented fundamental group and with this property on its cellular chain complex is called *finitely dominated*.[19]

Under such conditions, we have the covering translate $\eta \colon F \to F$, and a homotopy equivalence $M \to T(\eta)$ from M to the mapping torus. This mapping torus, $(T\eta)$, is a Poincaré complex (it is homotopy equivalent to M) and with some effort one can show that F is too (it's finitely dominated). Splitting this map is part of homotoping the map to a fibration.

Theorem 4.9 (Siebenmann, 1970a; Farrell, 1971b) *If $f \colon M \to S^1$ is a map which is a surjection on fundamental groups, with* $\dim M > 5$, *then f is homotopic to a fibration iff the following hold:*

(1) *the associated infinite cyclic cover of M is finitely dominated;*
(2) *an obstruction that lies in* $\mathrm{Wh}(\pi)$ *vanishes.*

If M has boundary and its boundary already fibers, the same result holds for the problem of extending the fibration to M.

The Whitehead group $\mathrm{Wh}(\pi)$ is defined purely algebraically. Let $\mathbb{Z}\pi$ be the integral group ring of the group π (it consists of finite formal sums of symbols of the form $a_g g$, where the a_g are integers, and the g are elements of π – made into a ring in the only sensible way imaginable).

$\mathrm{GL}_n(\mathbb{Z}\pi)$ is the group of invertible $n \times n$ matrices over this ring. We can

---

[18] This is reasonable given the special role that projective modules play in homological algebra: checking projectivity is often much simpler than freeness – cohomological vanishing suffices for the former, but not the latter.

[19] This is equivalent to being a retract of a finite complex, just like a projective module is retract (i.e., factor) of a free module.

stabilize by adding an identity in the bottom right to an invertible matrix:

$$GL_n(\mathbb{Z}\pi) \to GL_{n+1}(\mathbb{Z}\pi) \to \cdots \text{ whose limit is } GL(\mathbb{Z}\pi).$$

The first algebraic K-group is defined by $K_1(\mathbb{Z}\pi) = GL(\mathbb{Z}\pi)/E(\mathbb{Z}\pi)$, where $E(\mathbb{Z}\pi)$ is the group generated by elementary matrices. A lemma of J.H.C. Whitehead (which can be found in any introduction to K-theory) tells us that this quotient is abelian; indeed the product of matrices AB is equivalent in K_1 to $A \oplus B$ and $E(\mathbb{Z}\pi)$ is the commutator subgroups of $GL(\mathbb{Z}\pi)$.

Finally set

$$\text{Wh}(\pi) = K_1(\mathbb{Z}\pi)/(\pm\pi),$$

where we mod out by the obvious invertible 1×1 matrices $(\pm g)$ where, again, g denotes a group element in π.

Note that, if π is trivial, this group is trivial using row operations from linear algebra (and the Euclidean algorithm). When π is finite cyclic, $\text{Wh}(\pi)$ contains the obstruction that de Rham used to distinguish homotopy equivalent lens spaces (see Cohen, 1973. The group $\text{Wh}(\pi)$ has an important part to play in the Borel story and it is therefore important not to discuss it too early, since it also provides a possible obstruction to homotoping maps to homeomorphism. We will rely on a theorem of Bass *et al.* (1964) which tells us that $\text{Wh}(z^k) = 0$.

The fibering theorem is an analogue of an earlier important theorem:[20]

Theorem 4.10 (*h*-cobordism) *Let M be a compact manifold,* dim > 4. *Then there is a one-to-one correspondence:*

$$\tau: \{W \mid \partial W = M \cup ? \text{ and } W \text{ deform retracts to both } M \text{ and } ?\} \leftrightarrow \text{Wh}(\pi).$$

In particular, if $\text{Wh}(\pi) = 0$, then M and "?" are (Cat)-homeomorphic. In general, τ is called the torsion of the homotopy equivalence $M \to W$. A homotopy equivalence with zero torsion is called simple (see §5.5.3 for more discussion). The *h*-cobordism theorem asserts that *s*-cobordisms, i.e. *h*-cobordisms where the inclusion of one side is simple, are products, so finding *s*-cobordisms between M and "?" ends up being the same as finding (Cat)-homeomorphisms between these manifolds.

Almost all homeomorphisms constructed in high-dimensional topology make use of this theorem or ideas from its proof. In particular, the proof of the Borel conjecture for the torus $\mathbb{T}^k$ depends on the Bass–Heller–Swan calculation and the *h*-cobordism theorem in just this way.

---

[20] This is due to Smale in the simply connected case and is the backbone of his proof of the high-dimensional Poincaré conjecture. In general (for the PL and smooth categories) it is due to Barden, Mazur, and Stallings (see Kervaire, 1969); Rourke and Sanderson, 1982). The topological case is due to Kirby and Siebenmann.

The condition that W has its boundary components are deformation retracts is analogous to the finite domination of the infinite cyclic cover: it asserts the homotopical possibility of the geometric structure (i.e., a product structure or a fibering) we are seeking. In both cases, the obstruction lies in the same algebraic K-group, and they arise in both cases through "handlebody theory," the manipulation of handles to mimic geometrically that homotopy theory (which is a manipulation of cells) – except that one has to occasionally replace algebraic isomorphisms by geometric moves that require (freeness in the place of projectivity or) elementary matrices and their products.

In the presence of the h-cobordism theorem, the fibering theorem is then directly visible as a combination of two obstructions. The first is a splitting obstruction, which would give us a submanifold F' in M, homotopy equivalent to F. When we cut M open along F, we'd then get an h-cobordism from F to itself that also seems to involve a Wh obstruction to being a product. If it is a product, then we have exhibited M as a fiber bundle over S^1.

Actually, and this is an important point, the π–π theorem enables one to work in reverse and prove the splitting theorem from the fibering theorem – the theorem of Farrell. The advantage of this is that the splitting theorem is relevant to many more manifolds and submanifolds than the fibering theorem.

In general, there are splitting theorems unrelated to fibering problems. For simplicity we shall just state a version adequate for our current purposes[21] that incorporates the vanishing of Whitehead groups.

Theorem 4.11 *Suppose M is a manifold with free abelian fundamental group $\mathbb{Z}^k$, and $V \subset M$ is a codimension-1 submanifold with fundamental group $\mathbb{Z}^{k-1}$. Then if $\dim M > 5$, any homotopy equivalence $f\colon M' \to M$ can be split along V.*

We can now use this to prove the Novikov conjecture and try to prove the Borel conjecture.

If we are dealing with manifolds W with free abelian fundamental group $\mathbb{Z}^k$, then the Novikov conjecture is essentially the statement about the homotopy invariance of the signatures of the inverse images of subtori $\mathbb{T}^i \subset \mathbb{T}^k$ (for the classifying map $h\colon M \to \mathbb{T}^k$).

The tori are inductively stacked $\mathbb{T}^i \subset \mathbb{T}^{i+1} \subset \mathbb{T}^{i+2} \subset \cdots \subset \mathbb{T}^k$, so we only have to deal with the codimension-1 situation. An important but not difficult lemma (that is essentially the same one that arises in the three-dimensional topology of Haken manifolds) is that we can homotop the map $h\colon M \to \mathbb{T}^k$

---

[21] Cappell (1974a,b) gave an essentially complete theoretical analysis of this problem. In some cases, his analysis contains a non-K-theoretic obstruction that we will return to when we discuss group actions on aspherical manifolds and the Farrell–Jones conjecture.

to one where $h^{-1}(\mathbb{T}^{k-1})$ has fundamental group $\mathbb{Z}^{k-1}$. Then we can split the homotopy equivalence $f \colon M' \to M$ along $h^{-1}(\mathbb{T}^{k-1})$ till we get all the way down to the inverse image of $\mathbb{T}^i$.

So, by the homotopy invariance of signature, we have proven the Novikov conjecture.

Except for one little point: there is a dimension condition in the theorem. The fibering and splitting theorems do fail in low dimensions. So we should get stuck when the codimension gets high enough.

However, for the purposes of the Novikov conjecture, this is irrelevant, by the multiplicativity property of (higher) signatures. We can cross our manifold by $\mathbb{CP}^2$ a number of times to increase dimension as much as we need, without changing any invariants (sign $(\mathbb{CP}^2) = 1$) and then apply this argument.

For the Borel conjecture, it would be nice to argue inductively and get a decomposition of a homotopy torus as resembling homotopically the product of n copies of $(S^1, *)$ and then invoke the Poincaré conjecture (to handle these manifolds that inductively have boundary a sphere and are contractible). The dimension issue arises again but surgery theory gives a way to get around this. We will discuss the details of this later in this chapter. In the end, though, hidden behind the surgery method is a periodicity, which is an indirect application of the idea of crossing with $\mathbb{CP}^2$ still lurking behind the scenes.

4.4.2 Lusztig's Method

Recall that we had mentioned before that for any $\rho \colon \pi_1(M^{4k}) \to U(n)$ we can define a signature $\operatorname{sign}_\rho(M)$ using $H^{2k}(M; \rho)$ thinking of ρ as defining a flat bundle.

The simple idea in Lusztig's (1972) was to do this in a family. Concretely, Lusztig considered $n = 1$, so let $\Lambda = \operatorname{Hom}[\pi_1(M^{4k}), U(1)]$; it is a finite union of tori. (It is an abelian group under pointwise product.)

The idea is that, where before we thought of the signature as being an integer, one should reconsider it as a (virtual) vector space (and the integer as its dimension). Then, varying the construction over points of Λ, produces a vector bundle over Λ,[22] associated to M. So consider the bundle of $H^{2k}(M; \rho)$s over Λ; using an auxiliary Hermitian metric on this bundle, we can diagonalize the family of cup product pairings and obtain a virtual bundle: the difference between the positive and negative sub-bundles.

---

[22] There is an oversimplification here. The family of $H^{2k}(M; \rho)$ might not be a bundle, because of jumps in dimension. The same issue arises in the Atiyah–Singer index theorem for families where, for some values of the parameters, the dimension of kernel and cokernel might jump. One has to introduce some perturbations to the family to obtain genuine kernel and cokernel vector bundles. See Atiyah and Singer (1968b).

The Atiyah–Singer theorem for families gives a formula (for the Chern character of) $\mathrm{sign}_\pi(M) \in K^0(\Lambda)$. Note, by the way, that this invariant lies in a place that is covariant in π, as both Λ and K^0 are contravariantly functorial in π.

Now $\mathrm{sign}_\pi(M)$ *detects exactly the higher signatures associated to products of one-dimensional cohomology classes* (or dually the image of the higher-signature class from $\bigoplus H(K(\pi, 1); \mathbb{Q})$ in $\bigoplus(H(K(H_1(\pi, \mathbb{Z})/\mathrm{torsion}), 1); \mathbb{Q}))$. The numerology of $K^*(\Lambda)$ makes this at least believable. For $\mathbb{Z}^k$, the space of $U(1)$ representations, Λ is a torus $\mathbb{T}^k$, and K^* has a Künneth formula – giving us C_i^n copies of K^0, where C_i^n is the binomial coefficient; K^0 itself has a $\mathbb{Z}$ every fourth dimension (just the right size for a signature).[23]

This method is very nice in that not only does it prove that the relevant higher signature is a homotopy invariant, it also gives a formula that tells us why it's true. That is also true of the next variant that we will describe.

4.4.3 Using the Atiyah–Kodaira Fiber Bundle

A rather different verification of the Novikov conjecture is possible for surfaces of high genus[24] by making use of a different "representation." Atiyah and Kodaira have given a surface bundle over a surface with nonzero signature (see, for example, Atiyah, 1969).

The method is this: one takes a product of surfaces, and inside of it a subsurface that intersects each fiber the same way (i.e. in a fibered way). In this way the subsurface is a – perhaps disconnected – covering space of the base surface. If the subsurface is trivial as a class in mod 2 homology, then one can take the branched $\mathbb{Z}/2$ cover of the product along the surface. If the surface has nontrivial Euler class (e.g. its self-intersection is nontrivial integrally, or equivalently the cup square of its Poincaré dual is nontrivial), then it turns out that the signature of the total space of the branched cover is nontrivial.[25]

The details don't matter. What matters is this bundle whose total space has nonzero signature although the base does not. We denote the base of this bundle by Σ and the bundle itself by π.

Suppose now that we have $f\colon M^{4k+2} \to \Sigma^2$ a manifold with a map to a surface, we can *define*

$$\mathrm{sign}(f) := \mathrm{sign}(f * \pi).$$

[23] Actually, there's a $\mathbb{Z}$ every second dimension, but the ones that arise in $2 \bmod 4$ don't come up for these signature operators. Had we worked in KO, they wouldn't be observed.

[24] Note that, for any genus, we can reduce the Novikov conjecture to the special cases of $\mathbb{Z}$ and $\mathbb{Z}^2$ since all the cohomology of the surface is pulled back from one of these groups.

[25] These calculations can be done using the equivariant form of the signature theorem (Atiyah and Singer, 1968b).

Cobordism invariance of signature and its multiplicative properties show that this invariant of f only depends on the class that (M, f) represents in $\Omega_{4*+2}(\Sigma^2) \oplus_{\Omega^*(*)} \mathbb{Q}$, which is $f_*(L_k(M) \cap [M]) \in H_2(\Sigma; \mathbb{Q})$. In other words[26]

$$\text{sign}(f) = C\langle f^*[\Sigma] \cup L_{4k}(M), [M]\rangle,$$

where C is determined by setting $f = \text{id}$.

A by-product is that we now know the Novikov conjecture for high-genus surfaces by a homotopy-invariant formula (just as we had achieved for the case of $\mathbb{Z}$ in §4.3).

4.5 Novikov's Theorem

Gromov observed that the Atiyah–Kodaira example can be used to simplify[27] the proof of Novikov's theorem on topological invariance of rational Pontrjagin classes.

Theorem 4.12 *If $f \colon M' \to M$ is a homeomorphism between smooth manifolds, then $f^*(p(M)) = p(M') \in \bigoplus H^{4i}(M'; \mathbb{Q})$.*

Of course, we will prove the equivalent that $f * (L(M)) = L(M')$. We will first describe the argument if f is PL following Thom, Milnor, Rochlin, and Schwartz. Without loss of generality, we will assume that the dimension of M is odd, since we can cross M with an odd-dimensional sphere, without loss of information.

We would like to give a PL-invariant calculation of $\langle L(M), c \rangle$ for any homology class c. Note that its Poincaré dual is odd-dimensional. According to Serre's thesis, for every odd-dimensional cohomology class $\text{PD}(c)$, there is a nonzero multiple $N\text{PD}(c)$ and a map $f \colon M \to S^{2r-1}$ multiple (which is unique up to homotopy after a further multiple) so that $N\text{PD}(c) = f^*([S])$. We define $L(M)$ to be the unique cohomology class with

$$\langle L(M), c \rangle \text{sign}(f^{-1}(*))/N,$$

where f is the map to the sphere defined above associated to the Poincaré dual of c, and $*$ is a regular value for f. Using cobordism invariance, this is well defined and linear, defining a unique rational cohomology class – which, if M is smooth, the Hirzebruch formula identifies with the usual L-class.

---

[26] Atiyah deduces this and a stronger formula from the index theorem for families, but our point here is to point out that this example is somewhat different from the Lusztig example, although one can unify them.

[27] Although it still does make use of a key trick of Novikov, the audacious introduction of fundamental group into a simply connected problem.

Regular values exist by Sard's theorem in the smooth category; in the PL category, they exist for a less deep reason. Choosing triangulations so that f is locally affine, any point in the interior of a top simplex is a regular value. If we knew transversality in the topological category (which is indeed true, thanks to Kirby and Siebenmann), we could complete the argument in Top, as well, but that is a deeper result than Novikov's theorem.

What we have to prove is this:

Lemma 4.13 *If $h: U \to M \times \mathbb{R}^i$ is a homeomorphism between smooth manifolds, then* $\operatorname{sign}(M) = \langle L(U), h^*[M] \rangle$.

We shall actually make use of the fact that f is a hereditary homotopy equivalence: that is, f is a proper homotopy equivalence restricted to any open subset of the target.

Without loss of generality we will assume i is even, $i = 2l$. We note that there is a product of Atiyah–Kodaira bundles $\pi \times \pi \times \cdots \times \pi$ over $\Sigma \times \Sigma \times \cdots \times \Sigma$ the product of l-surfaces. This bundle can be used for proving the Novikov conjecture for the fundamental class of a product of these surfaces.

Note that the punctured manifold $(\Sigma \times \Sigma \times \cdots \times \Sigma - p)$ immerses[28] in $\mathbb{R}^{2l}$. Then $M \times (\Sigma \times \Sigma \times \cdots \times \Sigma - p)$ immerses in $M \times \mathbb{R}^{2l}$. We pull back the Atiyah–Kodaira bundle over this manifold and would like to take the signature of its total space (to recover $\operatorname{sign}(M)$). This is slightly tricky, because we are in a noncompact situation, so we have to see that the signature is what we expect it to be (now defined using signature where we mod out by the torsion). This is fairly easy because the bundle is trivialized at ∞ (which is the neighborhood of p – so we know that the homotopy type at infinity is that of $M \times S^{2l-1} \times$ Fiber, and can calculate the effect[29] of gluing or removing a plug of the form $M \times \mathcal{D}^{2l} \times$ Fiber.

Now for U we can pull back using the homeomorphism h to obtain a homeomorphic smooth manifold, and associated bundles, etc. Since everything is proper homotopy equivalent to the other side, we get the same total signature. However, on computing the signature of this total manifold we get (a nonzero multiple of) $\langle L(U), h^*[M] \rangle$.

The executive summary is that we find a codimension-2l signature by computing the signature of an associated 2l-dimensional bundle over the manifold!

---

[28] This is not at all obvious, but it follows from immersion theory (often called Smale–Hirsch theory): any parallelizable open manifold immerses in Euclidean space of the same dimension. This can be found in almost any treatment of h-principles, since it's the prototype of such a theorem.

[29] If one glues two manifolds with boundary together along their complete boundary, one obtains the sum of the signatures, and the signature of $M \times \mathcal{D}^{2l} \times$ Fiber is zero. This formula is called Novikov additivity.

This is the trick of §4.4 for some cases of the Novikov conjecture, and it suffices for the current application to Novikov's theorem.

4.6 Curvature, Tangentiality, and Controlled Topology

Out goal in this section is to introduce the idea of doing topology with control and explain the proof of the following theorem:[30]

Theorem 4.14 (Ferry and Weinberger, 1991) *Suppose W is a complete non-positively curved manifold and $f: W' \to W$ is a homotopy equivalence which is a homeomorphism outside of a compact set. Then f is tangential, i.e. f pulls back tangent bundles (in a way compatible with the identification given by the homeomorphism outside of some larger compact set).*

This implies[31] that $W' \times \mathbb{R}^3$ is homeomorphic to $W \to \mathbb{R}^3$ so it's definitely progress. In §4.7, we will see this, as well as why this result implies the Novikov conjecture for $\pi_1 W$.

The use of the rel∞ condition was discussed in the "morals" section (§3.8) of Chapter 3. Without it, the theorem would be highly false, as we've seen.

A key role in the proof is played by the following important theorem due to Ferry:[32]

Theorem 4.15 (Ferry (1979)) *Let M^n be a compact topological manifold, endowed with a metric. Then there is an $\varepsilon > 0$ such that if $f: M \to N$ is a continuous map to a connected manifold of dimension less than n, with diam$(f^{-1}(n)) < \varepsilon$ for all $n \in N$, then f is homotopic to a homeomorphism.*

The ε is related to the size of the smallest handle in a handle decomposition of M, so if M is noncompact, we can sometimes guarantee that the theorem holds anyway. There are also ε–δ statements that describe how far, in some sense, f has to be moved to make it into a homeomorphism. We'll need both kinds of refinement below when we apply the theorem.

---

[30] This was strongly inspired by earlier work of Kasparov proving the Novikov conjecture for $\pi_1 W$ (see §8.5) by analytic methods.

[31] This is a little white lie: the method of proof gives this improvement. Tangentiality by itself would not control what dimensional Euclidean space we'd need to cross with to obtain isomorphism. To get this dimension down to three, it's important that the tangentiality be "compatible with an identification of Spivak fibrations" so that one obtains vanishing normal invariant – not just the image of this under the map $[W/\infty : G/\mathrm{Top}] \to [W/\infty : B\mathrm{Top}]$, which is the assertion of the theorem. Once one has this, the π–π theorem quickly gives the homeomorphism.

[32] Actually, Ferry originally proved it for $n > 4$, but it's since been shown to be unconditional through advances in low-dimensional topology (the largest being the solutions of the three- and four-dimensional Poincaré conjectures by Perelman and Freedman).

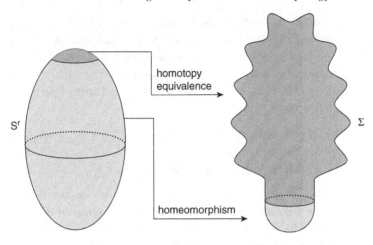

Figure 4.1 Proof of the Poincaré conjecture from Ferry's theorem.

This statement is of the form "an almost homeomorphism" is "almost a homeomorphism." Statements of this general type are sometimes trivial, sometimes trivially false, sometimes nontrivially false, and sometimes true, nontrivial, and useful. This theorem is of the fourth sort. To appreciate it, we shall give two examples:

Example 4.16 (Ferry $\Rightarrow$ Poincaré) Let S^n be the usual round n-sphere. And let ε be the epsilon guaranteed by Ferry's theorem. Let Σ be any homotopy n-sphere. I claim (assuming that Σ is a manifold!) that it is possible to build a map $f \colon S^n \to \Sigma$ where all point inverses have diameter less than ε.

Pick a point p and a neighborhood homeomorphic to $\mathbb{R}^n$. We map the complement of the $\varepsilon/2$ north polar ice cap in S^n homeomorphically to a ball in this neighborhood. The rest of Σ is contractible, so the map restricted to the $\varepsilon/2$-sphere around the north pole extends inwards (as a homotopy equivalence, although this is irrelevant to the application of Ferry's theorem) over the ice cap to Σ (lying entirely in the complement of the neighborhood of p; see Figure 4.1).

Let's examine the point inverses. If q lies in the neighborhood of p, then its inverse image $f^{-1}(q)$ is a single point, and has diameter equal to 0. If q lies outside the neighborhood, then its inverse image is constrained to lie in the $\varepsilon/2$ polar ice cap, and hence has diameter less than ε. Ferry's theorem then asserts that f is homotopic to a homeomorphism.

This proof shows the remarkable versatility of Ferry's theorem as a tool: the huge unexplored region in the range manifold is here shrunk to be in the $\varepsilon/2$

polar ice cap, while the small coordinate chart around one point is expanded to be almost the whole sphere. It feels like a talk by a weak student who spends almost his full hour explaining trivialities, leaving only a couple of minutes to the whole essence of the matter! Nevertheless, in topology, Ferry's theorem says that this works! Of course, there's a price – the continuity of the map from the domain to the range.

Example 4.17 (Ferry $\Rightarrow$ a virtual Borel conjecture)   Let V be a homotopy $\mathbb{T}^n$; then we shall see that every sufficiently large cover[33] of V, say with covering group $(\mathbb{Z}/k)^n$ for k large, is homeomorphic to the torus $\mathbb{T}^n$.

Let $f\colon \mathbb{T}^n \to V$ be a homotopy equivalence. Let's consider the map it induces between universal covers. Note that there is a universal bound C for all point inverses (for the map is automatically proper, and the bound for point inverses for one fundamental domain of the $\mathbb{Z}^n$-action works for all points by equivariance). Let ε be an ε appropriate to the torus $\mathbb{T}$. Suppose that $k > 2C/\varepsilon$, then we can identify the $(\mathbb{Z}/k)^n$ cover of $\mathbb{T}$ with $\mathbb{T}$, and we now have a map from $\mathbb{T}$ to a cover of V with point inverses of diameter less than ε. (The extra factor of 2 is to be in the range that the map from $\mathbb{R}^n \to \mathbb{T}^n$ is a local isometry.)

Remark 4.18 (On circularity)   Of course, if the Poincaré conjecture and a virtual Borel conjecture for the torus were used in the proof of Ferry's theorem, this would be a circular argument. (Even so, the above should convince that the theorem is not vacuous!)

There are probably three different arguments for this theorem. They all go via the α-approximation theorem of Chapman and Ferry, which I will not describe – but its essential difference is that it measures sizes over the target, not the domain.

The original proof (Chapman and Ferry, 1979) is based on modifying the proof of a weaker version of the theorem, Siebenmann's CE-approximation theorem (Siebenmann, 1972). That theorem asserts:

Theorem 4.19 (CE-approximation)   *A map $f\colon M \to X$ between manifolds is a limit of homeomorphisms (in the compact open topology) iff it is CE, i.e. if all $f^{-1}(x)$ are null-homotopic in arbitrarily small neighborhoods, i.e. iff f is a hereditary homotopy equivalence.*

This in turn used Kirby's torus trick, the basic tool in triangulation theory and requires a virtual Borel conjecture for tori. This argument would be indeed circular.

---

[33] That is, associated with any subgroup that intersects a metric ball of sufficiently large radius (depending on the original homotopy equivalence) only in the identity.

There is another proof, due to Quinn (1979, 1982b, 1982c, 1986), that is based on a controlled h-cobordism argument. This seems like it's essentially a generalization of the way the Poincaré conjecture was proved. However, the analogue of the fact that $\text{Wh}(e) = 0$ is more difficult in the controlled situation and Quinn's proof uses the torus trick. This can be avoided by more recent methods of calculating.

However, finally, there is a third proof that is based on combining two approaches. The first is an engulfing argument (based loosely on Stallings's proof of the Poincaré conjecture, rather than Smale's) due to Chapman (1981) that reduces the α-approximation theorem to the CE-approximation theorem.

The CE-approximation theorem itself has an amazing extension to the situation where X is not assumed to be a manifold (and is hence very useful for proving that spaces are manifolds!) due to R.D. Edwards (it is the goal – achieved – of Daverman, 2007). In any case, Edwards's proof is purely geometric and does not rely on any algebraic tools, neither torus trickery nor surgery. So, finally – using this combination – this argument for the second application does not have to be viewed as circular.

Thus, the least generous view one could have is that Ferry's theorem somehow *is* a form of the Poincaré conjecture, but in liberating that problem from the sphere, we have obtained an extremely useful tool and perspective on the problem of homotoping maps to homeomorphisms. And, indeed, there certainly are many other arguments in the spirit of the ones above based on Ferry's theorem that are far from circular; indeed, much of the work on the Borel conjecture since the 1980s has this flavor (but are much more involved; see Chapter 8).

Now let us return to the proof of the main theorem of this section. For simplicity, we will first sketch our argument for W compact, where the result is actually considerably simpler – although not much simpler from the point of view that we adopt![34] We first note that the salient feature of the tangent bundle to a manifold is that it is a bundle of $\mathbb{R}^n$s – given a section – over a manifold, so that around the 0-section, the fiber direction is the same as the base direction. (In the smooth case, the exponential map sets up such an isomorphism.)

As a result, for an aspherical manifold $W = \widehat{W}/\Gamma$, we can consider the following quotient as a model for the tangent bundle:

$$TW \approx (\widehat{W} \times \widehat{W})/\Gamma.$$

While we are used to saying "the universal cover" of a space, this notion actually

---

[34] The compact case was earlier proved by Farrell and Hsiang (1981) and doesn't need any version of Ferry's theorem.

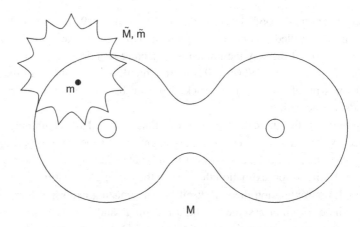

Figure 4.2 The moving family of universal covers as a tangent bundle.

requires a choice of base point, and as we vary the base point, this can indeed be a nontrivial bundle (see Figure 4.2).

In this model, the differential of the map $f: W \to W'$ (at the point w) is the lift of f to the universal cover $\widehat{W} \to \widehat{W}$ (based at the points w and $f(w)$, respectively). Notice that this map has bounded size point inverses (as in the case of the torus above). At this point let us use the Riemannian tangent bundle and the exponential map to a map $TW_w \to \widehat{W}$. The fibers are now Euclidean spaces, and, *using non-positive curvature*, the point inverses are still of bounded size: as geodesics spread apart in non-positive curvature, the inverse of the exponential map is Lipschitz.

Now we apply Ferry's theorem to the family of maps $TW_w \to \widehat{W'}$ to get a family of homeomorphisms.

In the closed case, one can actually "go all the way to ∞" and get an isomorphism between the ideal sphere bundles.[35] However, in the noncompact case we cannot do this, since we must interpolate between the "infinitesimal isomorphism" on tangent spaces coming from the homeomorphism outside of a compact, and the process we do in neighborhoods of points in $\widehat{W}$. As a result, Ferry and Weinberger (1991) instead argue about isotoping the family of balls of radius R, namely $\mathrm{B}(R, w)$, in TW_w to embeddings around $f(w)$ in $\widehat{W'}$.

Finally, as explained in §3.3, tangentiality is slightly less than we would want: we need the map to not change the identification of spherical fibrations guaranteed to us by Atiyah's theorem. This is essentially achieved by ensuring

---

[35] This method is due to Farrell and Hsiang (1981) and is reminiscent of one of the key steps in Mostow's (1968) proof of his rigidity theorem for closed hyperbolic manifolds.

that the isotopies are covered by homotopies that do not distort the function at ∞.

4.7 Surgery, Revisited

In Chapter 3, we discussed surgery in the special case where there are no obstructions, the π–π situation. In that case, the discussion ended up being essentially homotopy-theoretic, and the main results were the structure of classifying spaces, and therein lay the differences between the various categories.

We shall now discuss the case of closed manifolds and in a purely topological setting where the results seem to be in their most perfect form. The reader should treat these results as truths coming from on high: we shall not explain why they are true or take the form that they do.

On the other hand, since the moral of Chapter 3 was that functoriality is critical to our program, the reader should not object to a presentation of surgery theory in which functoriality plays a central role.

Let M be a compact manifold with ∂. We define the *structure set*:

$$S(M) =$$
$$\{(M', \partial M', f) \mid f : (M', \partial M') \to (M, \partial M) \text{ a simple homotopy}$$
equivalence which restricts to a homeomorphism on the $\partial\}/s$-cobordism.

For manifolds, s-cobordism, thanks to the h-cobordism theorem (see §4.4), is the same thing as being a product. However, to get the best properties of $S(M)$, it is convenient to allow in some non-manifolds in the definition of $S(M)$.

Definition 4.20 A homology manifold is a finite-dimensional absolute neighborhood retract, (ANR)[36] X so that, for all x, and all i,

$$H_i(X, X - x) \cong H_i(\mathbb{R}^n, \mathbb{R}^n - 0).$$

Such an X satisfies Poincaré duality (in the noncompact sense), and therefore so does every open subset. A good example is the suspension of a homology sphere.[37] The cone points are the only non-manifold points, but they satisfy the hypothesis of the definition as their links are homology spheres.

Definition 4.21 A space X has the disjoint disk property (DDP) if, for any

---

[36] we are including this hypothesis in the definition of homology manifold, which is not a fully standard decision.

[37] A homology sphere here is a closed manifold with the $\mathbb{Z}$-homology of the sphere. Thanks to the Poincaré conjecture, such a space is the sphere (if dim > 1) iff it is simply connected.

$f, g: \mathcal{D}^2 \to X$ and any $\varepsilon > 0$, there are perturbations, f' and g', such that $d(f, f') > \varepsilon$ and $d(g, g') < \varepsilon$ and $f'(\mathcal{D}^2) \cap g'(\mathcal{D}^2) = \varnothing$.

Note that manifolds of dimension 5 all satisfy DDP (as do many other spaces that are not homology manifolds). The DDP fails for the suspension example: any two $\mathcal{D}^2$s that go through a cone point, but whose boundaries map nontrivially in π_1(link) cannot be moved disjoint from the cone point.

Indeed DDP can fail quite dramatically: Daverman and Walsh gave an example (a *ghastly homology manifold*) of a homology manifold and a nice curve, so that every singular 2-disk it bounds contains an open set.

Edwards's theorem is a CE-approximation theorem for DDP homology manifolds:[38]

Theorem 4.22 (See Daverman, 2007)   *A map $f: M \to X$ from a manifold to a homology manifold with the DDP is a limit of homeomorphisms (in the compact open topology) iff it is CE, i.e. all $f^{-1}(x)$ are null-homotopic in arbitrarily small neighborhoods, i.e. iff f is a hereditary homotopy equivalence.*

A homology manifold X for which a CE map $f: M \to X$ from some manifold exists is called *resolvable*. Quinn (1982a, 1987b) showed that, if X is a connected homology manifold and it contains any resolvable open subset (e.g. it has a manifold point), then it is resolvable! Quinn defined an invariant $I(X) \in Z$, which is 1 iff X is resolvable. We can compute $I(X)$ locally from a neighborhood of any point: it is $1 \bmod 8$, and it has the property that $I(X \times Y) = I(X)I(Y)$.

It turns out that using DDP homology manifolds in the definition of $S(M)$ above is the "right" thing to do. First of all, if $\partial M \neq \varnothing$, by Quinn's theorem combined with Edwards's, we are not actually allowing any nonmanifolds in.

Secondly, we will see that even allowing in homology manifolds does not affect the Borel conjecture: if it is true for manifolds, it is true for homology manifolds.

And, finally, it is with this more elaborate definition that $S(M)$ achieves its strongest functorial properties. We begin with that last point:

Theorem 4.23 (Bryant *et al.*, 1996)   *If M is a manifold, $S(M) \cong S(M \times \mathcal{D}^4)$.*

This is an analogue of Bott periodicity. Actually, there is a version of the Thom isomorphism theorem: $S(M) \cong S(E)$ if E is an oriented $\mathcal{D}^{4k}$-bundle over M. However, even the periodicity statement has important consequences. Setting $M = S^n$ we see the right-hand side is $\mathbb{Z}$,[39] so we "need" the sphere

---

[38] But the version where both M and X are DDP homology manifolds would be much better!
[39] $[S^n \times \mathcal{D}^4/\partial: F/\mathrm{Top}]$ has a $\mathbb{Z}$ from $\pi^4(F/\mathrm{Top})$.

to have nontrivial structures – seeming "counterexamples" to the Poincaré conjecture.

This $\mathbb{Z}$ is given by $(I(X)-1)/8$. There is a $\mathbb{Z}$'s worth of homotopy spheres that are really determined by their local structure. Conjecturally, there is a unique homotopy sphere for each integer. This would make a lot of sense: the proof of the Poincaré conjecture by Smale uses the h-cobordism theorem, and starts by choosing a small ball neighborhood of a point in a coordinate chart, using the manifold hypothesis. For other local indices, the very first trivial step seems to be the one that is obstructed.

In addition, note that on the right-hand side of this equation there is an abelian group structure: the structures we are using are homeomorphisms on the boundary. Thinking of $\mathcal{D}^4$ as a cube, we can glue along faces to "add" elements, and the usual proof that π_2 is abelian shows that this is a commutative group structure. (The elements of $S(M)$ that correspond to manifolds are a subgroup. The map $(M',f) \rightarrow (I(M)-I(M'))/8$ is a group homomorphism $S(M) \rightarrow \mathbb{Z}$.)

Using this, we can easily finish off the proof of the Borel conjecture for the torus we had sketched in §4.4: using periodicity there is no "low dimension" to push past. Of course, in this view, the vanishing of $S(*)$ should not be viewed as a triviality: it is – in light of periodicity – the Poincaré conjecture[40] in dimensions a multiple of 4.

A next formal point is that it then becomes reasonable to have additional groups, namely $S(M \times \mathcal{D}^i)$ for any i. Doing this systematically leads to the following definition for arbitrary finite CW-complexes X.

Definition 4.24 Let X be a finite complex. We define $S_n(X)$ as $S(M)$ where M is any compact oriented n-manifold with boundary simple homotopy equivalent to X. If there is none, define $S_n(X) = S_{n+4k}(X)$ where $4k$ is large. If X is infinite, take the limit over the finite subcomplexes of X.

Note that, thanks to periodicity, the groups $S_n(?)$ are actually covariantly functorial.[41] After a periodicity if necessary, given $f: X \rightarrow Y$, we can embed the manifold for X into the manifold for Y, and get the "pushforward" of the structure to be obtained by gluing in the annular region. Such an embedding, after further stabilization, is unique up to isotopy. This turns the $S_n(?)$ into homotopy functors.

Moreover, it does not take much to imagine the meaning of $S_n(Y,X)$ for a pair. We have to consider manifolds with boundary whose interiors are homotopy

---

[40] Here viewed as the vanishing of structures of the disk, which tacitly has the boundary condition – and therefore the manifold hypothesis – built in.

[41] This is completely analogous to the wrong way maps defined at the beginning of Atiyah and Singer (1968a) using Bott periodicity (interpreted as a Thom isomorphism kind of statement).

equivalent to Y and whose boundaries contain a piece homotopy equivalent to X (and we work relative to the rest of the boundary).

Theorem 4.25 *The $S_n(?,?)$ are a sequence of covariant homotopy functors. They are 4-periodic and fit into an exact sequence of abelian groups:*

$$\cdots \to S_n(X) \to S_n(Y) \to S_n(Y,X) \to S_{n-1}(X) \to S_{n-1}(X) \to \cdots .$$

If M is an oriented n-manifold, then $S_n(M) \cong S(M)$.

(There is no difficulty in setting up an analogous theory for nonorientable manifolds, and a proper theory for the noncompact situation.)

Note that, from this perspective, the "cohomological term" that we had looked at in the $\pi-\pi$ theorem actually is naturally a homology theory (the functoriality flipped!); the perspectives in these two approaches to the theory are Poincaré dual: $[M^n : \mathbb{Z} \times F/\text{Top}] \cong H_n(M,\partial M; L(e))$, where $L(e)$ is the homology theory associated to $\mathbb{Z} \times F/\text{Top}$.

Theorem 4.26 *If $K(\pi,1)$ is a finite complex, then the statement that, for all n, $S_n(K(\pi,1)) = 0$, follows from the Borel conjecture.*[42]

(If it is a manifold, then the vanishing of all the $S_n(K(\pi,1))$ follows from the vanishing of all of the $S(K(\pi,1) \times \mathbb{T}^k)$. Note that $S(K(\pi,1) \times D^4)$ is a summand of $S(K(\pi,1) \times \mathbb{T}^4)$, so if there is an extra $\mathbb{Z}$ arising from a homology manifold, it arises for manifolds four dimensions higher.)

The theorem follows immediately from functoriality together with the Davis construction:[43] it produces from $K(\pi,1)$ an aspherical manifold M (of any large enough dimension) that has $K(\pi,1)$ as a retract, and therefore $S_n(K(\pi,1))$ is a summand of $S_n(M) = S(M)$ (or if one is opposed to homology manifolds, a factor of $S_n(K(\pi,1) \times \mathbb{T}^4)$).

We can now think of the Borel conjecture as the statement for π torsion-free, $S(K(\pi,1)) = 0$ (in all dimensions).[44]

We now note that if the Borel conjecture is true for π, and $\pi_1(M^n) \to \pi$ is an isomorphism, then we get an exact sequence:

$$\cdots \to S_{n+1}\big(K(\pi,1)\big) \to S_{n+1}\big(K(\pi,1),M\big) \to S_n(M) \to S_{n-1}(K(\pi,1)) \to \cdots ,$$

which becomes a homological calculation of $S(M)$:

$$S(M) \cong S_n(M) \cong S_{n+1}(K(\pi,1),M) \to H_{n+1}(K(\pi,1),M; L(e)),$$

---

[42] The converse requires an additional statement about vanishing of Whitehead groups. We will discuss it in Chapter 5.

[43] As noticed by Davis.

[44] We note that groups with torsion do not have finite-dimensional $K(\pi,1)$s. Here we are taking a leap of faith that all torsion free groups behave the same way as those with a finiteness condition.

the final isomorphism coming from the π–π theorem.

This isomorphism means that (when we take ∂) the (relevant) data comparing $M\prime$ to M, namely the difference of the L-classes, if we were working rationally, vanishes by the time we push further to $H_n(K(\pi, 1); L(e))$. This is exactly (an integral form of) the Novikov conjecture.

Note then the philosophy that emerges from functoriality (conditionally on the Borel conjecture): *A manifold is exactly as rigid as it is homologically similar to $K(\pi, 1)$.*

Note also, by a diagram chase, unconditionally, no homology in $H_n(M; L(e))$ that dies in $H_n(K(\pi, 1); L(e))$ contributes to $S(M)$. (The cokernel in degree $n + 1$ only contributes conditionally on Novikov-type statements.)

Let us be a bit more explicit and go back to a more classical view of surgery. It is high time that we mention the surgery exact sequence![45] This is an exact sequence that looks like:

$$\cdots \to L_{n+1}(\pi) \to S_n(M) \to H_n(M; L(e)) \to L_n(\pi) \to \cdots .$$

These groups, $L_n(\pi)$, are called "L-groups" or "Wall groups" and have a purely algebraic definition. They are 4-periodic, and describe the "obstruction to doing surgery to convert a degree-1 normal map into a (simple) homotopy equivalence."[46]

Classically,[47] this all was described as an exact sequence, valid in all categories:

$$\cdots \to L_{n+1}(\pi) \to S^{\text{Cat}}(M) \to [M : F/\text{Cat}] \to L_n(\pi),$$

with the following understanding: $S(M)$ is just a set with a distinguished element (the identity); and $[M : F/\text{Cat}]$ is just a set.[48] Thus, exactness has to be interpreted appropriately, with "distinguished element" taking the place of 0. The map $L_{n+1}(\pi) \to S^{\text{Cat}}(M)$ is then a group action. Given M and an element $\alpha \in L_{n+1}(\pi)$, there is a degree-1 normal map $W \to M \times [0, 1]$, such that, on the bottom ∂ of W, one has a homeomorphism to M, and on the top, one has a homotopy equivalence. In that circumstance, there is a rel ∂ surgery obstruction

---

[45] Ordinarily (and very properly) the centerpiece of presentations of surgery theory.

[46] There are different theories of surgery based on whether one wants to obtain homotopy equivalences, and then the equivalence relation is h-cobordism; or whether one wants a simple homotopy equivalence, and then s-cobordism is the equivalence relation. The L-groups differ by 2-torsion in a way described by the Rothenberg sequence; see Shaneson (1969).

[47] We refer to Wall (1968), Lees (1973), and Lück (2002a) for some expositions of the classical theory.

[48] Although for Cat = PL or Top, Sullivan's H-space structure turns this into an abelian group and the map $[M : F/\text{Cat}] \to L_n(\pi)$ is a homomorphism.

of the map $W \to M \times [0,1]$, which is α. The action then assigns to α the upper boundary homotopy equivalence.[49] One can show that this is well defined.

The groups $L_n(\pi)$ have purely algebraic definitions. The classical definition is given by Wall (1968), which makes it clear that $L_{2k}(\pi)$ is associated to $(-1)^k$-symmetric quadratic (or, better, Hermitian) forms over $\mathbb{Z}\pi$ and that $L_{2k+1}(\pi)$ are associated to their automorphisms. On the other hand, Ranicki (1980a,b) gave a very nice definition[50] in terms of chain complexes with duality, and their cobordism that makes the algebraic treatment "dimension-independent" and that is quite useful, because many more things end up directly defining elements in L-groups, and also it is more flexible for making constructions. Note that, by the algebraic definition, L-groups are 4-periodic. Indeed, this is the *source* of the periodicity of the S_n functors.

In any case, I should describe at least what happens in the simply connected case:

For $\pi = e$, then for n odd, $L_n(e) = 0$. For $4|n$, we have $L_n(e) \cong \mathbb{Z}$. The invariant is this: If $M \to X$ is a degree-1 normal map (ignoring the bundle data), then sign M = sign X is a necessary condition to be able to normally cobord f to a homotopy equivalence, since signature is both cobordism-invariant and homotopy-invariant. The isomorphism $L_0(e) \cong \mathbb{Z}$ is given by $1/8(\text{sign } M - \text{sign } X)$, the divisibility being a consequence of the bundle data that we suppressed: it ultimately forces the quadratic form on the cok $f*$ to have even numbers on the diagonal, which makes divisibility by eight automatic.

In dimensions that are $2 \bmod 4$, $L_2(e) \cong \mathbb{Z}/2$ with the isomorphism provided by the "Arf invariant." Here the antisymmetric bilinear form over $\mathbb{Z}$ is standard, but the refinement that a quadratic form has for $\lambda(x,x) = 2\mu(x)$ gives rise to an invariant $L_2(e) \cong L_{0,2}(F_2)$ (in the target, being characteristic 2, there is no difference between 0 and $2 \bmod 4$).

The homotopy group isomorphism $\pi_n(F/\text{Top}) \cong L_n(e)$ is essentially a consequence of the Poincaré conjecture (although one needs special low-dimensional arguments for $n < 5$), and perhaps makes calling the spectrum whose homology theory we said was dual to $[?: \mathbb{Z} \times F/\text{Top}]$ by the name $L(e)$ seem less peculiar.[51]

The classical surgery exact sequence continues infinitely to the left, but not

---

[49] This is an action because, for any homotopy equivalence $M' \to M$, we can build such a W associated to α, with the given homotopy equivalence being the bottom boundary.

[50] Strongly motivated by Chapter 9 of Wall (1968) that gives a cobordism treatment of relative L-groups that are complicated by the fact that manifolds with boundary are always both odd- and even-dimensional! An algebraic cobordism approach to L-groups was given first by Mischenko, but it was somewhat buggy at the prime 2.

[51] The truth lies somewhat deeper than this – and arises either from blocked surgery or from controlled topology. This might already be clearer in the coming section. For more information about all the classifying spaces arising in surgery theory, see Madsen and Milgram (1979)).

to the right (unless one restricts to the setting we described: Top and including homology manifolds).

Using the π–π theorem, we can describe what we have written as an analysis of the map obtained by crossing with $\mathcal{D}^3$:

$$S_n(M) \to S_n(M \times \mathcal{D}^3, M \times \mathcal{S}^2) \cong H_{n+3}(M \times \mathcal{D}^3, M \times \mathcal{S}^2; L(e)) \cong H_n(M; L(e))$$

(the last isomorphism is a suspension isomorphism in a generalized homology theory) and, in these terms, $L_n(\pi)$ occurs as measuring the obstruction to solving a splitting problem. In any case, this perspective gives the normal invariants functoriality as well, and the whole surgery exact sequence becomes functorial. We note the special case:

$$
\begin{array}{ccccccccc}
\cdots \to & L_{n+1}(\pi) & \to & S_n(M) & \to & H_n\big(M; L(e)\big) & \to & L_n(\pi) & \to \cdots \\
& \downarrow & & \downarrow & & \downarrow & & \downarrow & \\
\cdots \to & L_{n+1}(\pi) & \to & S_n\big(K(\pi,1)\big) & \to & H_n\big(K(\pi,1); L(e)\big) & \to & L_n(\pi) & \to \cdots .
\end{array}
$$

This factors the surgery obstruction map through a universal functorial map between two group-theoretic objects $H_n(K(\pi,1); L(e)) \to L_n(\pi)$.

This map is called the assembly map. It has several interpretations, one of which has to do with assembling things. In §4.8 we will explain that it has another interpretation in terms of "forgetting control" in the sense of controlled topology. There are similar maps in algebraic K-theory and operator K-theory, and they will occupy us in Chapter 5.

Note that the Borel conjecture can be rephrased using the assembly map as follows:

Conjecture 4.27 *If π is a torsion-free group, then the assembly map is an isomorphism.*

The Novikov conjecture then is the following:

Conjecture 4.28 *If π is any group, then the assembly map is rationally an injection.*

We leave the verification that this is equivalent to the version involving higher signatures as an exercise. (Consider the map $[X : F/\text{Top}] \to \Omega(X) \otimes_{\Omega_*(*)} \mathbb{Q}$ given by $[M \to X] \to f_*\big(L(M) \cap [M]\big) - L(X) \cap [X].^{52}$)

---

[52] We note that the right-hand side of the equation commutes with the periodicity isomorphism.

4.8 Controlled Topology, Revisited

Having discussed briefly one result in controlled topology and then classical surgery theory, we would be remiss if we did not discuss their marriage. In general, the theme of controlled topology is to redo the problems solved in classical topology, but now with attention paid to the size of the constructions.

Size can be measured in various ways, and this theme has many incarnations and variants: indeed, we have tacitly already used two different kinds of controls. When discussing the Novikov conjecture, we used the fact that we had a uniform bound on the sizes of things, but this size was not made available – we were obliged just not to leave the category of maps that maintain that same property of uniform boundedness: this is called *bounded control*. Our discussion of Novikov's theorem was based on *epsilon control* because we used the fact thatc all point inverses become contractible in a small neighborhood of themselves.[53]

For stratified spaces, it is useful to use *continuously controlled at* ∞ techniques, and in Chapter 8 we will discuss the beautiful idea of *foliated control* introduced by Farrell and Jones.

In all settings, the basic idea is to:

(1) do what is classically done in topology, i.e. reduce the geometric problem to one of algebra – so we have already seen $\mathrm{Wh}(\pi)$, $L(\pi)$, etc. – so there should be some such structure associated to the problem and it will now be algebra associated to the "control space" as well as the fundamental groups involved; and then

(2) actually *do* the algebra. This point of view is mainly due to Quinn, who developed many consequences of it; the first nontrivial cases of the theory were already in place in advance: notably the work of Anderson and Hsiang (1976, 1977, 1980) and Chapman and Ferry (1979).

A simple example is this. The space of proper maps from $\mathbb{R}^n \to \mathbb{R}^n$ is highly nontrivial – for example, it has $\mathbb{Z}$ components given by degree. If we add the condition that $d(x, f(x)) < C$ (where C is allowed to vary), then the function space becomes contractible.

Another example is this (for the impatient, skip to the formal definition a few paragraphs from here). If we take a homotopy equivalence $f: X \to Y$, with

---

[53] There is room for a distinction here: one can also study *approximate control* where one tries to prove an ε–δ theorem, where one wants to move things by at most δ to a solution, willing to assume initial data which are "ε-controlled." Such results are called "squeezing theorems" and the prototype might be Chapman's proof of the α-approximation theorem: a squeezing theorem reduces an approximate problem to an ε-controlled one – in our terminology. Quinn's (1979, 1982b, 1982c, 1986) papers deal with both issues, the squeezing and the ε-controlled, simultaneously.

X and Y polyhedral, and ask that, after taking an open cone,[54] Cf is still a homotopy equivalence in the category of maps where no point is moved more than a bounded amount (measured in CY), then we can deduce that, for every open set U in Y, $f^{-1}U \to U$ is a proper homotopy equivalence. If X and Y were manifolds of the same dimension, then f would be a uniform limit of homeomorphisms!

Of course, the benefits of having control are not always this dramatic. Having various controlled categories provides us with language and tools to scaffold incremental progress towards building homeomorphisms (or other useful geometric maps).

To give the idea and stay close to our roots, we will focus on the bounded category.

Definition 4.29 Let X be a metric space. Then the bounded category Bdd(X) is the category whose objects are spaces with maps, $(Z, f: Z \to X)$, and morphisms consisting of continuous maps $g: Z \to Z'$ so that $d(gf', f) < C$ for some C. Note that (Z, f) and (Z, f') are canonically equivalent in this category by the "identity map" if $d(f, f') < C$ via the identity, and that we do *not* insist that f be continuous.

Because of the last point, many metric spaces give rise to equivalent bounded categories.

Definition 4.30 If X and Y and are metric spaces, then a map $\varphi: X \to Y$ is a *coarse quasi-isometry* if (1) there are constants such that $A^{-1}d(x, x') - B < d(\varphi(x), \varphi(x') < Ad(x, x') + B)$; and (2) $\varphi(X)$ is C-dense in Y, i.e. every point in Y is within C of some point of X.

Coarse quasi-isometric metric spaces clearly have equivalent bounded categories.[55] As a trivial example, any bounded diameter metric space is equivalent (i.e., coarse quasi-isometric) to a point.

A very important example is that the universal cover of a compact manifold is coarse quasi-isometric to its fundamental group (with the word metric). The subject of geometric group theory is largely the study of this equivalence relation on finitely generated (or finitely presented) groups.

With the above terminology in place, it makes sense to raise questions such as the h-cobordism problem or the surgery question in Bdd(X). If X is a point, then this is just the old question and the obstructions involve Wh(π) and $L_n(\pi)$,

---

[54] We metrize so that the cone on a simplex is a Euclidean octant.

[55] Actually, there is a looser equivalence relation that also gives isomorphisms of bounded categories, wherein one replaces the linear upper and lower bounds by non-decreasing functions that go to ∞ (such as log and exponential).

etc. So we will need to take the fundamental groups of the objects into account, in particular the system of fundamental groups of inverse images of large balls of X.[56] But, for simplicity at this point, let's stick to the simply connected case.

Given that there is freedom in choosing the X when considering the category Bdd(X), we can try to choose the best possible model for X. For example $\mathbb{R}^n$ is in some ways a better model than $\mathbb{Z}^n$, because every object over the former can be replaced by one where the reference map f is *continuous*. We leave this an exercise, but observe that the key property that allows this (when the underlying space of the object is finite-dimensional) is the following:

Definition 4.31 A metric space X is *uniformly contractible* if there is a function $u(R)$ from $\mathbb{R}^+ \to \mathbb{R}^+$ such that any point $x \in X$, the ball $B(x, R)$ is *null-homotopic* in a larger concentric ball, given by $B(x, R)$ included in $B(x, u(R))$.

One example is the open cone of a compact ANR, as we leave to the reader.

Another very good (and *important*) example is the universal cover of a compact $K(\pi, 1)$.

Indeed, uniformly contractible metric spaces generally seem like good analogues of $K(\pi, 1)$s. They are terminal objects in the bounded category of spaces over X that are coarse quasi-isometric to X – just as $K(\pi, 1)$s are the terminal objects in the category of spaces that have the same 1-type as X.

Given the philosophy we have espoused in the previous chapter's morals section (§3.8), we should conjecture some type of bounded rigidity for uniformly contractible manifolds.[57]

Conjecture 4.32 (Bounded Borel) *If M is a uniformly contractible manifold, and $f : M' \to M$ is a homotopy equivalence in the bounded category over M, then f is homotopic (in this category) to a homeomorphism.*

---

[56] As a pro-system: the fundamental groups themselves have no real meaning, but 'system', as we allow larger and larger balls, does make sense.

[57] Unlike the Borel conjecture itself, the following conjecture is known to be false (Dranishnikov *et al.*, 2003). The example is a rather pathological Riemannian manifold that is abstractly a Euclidean space. It is based on an amazing example of Dranishnikov of a space of finite cohomological dimension but infinite covering dimension, and requires a violation of "bounded geometry." For example, if M has a triangulation with all simplices of bounded size, *and* a uniform bound on the valence of any vertex (or even a lower bound on injectivity radius and bounds on curvature), such as the universal cover of a finite $K(\pi, 1)$-complex then the methods of that paper do not apply.

This example could suggest that the Borel and allied conjectures are not as well founded for groups of infinite cohomological dimension. However, there are many cases where the conjectures do seem to be correct even in this setting and the question requires a lot more thought.

Remark 4.33 (1) With a bit of care, one can show that the CE-approximation theorem is essentially a verification of a special case of this conjecture.

(2) Actually taking the fundamental group into account suggests a unification of these conjectures, the bounded rigidity of a *uniformly aspherical* manifold. Good examples of these are $K \backslash G / \Gamma$ for lattices – and they are indeed rigid (in high dimensions as observed in Chang and Weinberger, 2007).

Let us consider what the surgery exact sequence should suggest for this problem:

$$\cdots S_{n+1}(M \downarrow M) \to H^{\mathrm{lf}}_{n+1}(M; L(e)) \to L_{n+1}(M \downarrow M) \to S_n(M \downarrow M) \to \cdots .$$
$$\quad = 0 \qquad\qquad\qquad\qquad\qquad\qquad\qquad\qquad = 0$$

So the bounded Borel conjecture asserts that *bounded L-groups are a homology theory of the control space* in the uniformly contractible case.

And, indeed frequently this is the case. For example, for the cone on a finite polyhedron this can be verified in a couple of ways. One can deduce this from the α-approximation theorem – rather like the way the L-groups of the trivial group are identified with the homotopy of F/Top via the Poincaré conjecture[58] (Ferry, 2010). Or one can show that this is a homology theory in the space being coned – with a codimension-1 splitting argument being used for the critical verification of the excision (or, equivalently, the Mayer–Vietoris) axiom (Carlsson and Pederson, 1995; Ferry and Pederson, 1995).

Versions of this principle are true for all the types of control we had mentioned and such results are central to controlled topology's many geometric consequences.

We end this section with a discussion of the following points.

Remark 4.34 (1) What to do when we don't have (or know) a uniformly contractible model.

(2) How to formulate a bounded Novikov conjecture.

(3) Then observe that Novikov's theorem on topological invariance of rational Pontrjagin classes is a special case of the latter.

(1) What can we do when we don't have (or know) a uniformly contractible model? This is completely analogous to the situation in ordinary surgery when we don't have a finite complex $K(\Gamma, 1)$.[59] What we do there is choose a sequence of finite complexes that approximates it. The limit is a space that we call $K(\Gamma, 1)$,

---

[58] With the opposite goal in mind. When we made this identification before, the goal was to learn about the homotopy type of F/Top, because the groups $L(e)$ were under control. Here, the bounded L-groups are the objects we are interested in learning about.

[59] Or, indeed, whenever a category doesn't have a terminal object. One takes a limit in a pro-category.

but had our ideology allowed only finite complexes we wouldn't be able to call it a space. The homology that we use is cognizant of this fact: we take the homology with finite chains, not the locally finite chains.

So if Z is a metric space, we can form a simplicial complex $R_1 Z$ by having the points of X (or, for technical convenience, a discrete, 1/2-dense subset of Z) be vertices, and then adding edges when both points have distance ≤ 1, adding 2-simplices when all three vertices have distance ≤ 1, and so on. Then we can include this complex as a subcomplex of the same construction with 2 replacing 1, $R_2 Z$, and so on. We take the direct limit of these complexes. We also take the limit of the locally finite homologies:

$$\mathrm{HX}_i(Z; L(e)) = \lim H_i^{lf}(R_n Z; L(e)).$$

The X in HX reminds us that we are working at a large scale: all finite-scale phenomena have been wiped away (e.g. a cycle present at size t is killed in $R_{t+1}Z$).

(2) In these terms, which higher signatures should we expect to be homotopy invariant in Bdd(Z)? Exactly the pushforwards in $\bigoplus HX_{n-4i}^{lf}(Z; \mathbb{Q})$ of $L(M) \cap [M]$, of course.

(Moreover, we can conjecture that there are integral versions in more refined theories, just like the Novikov conjecture has integral refinements for torsion free groups, as we will emphasize in Chapter 5. We will see that, frequently, these conjectures are indeed correct.)

(3) As for Novikov's theorem? Note that the bounded Novikov conjecture for $\mathbb{R}^n$ implies Novikov's theorem as we explained in §4.5. The map $U \to V \times \mathbb{R}^n$ considered there is surely a bounded homotopy equivalence over $\mathbb{R}^n$ (as it's a homeomorphism). The bounded Novikov conjecture implies that the transverse inverse image of V has the same signature as V.

4.9 The Principle of Descent

We can now reformulate the argument in §4.6 more abstractly. It shares with the Lusztig argument the idea of taking a naive homotopy invariant and then varying it in a family to get more information out of it.

For $\mathbb{Z}^n$ there is an important map:

$$L_i(\mathbb{Z}^n) \to L_i^{\mathrm{Bdd}}(e; \mathbb{Z}^n).$$

The left-hand side is isomorphic to a sum of $n!/j!(n-j)!$ copies of $L_{i-j}(\mathbb{Z}^n)$ (varying j). These are essentially the simply connected parts of the surgery obstructions you'd see along all of the codimension-j tori. The right-hand side we discussed in the previous section; it is just the codimension-n surgery obstruction, and we have seen that one way to obtain this is by Ferry's theorem. (It is $HX_i(\mathbb{Z}^n; L(e))$.)

This map is a "transfer."[60] If we have a normal invariant over a space with fundamental group $\mathbb{Z}^n$, then when we take the universal cover it is a normal invariant of the cover, and we can try to perform surgery on it to make it into a bounded homotopy equivalence over $\mathbb{R}^n$. This obstruction is powerful: it gave us Novikov's theorem in Remark 4.33(3).

Using it and projection to all of the smaller tori gives us a proof of the Novikov conjecture for free abelian groups (although not essentially more elementary than the ones we've already discussed).

Now, if W is a complete simply connected manifold with nonnegative sectional curvature, the inverse of the exponential map gives a Lipschitz diffeomorphism:

$$\mathrm{Log} \colon W \to \mathbb{R}^n,$$

and hence a map $L^{\mathrm{Bdd}}(W) \to L^{\mathrm{Bdd}}(\mathbb{R}^n)$. Direct application of this map is clearly strong enough to prove the Novikov conjecture for the fundamental class of W/Γ for any cocompact group of isometries of W.

What we did in §4.6, though, was much stronger in that we made use of the fact that there were "logarithm maps" at *all* points w of W, and made a family of bounded surgery obstructions. These were strong enough to detect the whole $H_W(W/\Gamma; L(e))$.

Setting up the formalism of these families can be done in more than one way. Two of the most popular are: (1) blocked surgery; and (2) homotopy fixed-point sets. Both of these depend on "spacification," which means finding spaces whose homotopy groups are the L-groups, normal invariants, and structure sets, and so that the surgery exact sequence becomes the exact sequence of a fibration. This process is similar to viewing indices as vector spaces and then being able to associate bundles to families of operators.

There is an important difference between index theory and manifold theory in this "spacification." In index theory, one uses genuine families of operators. In surgery, we do not need to. Genuinely parameterized surgery is a much more complicated subject than we need for these purposes.

Recall that, when we discussed the Farrell fibering theorem, although we fibered a manifold over the circle, the actual process was different: we found a single fiber (via a splitting theorem, but never mind). In other words, we solved a problem over a vertex in a triangulation of S^1. (Solving it over other vertices

---

[60] Transfer (for covariant functors) is generally a map that passes from an invariant of a quotient to one of the original space (or perhaps an intermediate quotient).

is no additional problem.) Then we cut open the circle and had to see that we were OK over the resulting interval.

In general, in spacification, we want the families to not involve more complicated objects than arise in the vertices.[61] if we can. We are being conservative in the type of difficulties that ever need to be considered. This can achieved because when we did surgery there was no difference in the theory of closed manifolds and manifolds with boundary if we work relative to the boundary.

And a similar method can be done with manifolds with corners.

One is thus led to consider a simplicial complex, where the k-simplices are surgery problems that are "modeled" on Δ^k. In this space, $L_n(\pi)$, vertices are n-dimensional surgery problems with fundamental group π; 1-simplices are cobordisms between such, i.e. they are $(n + 1)$-dimensional surgery problems with two boundary components (labeled by 0 and 1), everything with fundamental group π, and the boundary of the problem determines the boundary of the 1-simplex. And so on for defining the higher simplices. Doing this gives a space, and $\pi_i\big(L_n(\pi)\big) = L_{i+n}(\pi)$. Moreover, there are homotopy equivalences $\Omega^4\big(L_n(\pi)\big) \cong L_{n+4}(\pi)$ that are an analogue of the 4-fold periodicity of L-groups.

It is worth saying a word about this equivalence, since it is the prototype of the notion of an assembly map, and it explains the use of the word "assembly," which often mystifies people who see other maps that are also called assembly maps because they somehow resemble this one. An element of $\Omega^4(L_n(\pi))$ (or a vertex in a simplicial model of it) is a map from S^4 into $L_n(\pi)$). So, we can think of S^4 as being triangulated, and each simplex in that triangulation being assigned one of the defining simplices in the space $L_n(\pi)$. We can "assemble" all of these together to define a vertex in the space $L_{n+4}(\pi)$.

Of course, if one is fibered over a base, one has this situation, but this "blocked" theory is much simpler: it satisfies our desideratum that no object is more complicated than which occurs already over a vertex.[62] So, above, we exactly have this from a surgery problem:

$$N' \to N \to W/\Gamma.$$

We can take universal covers and lift

$$\tilde{N}' \to \tilde{N} \to W,$$

giving the family $(\tilde{N}' \to \tilde{N}) \times_\Gamma W$, which fibers over W/Γ. Over each Δ in W/Γ one has the product $(\tilde{N}' \to \tilde{N}) \times \Delta$ (although identifications change as one

---

[61] Except combinatorially. An object over a simplex will need the combinatorial complexity of a simplex (at least).

[62] That is, that surgery on manifolds with boundary relative to a solution on the boundary has exactly the same type of obstruction that already arises on closed manifolds.

moves around). This is a simplex is a space of surgery problems of the form $L^{Bdd}(W)$. If one thinks through all the identifications made, then one realizes that one has detected $H_i(W/\Gamma; L(e))$ in $L_i(\Gamma)$ via the composite

$$L_i(\Gamma) \to H^0(W/\Gamma; L^{Bdd}(W)) \to H^0(W/\Gamma; L^{Bdd}(T_w W)) \cong H_i(W/\Gamma; L(e)),$$

where $H^0(W/\Gamma; L^{Bdd}(W))$ denotes the *twisted* cohomology of the spectrum $L^{Bdd}(W)$ over W/Γ and TW denotes the tangent bundle of W, called the *family bounded transfer*.[63]

A very nice alternative description of this method is to recall the notion of homotopy fixed set. If G is a group acting on a space X, then X^G, the fixed-point set of the action, can be thought of as the equivariant mapping space $\text{Map}_G[pt: X]$. There is a map of this space into a more homotopical object[64]

$$X^{hG} = \text{Map}_G[EG: X].$$

Unlike fixed points, that are quite sensitive to a space being acted upon, if $f: X \to Y$ is an equivariant map that is a homotopy equivalence, then f induces a homotopy equivalence $X^{hG} \to Y^{hG}$.

Moreover, there is clearly a map $X^G \to X^{hG}$.

If one unravels the notation in the family bounded transfer, one sees that one has the map

$$L_i(\Gamma) \to L_i^{Bdd}(\Gamma)^{h\Gamma}$$

and thus one can interpret our proof in the non-positively curved situation as using a *homotopy fixed set* for the purpose of splitting an assembly map.

In any case, these ideas lead to a method of descent, wherein a suitable Borel-type conjecture (or maybe a little less) for Γ as a metric space gives rise to the Novikov conjecturec for Γ itself as a group – again, these are little different from what we did by hand in §4.6, but this interpretation, for example, makes sense for many functors other than L, and also is now suitable for situations where we get our bounded information from any source, not only from the Ferry ε-map theorem.

---

[63] This discussion ignores the aspect where we put support conditions on the homotopy equivalence and changed the cohomological term to one with compact supports.

[64] Below, EG is the universal contractible space on which G acts freely, and hG is homotopy fixed set of the G (or Γ) action.

4.10 Secondary Invariants

... And a Little More Surgery

It is time to return to lens spaces, those remarkable explicit manifolds which, while homotopy equivalent, can frequently not be distinguished by their tangential information (say in dimension 3). Recall de Rham's theorem that lens spaces are only diffeomorphic[65] when they are linearly so – it is time to understand why this is.

This discussion contains embryonically one of the main keys to understanding closed manifolds whose fundamental groups have torsion. If we think functorially, the key question is:

Problem 4.35 What does $S(K(\pi, 1))$ look like when π has torsion?

(Indeed, *when the Novikov conjecture is true*,[66] we always have, rationally, a decomposition of

$$S_n(M) \otimes \mathbb{Q} \cong S_n(K(\pi, 1)) \otimes \mathbb{Q} \times S_{n+1}(K(\pi, 1), M) \otimes \mathbb{Q}$$
$$\cong S_n(K(\pi, 1)) \otimes \mathbb{Q} \times H_{n+1}(K(\pi, 1)), (M; L(e)) \otimes \mathbb{Q}$$
$$\cong S_n(K(\pi, 1) \otimes \mathbb{Q} \times \bigoplus H_{n \pm 4i+1}(K(\pi, 1)M; \mathbb{Q}).)$$

We shall see[67] that $S_3(K(\pi, 1)) \otimes \mathbb{Q}$ is never 0 for groups with torsion[68] (and that, furthermore, $S_3(M) \otimes \mathbb{Q}$ is nonzero for any closed orientable manifold whose fundamental group has torsion[69]). For general groups, the Farrell–Jones conjecture (to be discussed in Chapter 8) gives a conjectural answer.[70]

To begin answering this, we must first consider the important case of π finite.

For finite groups, the homology term is irrelevant (rationally), so we need to think about the L-groups. Wall (1974, 1976a) showed that, for finite groups π, the $L_n(\pi)$ are finitely generated abelian groups, with 2-primary torsion. Moreover, the groups for n odd are finite, so we shall concentrate on n even. Indeed, for future reference, let me go so far as to actually define these groups.[71]

---

[65] Indeed, homeomorphic, although de Rham could not have known that!

[66] This is a diagram chase, when one takes into account that the assembly map, as a map of spectra, being rationally injective on homotopy groups, has a rational splitting as a map. For integral analogues see, e.g., Weinberger *et al.* (2020).

[67] Following Chang and Weinberger (2003).

[68] But it can be 0 in the setting of groups with nontrivial orientation character.

[69] This is not a formal consequence of the previous remark, as it was conditional.

[70] When we discuss the equivariant version of the Borel conjecture, we will be led to a more straightforward geometric conjecture for $S(K(\pi, 1)) \otimes \mathbb{Z}[1/2]$. However, the "correct formula" for $S(M)$ will then be a purely homological variation of the first summand in the decomposition above.

[71] I have always been amazed at how much is possible to do in surgery theory without a definition of the obstruction groups, and only a modicum of their properties. However, for

Definition 4.36 Let π be a group, and $w: \pi \to \mathbb{Z}/2$ a homomorphism (called the orientation character). Then $L_{2k}(\pi, w)$ is the group generated by 3-tuples (M, λ, μ), where M is a free finitely generated $\mathbb{Z}\pi$-module, $\lambda: M \times M \to \mathbb{Z}\pi$ is bilinear and nonsingular[72] over $\mathbb{Z}\pi$, Λ is $(-1)^k$ Hermitian (the conjugacy on $\mathbb{Z}\pi$ generated by sending g to $w(g)g^{-1}$), and $\mu: M \to \mathbb{Z}\pi/(u - (-1)^k \bar{u})$ is a quadratic refinement of λ so that $\mu(x + y) - \mu(x) - \mu(y) = \lambda(x, y)$ and $\mu(ux) = \bar{u}\mu(x)u$ for u a multiple of a group element.

An element is trivial if M contains a subspace K such that λ and μ restrict trivially to K, and $\lambda: K \to M/K$ is an isomorphism.

One can define variants using projective modules rather than free, or free-based modules, and then impose a condition on $\det(\lambda)$. All of these just affect L at the prime 2, so we will not worry about them here.

Ranicki (1979a) showed that the map $L_n(\mathbb{Z}\pi) \to L_n(\mathbb{Q}\pi)$ is an isomorphism away from 2 for *any* group π. Moreover, with 2 inverted in the coefficient ring, the μ is irrelevant (i.e. determined by the λ).

So we now have a stripped-down picture of the kind of invariant we are seeking: a Witt class of a quadratic form over $\mathbb{Q}\pi$.

That is a straightforward invariant to try to get: whenever the finite group π acts on an oriented space[73] X^{2k} satisfying Poincaré duality (with orientation properties given by w), we get such a structure on $H^k(X; \mathbb{Q})$. Let $\langle x, y \rangle = \sum \langle (x \cup g_* y), [X] \rangle g \in \mathbb{Q}\pi$. We shall call this invariant π-sign $(X) \in L_{2k}(\mathbb{Q}\pi) \otimes \mathbb{Q}$.

For π trivial, this invariant is trivial for k odd (every skew-symmetric form over $\mathbb{Q}$ is determined by its dimension as a vector space: it is a symplectic vector space, and every Lagrangian in it defines an equivalence to the trivial element). For k even, $L_{2k}(\mathbb{Q}) \cong \mathbb{Z} \oplus T$, where T is an infinite sum of $\mathbb{Z}/2$s and $\mathbb{Z}/4$s. The $\mathbb{Z}$ is just the signature of the quadratic form.

In general, we can analyze $L_{2k}(\mathbb{Q}\pi)$ in a few equivalent ways. The invariants we will be discussing are representations, and therefore can be thought of as characters – which means that we only need pay attention to cyclic groups.[74] For $k = 0$, we can diagonalize the quadratic form, and then consider the difference of positive and negative definite parts $[H^+] - [H^-]$ in $RO(\pi)$ as an invariant. When k is odd, and π is cyclic, we can take a complex representation and (after multiplying by i) get a Hermitian inner product. It has a signature.

---

almost anything involving groups with torsion, the definitions are necessary, and hands made dirty by calculation cannot be avoided.

[72] That is, λ defines an isomorphism $M \to M^*$ (where $M^* = \text{Hom}(M: \mathbb{Z}\pi)$).

[73] We are reserving the right to allow X not to be a manifold, and the action not to be free – since these affect nothing. Moreover, by allowing the modules to be projective in the definition of the L-group, we really have a very transparent invariant.

[74] Of course, the L-groups themselves are more complicated than this. The reader might wish to think about the cases of Q_8 and the symmetric groups to see various phenomena.

Proposition 4.37   *If π is a finite group that acts freely on M^{4k}, then π-sign (M) is a multiple of the regular representation (i.e. its character is trivial for all $g \neq e$). If π acts freely on M^{4k+2}, then π-sign (M) vanishes.*

This can be proved in several ways. First of all, it is a consequence of the Atiyah–Singer G-signature theorem (Atiyah and Singer, 1968b). It can also be easily proved by cobordism considerations: bordism of free π-actions is equivalent to $\Omega(K(\pi, 1))$, but after $\otimes \mathbb{Q}$ this is the image of $\Omega(*)$, i.e. every bordism class is induced from a trivial action, immediately giving the result.

Note that this proposition includes our earlier observation that signature is multiplicative for finite covers of closed manifolds.

Now we can define a basic invariant of an odd-dimensional manifold with finite-order fundamental group.

Definition 4.38   Suppose M^{2k-1} is a closed manifold with finite fundamental group π. Let W be such that $km = $ boundary of W; we define $\rho(M) \in L_{2k}(\mathbb{Q}\pi)/L_{2k}(\mathbb{Q}) \otimes \mathbb{Q}$ as follows. Some multiple, kM, of M bounds with fundamental group $\pi \cdot kM = \partial W$:

$$\rho(M) = (1/k) \, \pi\text{-sign}\,(W) \in L_{2k}(\mathbb{Q}\pi)/L_{2k}(\mathbb{Q}) \otimes \mathbb{Q}.$$

Remark 4.39   We have been quite cavalier in ignoring integrality and torsion issues. With more care,[75] one need not be.

For lens spaces we can make this completely explicit.

Start with the following $\mathbb{Z}/n$ action of a surface. Take the branched cover of S^2 branched at n points so that one gets a surface with a semifree $\mathbb{Z}/n$ action, so that all n fixed points have the same tangential representation – say t which equals rotation by $2\pi/n$. Changing the generator gives an analogous $\mathbb{Z}/n$ equivariant surface but with any rotation number one wishes. A product of k such surfaces will give a manifold of dimension $2k$ with a semi-free $\mathbb{Z}/n$ action, with n^k isolated fixed points, all with the same given normal representation. If we removed a deleted neighborhood of the fixed point set, and take quotients, we obtain n^k lens spaces all with our chosen set of rotation numbers. Moreover, the ρ-invariant can be computed from the calculation for the original branched cover using Galois invariance and multiplicativity of signatures. In any case, the calculation is done (by another method) in Atiyah and Bott (1968) and furthermore they explain the quite nontrivial and interesting proof that this invariant is strong enough to distinguish the lens spaces from one another.[76]

---

[75] Using the technology of assembly maps (and using calculations of equivariant Witt groups).

[76] It is interesting that both this proof and de Rham's original proof both rely on the same number-theoretic fact: the Franz independence lemma (see Milnor, 1966; Atiyah and Bott, 1968; Cohen, 1973).

In any case, this invariant ρ now presents a challenge to the Borel conjecture. Using our surgical description of ρ, however, we have:

Conjecture 4.40 *If Γ is torsion-free and π is finite, then $L(\Gamma) \to L(\pi)/L(e)$ has finite image.*[77]

Theorem 4.41 *The Borel conjecture[78] implies the above conjecture.*

The theory of ρ-invariants is susceptible to a nice generalization by Atiyah *et al.* (1975a,b) where one assigns an invariant for every finite-dimensional unitary representation of Γ (whether finite, torsion-free, or anything at all!).

This analytic method does not have, as far as I know, a purely topological approach. It is a descendent of the above-mentioned fact that, for closed manifolds (but not for manifolds with boundary), the signature with coefficients in any flat unitary bundle is the same as the ordinary signature (times the dimension of the representation).

Definition 4.42 Let D be a self-adjoint elliptic operator on an odd-dimensional manifold. Associated to D we form the series

$$\eta(s) = \sum |\text{sign}\,\lambda| \lambda^{-s}$$

(summed over the *nonzero* eigenvalues λ of D) and, via analytic continuation, form the real number $\eta(0)$.

The η-invariant enters as a correction term from the boundary in an Atiyah–Singer theorem for manifolds with boundary. Therefore, as we are in a situation where relationships that hold for closed manifolds do not apply to manifolds with boundary, the η-invariant arises – without a choice of cobounding manifold – to give an invariant of M itself.

For M^{2k-1} there is a "signature operator" B on forms of even degrees $(2p)$ given by

$$B\phi = i^k(-1)^{p+1}(*d - d*)\phi.$$

If $\alpha: \pi_1 M \to U(n)$ is a unitary representation, then we can also consider B with coefficients in the flat bundle determined by α.

Definition/Theorem 4.43 (Atiyah *et al.*, 1975b) *The invariant $\rho_\alpha(M)$ is*

---

[77] Actually, most of the 2-torsion should also not be hit, as one can see using more detailed information about the assembly map for finite groups, i.e. the problem of which surgery obstructions arise from problems involving closed manifolds. This is called the "oozing problem" for historical reasons, with important contributions being Wall (1976b), Cappell and Shaneson (1979), Morgan and Pardon (unpublished), and Hambleton *et al.* (1988).

[78] As well as the Baum–Connes conjecture.

defined as the difference of the η-invariants for the signature operator with coefficients in the trivial bundle and that with coefficients in α:

$$\rho_\alpha(M) = n\eta_B(0) - \eta_{B\alpha}(0).$$

This invariant is independent of the Riemannian metric on M. If $M = \partial W$ so that the flat bundle extends, then

$$\rho_\alpha(M) = n\text{sign}(W) - \text{sign}_\alpha(W).$$

So, of course, all the flat bundles and APS invariants give potential obstructions to the Borel conjecture. We can turn this around and make a theorem:

Theorem 4.44   *If the Borel conjecture is true for (the torsion-free group) Γ, then, for all α, ρ_α is a homotopy invariant.*

Remark 4.45   See Weinberger (1988b, 1989) for the details.

(1) While the numbers $\rho_\alpha(M)$ can be arbitrary real numbers if we make no assumption about the unitary representation α (even for the circle, this is a non-constant continuous function of the representation), the non-homotopy invariance $\rho_\alpha(M') - \rho_\alpha(M)$, for homotopy-equivalent manifolds, is always an element of $\mathbb{Q}$. Note that for cobordant manifolds for which the flat bundle extends, this difference is an integer. It stands to reason, then, that the cobordism information implicit in the Novikov conjecture could lead to this rationality at the level of conjecture. That it is true unconditionally is based on ideas developed through work on the Novikov conjecture.

(2) It is not hard to see that if Γ is residually finite and has torsion, then there is an α which detects the infinitude of $S_3(K(\Gamma, 1))$.[79] This method actually gives more information, because ρ_α is an invariant of manifolds, so it can be used to implies that the manifolds are different from each other, not only that some given homotopy equivalence is not homotopic to a homeomorphism.

This theorem is related, but not quite equivalent, to the following obstacle to the surjectivity of the assembly map.

If $\alpha\colon \Gamma \to U(n)$ is a unitary representation, then it induces a homomorphism $\mathbb{R}\Gamma \to GL_n(\mathbb{C})$, compatible with conjugation, and thus a map (by Morita invariance – i.e., viewing a matrix of matrices as a larger matrix with ordinary entries) $\text{sign}_\alpha\colon L_{2k}(\Gamma) \to L_{2k}(\mathbb{C}) = \mathbb{Z}$. By the index theorem, if the assembly map is surjective (so that every element of $L_{2k}(\Gamma)$ comes from a closed manifold), we must have $\text{sign}_\alpha \equiv n\text{sign}$ on $L_{2k}(\Gamma)$. Note that if this weren't true,

---

[79] This infinitude is true even if Γ is not residually finite, as can be seen (Chang and Weinberger, 2006) using an L^2-variant of the η-invariant introduced by Cheeger and Gromov (1986).

then we could use such an element and the Wall realization theorem to give a counterexample to the Borel conjecture.

The proof of the theorem also used the Novikov (i.e. injectivity) half of the Borel conjecture.

In the following notes we describe a less classical invariant based on these considerations.

4.11 Notes

In §4.1, the proof of de Rham's theorem was based on calculations of Reidemeister torsions of the lens spaces. The Reidemeister torsion can be defined for any space that has an acyclic flat bundle on it. Torsions are definable more generally when one has a situation where a finitely generated free chain complex is acyclic (and, crucially, the chain groups have given bases): such as a homotopy equivalence between finite complexes (where the chain complex of the mapping cylinder has this nature).

The torsion measures the determinant of the underlying geometric chain complexes. For a finite $\mathbb{Q}$-acyclic complex, the torsion is essentially the alternating product of the orders of the integral homology groups. Importantly and by contrast, in non-simply connected situations the torsion is *not* a homotopy invariant. Of course, the non-homotopy invariance is often an obstacle to calculation.

The torsion also occurs in the h-cobordism theorem (see §4.4) as the obstruction of an h-cobordism being a product. An h-cobordism is a manifold that deform retracts to its boundary components. The basic obstruction to being $\partial \times [0, 1]$ is that the torsion of the inclusion of (either) one of the ∂ components is trivial and the h-cobordism theorem asserts that (in dimension greater than 5) this is the only obstruction.

See §5.5.3 for more discussion.

As mentioned in the body of §4.2, the Hirzebruch signature theorem has two rather different proofs. There is the cobordism-theoretic proof and the index-theoretic proof. Both of these are subject to extensive and important generalization.

The cobordism-theoretic proof can be modified to allow cobordism of more algebraic or singular objects than merely manifolds. Doing so, the Hirzebruch theorem then becomes a calculation of where smooth manifolds fit into those cobordism theories. One such version is to consider the bordism of "controlled algebraic Poincaré complexes" (Yamasaki, 1987) – where control here is as in controlled topology introduced in §§4.6 and 4.8. This turns out to be the

$H_*(X; L(R))$, where X is the control space, and R is the ring (and L is the spectrum discussed in §4.9). Doing this gives a most satisfying proof of Novikov's theorem – the L-classes have been topologically defined as "the controlled symmetric chain complex of M over M."

The work has been hidden. As we discussed in §4.8, proving that a controlled algebraic functor is a homology theory boils down to a statement like the α-approximation theorem and we've already seen such a result of the correct level of depth for this type of applications.

The Atiyah–Singer index theorem has had numerous extensions and variants. Many of these are subsumed by very some general theorems (and even more by philosophies) in the setting of noncommutative geometry and C^*-algebras. A reference to the Connes (1994) book is surely necessary, but not sufficient. An excellent introduction is Higson and Roe (2000).

Broadening one's viewpoint in this way, besides enabling the proofs of the most advanced known results on the Novikov conjecture, also significantly expands its scope of application – as I hope will become clearer as we continue.

Section 4.4 explains two of the early approaches to the Novikov conjecture.

Probably the most misleading aspect of my treatment is viewing the vanishing of the Whitehead group of free abelian groups as an exogenous fact, that we just exploit, rather than an integral part of the "Borel package" of conjectures. Indeed, for the Borel conjecture to be true, one must have such vanishing, and one should view the vanishing of the Whitehead groups of torsion-free groups as being completely analogous to the conjectured isomorphism statement for L-groups that we take as the algebraic version of the conjecture in §4.7. We will rectify this failure in Chapter 5.

Besides the h-cobordism theorem that involved $K_1(\mathbb{Z}\pi)$, one had Wall's 1965 paper which showed how $K_0(\mathbb{Z}\pi)$ regulates whether a finitely dominated complex is homotopy equivalent to a finite complex. Siebenmann's 1965 thesis showed the bearing of Wall's work on the question of when noncompact manifolds are the interiors of manifolds with boundary (i.e. the compactification problem). The first approach to the fibering problem by Farrell (1996) had multiple obstructions involving various K-groups and a nil-type group.[80] Siebenmann (1965) gave another approach to the problem where all the obstructions were unified into one – the connections between the pieces in Farrell's approach being given by a nonabelian generalization of the Bass–Heller–Swan theorem (Farrell and Hsiang, 1970). Farrell (1971b) gave another very elegant approach to fibering in his ICM talk.

---

[80] Nil groups are Grothendieck groups of nilpotent matrices. The connection to the K-theory of Laurent series is straightforward: if N is nilpotent over $\mathbb{R}$, then $I + t^{\pm 1}N$ are invertible over $\mathbb{R}[t, t^{-1}]$. See Bass and Murthy (1967) for some calculations of these for abelian groups.

More general splitting was developed by Cappell and applied to the Novikov conjecture by him (Cappell, 1976a,b). It turns out that it has aspects that are not attributable to algebraic K-groups. A consequence of Cappell's theorem is that the problem of being a connected sum is homotopy invariant (in dimension greater than 4) if the fundamental group has no 2-torsion (but not in general).

Developing the relevant algebraic K-theory for amalgamated free products by Waldhausen (1968) – motivated by the work he did on 3-manifolds, as one of the authors of "Haken–Waldhausen theory" – was an important step in the development of higher algebraic K-theory. In any case, this work led to consideration of the "Cappell–Waldhausen class" of groups, which are accessible from the trivial group by amalgamated free products and HNN extensions any number (including transfinite) of times. For these, the assembly map is an isomorphism after $\otimes \mathbb{Z}[1/2]$. In low dimensions, this includes many of the fundamental groups that seem important, but, in light of Property (T), no high-rank lattices in simple groups lie in this class.

On the other hand, after introducing the ideas of bounded and controlled topology, the splitting methods return, as we split the control spaces (spaces can be broken up into pieces much more easily than groups can) and thus this method is implicit in many of the subsequent topological (and many of the analytic) approaches.

An exception is Lusztig's method. Extension of this to non-positively curved situations (and beyond) was taken up fairly soon after by Mischenko and Kasparov. Mischenko, besides using infinite-dimensional bundles, also introduced the formalism of algebraic Poincaré complexes and their cobordism to get invariants of manifolds (essentially elements of $L(\mathbb{Q}\pi)$). A useful exposition of Mischenko's work can be found in Hsiang and Rees (1982).

Kasparov (1988) developed an extensive new technology, *KK*-theory, for the problem, which he applied to give the first proof of the Novikov conjecture for fundamental groups of complete non-positively curved manifolds (we gave a geometric approach in §4.6). It is fundamental for most of the subsequent analytic results. A useful reference for Kasparov theory is Blackadar (1998).

(From a noncommutative geometry perspective, Lusztig's method uses a family of operators parameterized by a commutative space, and one can look for families parameterized by a noncommutative space.)

In §4.5, the survey paper of Ferry *et al.* (1995) translates one of Novikov's papers and helps track his train of thought. Novikov's master stroke of using non-simply connected manifolds to get information about the topology of $\mathbb{R}^n$ was commented on, for example, in Atiyah's citation of Novikov for his Fields Medal. The approach I took is based on the idea of Kirby's torus trick (a somewhat different trick that has the same crude description) and is a variation of one of Gromov (1993).

In the original version of the torus trick, the matter of filling in the "hole" was accomplished using Siebenmann's completion (or end, or boundary) theorem in his thesis (Siebenmann, 1965). A nice aspect of using the signature of fiber bundle approach is that this is unnecessary.

The torus trick, or alternatively controlled topology, is used in proving the annulus conjecture and the other foundational theorems for the topological category (see Kirby, 1969; Kirby and Siebenmann, 1977; Quinn, 2010).

It is important to realize that in the equivariant setting all of the basic tools of the topological category that the above work fashions, dramatically fail. Handlebody structures neither exist nor are unique; equivariant Whitehead torsion is not topologically invariant; and transversality fails. Nevertheless, the equivariant signature operator is a topological invariant. We will discuss these matters in Chapter 6.

In §4.6 the Novikov conjecture for closed non-positively curved manifolds was first proved by Mischenko (1974). Farrell and Hsiang (1981) gave a direct geometrical proof (that includes the stable homeomorphism statement). Their method uses the compactification of the fibers and Alexander tricks rather than the use of the α-approximation theorem that we do. The result of this section is the main result of Ferry and Weinberger (1991) and is a slight improvement – from the topological perspective – of Kasparov's result.

Kasparov's theorem was important for the philosophical reason that it did not *seem* to require a hypothesis on the quasi-isometry type of the group, while in the closed case one immediately sees the "sphere at ∞" implicated in the solution: an idea already present explicitly in Mostow's work. When one has infinite volume, there was a strong psychological presentiment that "almost anything is possible." Indeed, we will take up this theme in Chapter 8.

Ferry's theorem was the solution to a problem of Siebenmann from his CE-approximation paper. I consider it one of the high points of twentieth-century topology. It is based on his joint result with Chapman on the α-approximation theorem, which says, in modern terminology, that a homotopy equivalence $M' \to M$ that is controlled over M is controlled homotopic to a homeomorphism.[81]

Quinn's (1979, 1982b, 1982c, 1986) papers essentially liberated the place where the control was being measured from the space where the problem was

---

[81] And, here, we can have ε-δ control in this theorem. In fact, the name α-approximation means that one can use any cover α for an open manifold, and then refine it to a β, so that β-homotopy equivalences are α-homotopic to homeomorphisms. So, oddly enough, the α is just the name chosen for a variable.

solved. This simple idea has proved to be enormously important, as the many applications in that series already showed. And there have been very many more; we use this type of reformulation in §§4.8 and 4.9.

A homology manifold can be thought of as a space that is a controlled Poincaré duality space, controlled over itself. This was a critical insight that led to Quinn's obstruction to resolution (Quinn, 1982a, 1987b), and the construction of nonresolvable homology manifolds and their classification (Bryant *et al.*, 1993).

In §4.7, the topological form of surgery is decidedly less elementary than the smooth theory, but it has the much better features described in this section.

The use of periodicity is an elaboration of the idea, which occurred first in Shaneson (1969), of using the periodicity of *L*-groups to do an end-run around the problem that low-dimensional problems cause for studying high-dimensional problems. (He showed that there is a smooth manifold homotopy equivalent to $\mathbb{T}^2 \times S^3$ that looks like a product of $\mathbb{T}^2$ with a counterexample to the Poincaré conjecture despite the fact that one can't really unwrap those circles via Farrell's theorem.) In Wall (1968) and Hsiang and Shaneson (1970), this idea is used to prove the Borel conjecture for the torus.

Nowadays, the inclusion of periodicity into the functoriality means that we do not have to consciously think about these issues. On the other hand, having included homology manifolds into our structure sets, the objects we study are even less elementary than topological manifolds. In Bryant *et al.* (1993), the paper in which they were constructed, they are also classified up to *s*-cobordism (under some technical conditions)[82].

Homology manifolds were initially studied as places where sheaf theory behaved similarly to the theory of manifolds, and then, later, in the Bing school as cousins to manifolds with interesting topological properties in their own right (e.g. being manifold factors or fixed sets of group actions). Edwards's theorem (and the earlier work of Cannon and Edwards on the double suspension problem[83]), and Sullivan's observation that Novikov's theorem applies to CE maps, made their study central to geometric topology, and Freedman's proof of the four-dimensional Poincaré conjecture was perhaps their crowning triumph – showing that even those tame souls just interested in manifolds could not ignore these spaces as pathological.

While the resolution conjecture, asserting that high-dimensional homology

---

[82] The argument in that paper is only correct as given for homology manifolds that are $L^*(\mathbb{Z})$-orientable, which includes any that are homotopy equivalent to a manifold. An erratum (in preparation) will show how to deal with more general homology manifolds.

[83] Is the suspension of the suspension of a homology sphere a manifold (and therefore the sphere)? Yes.

manifolds are all resolvable, is false (and therefore the characterization of manifolds cannot be expressed just in terms of DDP), it is conceivable that DDP homology manifolds are in all regards just as beautiful as manifolds. However, the most basic properties of these spaces, e.g. whether they are topologically homogenous, whether the h-cobordism theorem is true for them, etc., remain open (see, e.g., Weinberger, 1995).

An ideal situation would be the true extension of the CE-approximation theorem to the setting of DDP homology manifolds: any CE map between ANR homology manifolds with the DDP should be a uniform limit of homeomorphisms – this would lead to homogeneity and the s-cobordism theorem, but can it really be that Edwards's theorem is around the same depth as homogeneity? (Surely for manifolds this isn't the case.)

Section 4.8 had as its goal to explain in more detail how controlled topology works, more formally and systematically than by example. I had tried once before in Weinberger (1994). Other (more) useful references are Chapman (1981, 1983), Quinn (1987a), Anderson *et al.* (1994), and Ferry and Pederson (1995). It is possibly fair to say that the prehistory of controlled topology began with the work of Kirby and Siebenmann on topological manifolds, and Chapman and Ferry on the α-approximation theorem and metric criteria for simplicity of a homotopy equivalence, but was consciously and effectively developed by Quinn (1979, 1982b,c, 1986) and turned into a systematic tool (wherein the control space gained its independence from the formulation of the problem).

The bounded Borel conjecture was, I think, in the air with controlled topology and this whole circle of problems. Its formulation using the Rips complex (i.e. the direct limit of the nerve of coverings by bigger and bigger balls) was natural given that uniformly contractible models do not always seem to exist. See Block and Weinberger (1992, 1997), Gersten (1993), and Roe (1993) for early uses of the Rips construction and its homology.

That this substitute should work out better than a uniformly contractible model was a great surprise to me, and this was the source of the example in Dranishnikov *et al.* (2008). (Although, I guess, the moral is that functorial constructions that work in great generality, substituting for objects that don't necessarily exist, will occasionally beat those objects, even when they do exist.) The phenomenon itself was a derivative of Dranishnikov's (1988) discovery that if $X \to Y$ is a CE map, then, while it induces an isomorphism on ordinary homology, it does not necessarily induce one on non-connective (and, in particular, periodic) homology theories, when Y has infinite covering dimension.

In the C^*-algebra setting, Yu (1998) gave other more dramatic failures of even the coarse Novikov conjecture (discussed in Chapter 5 in the setting of positive scalar curvature). And, even in the presence of bounded geometry, expander graphs give rise to other examples in this setting, as we will discuss in Chapter 8.

In §4.9, the principle of descent seems to have been developed and redis-covered multiple times. Its job is to explain the miracle (not present in our treatment of the torus) of how understanding, say, hyperbolic space, extremely well is enough for the understanding of how *every* cohomology class of every hyperbolic manifold (and we do not really understand very well this cohomol-ogy!) enters as a potential obstruction to homotopy equivalence.

Besides the work of Kasparov on the Novikov conjecture mentioned above, Gromov and Lawson (1983) used a variant in their beautiful paper on positive scalar curvature (see Section 13 therein). Our treatment here is based on Ferry and Weinberger's 1995 reformulation of Ferry and Weinberger (1991); Carlsson (1995) developed it in the guise of homotopy fixed sets (and extended its reach in papers with Pedersen, for example Carlsson and Pederson, 1995). An excellent explanation of its C^*-algebra version appears in Roe (1993).

The technique used here can be used to prove the Novikov conjecture for groups of finite asymptotic dimension. This was first done by analytic methods by Yu (1998). But methods based on the squeezing properties of a finite complex (i.e. α-approximation type results) together with descent have been successfully applied to give this result in Bartels (2003), Carlsson and Goldfarb (2004), and Chang *et al.* (2008). A completely different topological approach (based on the existence of appropriate acyclic completions of the EΓ) is given in Dranishnikov *et al.* (2008).[84]

Spacification was introduced by Casson (1967) to get information about fibering a manifold over S^2. Quinn's thesis (see Quinn, 1970) developed it sys-tematically (see also Nicas, 1982). Other treatments can be found in Burghelea *et al.* (1975), Weinberger (1994), and Cappell and Weinberger (1995).

Finally, in §4.10, W. Neumann (1979) was the first to show that, for the case of $\mathbb{Z}^n$, Atiyah–Patodi–Singer invariants are homotopy invariants by means of an explicit homotopy-invariant formula for them. In Weinberger (1985a) I explained how the Borel conjecture implies that for torsion-free groups these are homotopy invariant. Keswani (2000) showed how a version of the Baum–Connes conjecture implies this as well, and therefore, for torsion-free amenable groups, the work of Higson and Kasparov implies this conclusion.

The fact that APS invariants are homotopy invariant up to rational numbers

---

[84] The Higson corona used in that paper is a variant of the Stone–Čech compactification. The utility of a generalization of the boundary of hyperbolic space for rigidity purposes is a central theme in Mostow, Tits, Gromov, and through to the present.

was something I had worked on, off and on, for almost a decade. My final proof (Weinberger, 1988a) used a deformation argument (whose key point was a calculation of Farber and Levine) to reduce to subgroups of GL_n with algebraic entries. This argument ended up being only a couple of pages long. Later, Higson and Roe (2005a,b,c, 2010) gave a more direct argument.

Mathai (1992) was the first to study the homotopy invariance properties of the Cheeger–Gromov reduced L^2-η invariant. Keswani related this to a variant of the Baum–Connes conjecture (one true for all amenable groups, but false for groups with Property (T)). Chang (2004) showed that the Borel conjecture implies homotopy invariance in the torsion-free case. Chang and Weinberger (2003) showed the non-homotopy invariance for all groups with torsion.

Given that any nontrivial torsion in π gives rise to the infinitude of $S_3(M)$, it seems reasonable to believe that the size of $S_3(M)$ (and of similar invariants) should be larger when the fundamental group of M has more torsion. This has not been shown unconditionally, but, for very many fundamental groups, lower bounds in terms of the number of orders of torsion elements (or even on the number of conjugacy classes of torsion elements) have been given in Weinberger and Yu (2015).

There is a general philosophy of secondary invariants that comes out of the Novikov conjecture and, complementary to these, homotopy-invariant secondary invariants. These were explicitly introduced in Weinberger (1999b) and are "higher ρ-invariants" (although they were implicitly used already in Weinberger (1988a)). Like Reidemeister torsion, they require some amount of acyclicity[85] to define (and examples show that this is actually necessary). Typical places that they take values in is a quotient of $S(K(\Gamma, 1))$ or of an L-group or a some kind of homological (or K-theory) invariant related to the fundamental group – where the quotient is determined by the type of Novikov technology that we will discuss in Chapter 6. I am being vague about this because there isn't yet an overarching general theory that includes all others. Weinberger (1999b), for example, does not deal at all with torsion issues – although some of the later literature (Higson and Roe, 2005a,b,c; Piazza and Zenobi, 2016; Weinberger *et al.*, 2020) does – and the context of their definition is "up for negotiation," essentially in terms of what kind of acyclicity hypothesis is necessary, or what is its source.

---

[85] Actually the relevant acyclicity is only necessary around the middle dimension. That manifolds with this property are special and can be more easily understood than general manifolds was first realized by Jean-Claude Hausmann, who studied (in unpublished work sometime in the 1970s) them under the slightly less general but more geometric condition of having no middle-dimensional handles in a handle decomposition.

This invariant is adequate for distinguishing lens spaces after crossing with aspherical manifolds for which the Novikov conjecture is known.

The reason that they are somewhat subtle is that, unlike higher signatures that are signatures of appropriate submanifolds associated to cycles in $K(\Gamma, 1)$-, ρ-invariants are very definitely *not* cobordism invariant, and therefore it is *prima facie* unclear that any higher version of an such invariant should be definable. Interestingly, acyclicity solves this and this makes some cycles more canonical than others, in a way that is not apparent to straightforward transversality.

The precise definition uses the fact that sufficiently acyclic manifolds are both algebraically (tautologously) and geometrically null-cobordant[86] (using the Novikov conjecture), and that one gets an interesting invariant by comparing these two nullcobordisms.

These invariants have a number of interesting applications. The first is that it gives a way of showing that manifolds are not homeomorphic, not only that a certain map is not homotopic to a homeomorphism. Nabutovsky and I used this to show that, even among homotopy equivalent manifolds, the homeomorphism problem can be algorithmically undecidable (Nabutovsky and Weinberger, 1999).

Another, more recent, application is to Gromov–Hausdorff space. Recall (Gromov, 1999) that Gromov–Hausdorff space is a compact metric space of compact metric spaces, and that spaces are close if they can be approximately "aligned" like two fairly dense subsets of a third metric space. Gromov–Hausdorff space and limits in it have become an important tool in comparison differential geometry.

One can hope to find strong geometric restrictions on sets of manifolds in Gromov–Hausdorff space that have a contractibility function.[87] It turns out, for example, that sufficiently close manifolds of this sort have the same simple homotopy type and rational Pontrjagin classes (Ferry, 1994). However, nevertheless, there are infinite families of manifolds M_i that are pairwise distinct, but which can all be made arbitrarily close to each other. See Dranishnikov *et al.* (2020) for how this goes, and how higher ρ-invariants are used.

In this example, it is important that the contractibility function (including the ε that describes a threshold at which balls are null-homotopic) is allowed to vary with i. For a fixed contractibility function f, Ferry (1994) proved a

---

[86] In some sense; for example, in the Witt cobordism sense.

[87] A contractibility function for X is a function $f : [0, \varepsilon) \to \mathbb{R}$ such that f is continuous, $f(t) \leq t$ and $f(0) = 0$, so that for each point x in X, the ball around x of radius t in null-homotopic in the ball of radius $f(t)$. It is a generalization of the notion of injectivity radius for a manifold, which corresponds to the case of the $f(x) = x$ on [0, inj], where inj is the injectivity radius. It is an easy exercise that, given a local contractibility function f and a dimension n, there is a δ such that δ-close n-dimensional ANRs are homotopy equivalent.

contrasting finiteness theorem: the number of manifolds in a precompact part of Gromov–Hausdorff space with any specified contractibility function is finite.

The same technology that defines the higher ρ-invariant is used in Leichtnam *et al.* (2000) to define higher signatures for noncompact complete manifolds under the same type of middle-acyclicity condition at ∞. (These higher signatures involve the cohomology of the fundamental group of the manifold, and nothing further about ∞. Assuming the Novikov conjecture, they are proper homotopy invariant.) Leichtnam *et al.* (2002) used a variant of this idea to study how higher signatures of closed manifolds change if one cuts them open along submanifolds and glues back differently.

5
Playing the Novikov Game

5.1 Overview

It turns out that the topological Novikov conjecture is only the first example of a more general phenomenon wherein the fundamental group of a manifold (or variety or . . .) plays an extremely large role on the geometry of the manifold – often mediated through analysis. And, as is clear from Chapter 4.1, this theme also extends to noncompact manifolds where the role of the fundamental group is supplemented by the quasi-isometry type of the manifold.

This chapter is about the "Novikov game": what it is, how to play, and what are the typical things that happen when you play.

One starts with a theorem about characteristic classes (or an index) true for all closed manifolds and interprets as being merely the simply connected version of a more general statement, hopefully true for all groups, where one augments the simply connected statement by the cohomology of the fundamental group.

As far as I can tell, the first player of this game was Reinhardt Schultz, in the mid-1970s. One of the nice topological applications of the index theorem is:

Theorem 5.1 (Atiyah and Hirzebruch (1970)) *If M is a closed smooth spin manifold, and M admits a nontrivial smooth S^1-action, then*

$$\langle A(M), [M] \rangle = 0.$$

For the definition of the A-genus of a spin manifold, see Borel and Hirzebruch (1959a,b, 1960): it is an analogue of the L-genus that played such an important role in Chapter 4.

Schultz asked whether this theorem is true for the "higher A-genus" for non-simply connected manifolds that have smooth circle actions.

One has to be a little careful with this statement. After all, the higher A-number associated to the fundamental class of any $K(\pi, 1)$-manifold is nonzero, but the torus $\mathbb{T}^n$ has a circle action! The way around this issue is to not consider all of $\pi_1(M)$, but rather the part that is "orthogonal to the circle," i.e. the following:

Theorem 5.2 (Schultz's conjecture; see Browder and Hsiang, 1982) *If M is a smooth spin manifold admitting a smooth S^1-action, and $f: M \to K(\pi, 1)$ is any map, then for any $\alpha \in H * (K(\pi/\text{orb}, 1); \mathbb{Q})$, the higher A-genus vanishes; that is $\langle f * (\alpha) \cup A(M), [M] \rangle = 0$.*

Here orb is the class in the fundamental group represented by any orbit. (This class is clearly independent of the orbit. It always lies in the center of the fundamental group so we can quotient by the subgroup it generates.)

Similarly, there is another result for (smooth) S^1-actions:

Theorem 5.3 (Atiyah–Singer) *If S^1 acts on a compact manifold M, then* $\text{sign}(M) = \text{sign}(F)$, *where F is the fixed set of the action (if F is suitably oriented).*[1]

This has the expected generalization to the non-simply connected case.

Theorem 5.4 (Weinberger, 1985b, 1987) *If S^1 acts on a compact manifold M, $f: M \to K(\pi, 1)$ is any map, and F is the fixed set of the action (if F is suitably oriented), then*

$$\langle f^*(\alpha) \cup L(M), [M] \rangle = \langle f^*(\alpha) \cup L(F), [F] \rangle$$

for all $\alpha \in H^*(K(\pi/\text{orb}, 1); \mathbb{Q}$ (which equals $H^*(K(\pi, 1); \mathbb{Q})$ if $F \neq \varnothing$).

However, when we begin examining the same story for $\mathbb{Z}/n$ actions, the situation is more complicated.

Theorem 5.5 (Consequence of the G-signature theorem) *Suppose that $\mathbb{Z}/n$ action acts homologically trivially on M, that $f: M \to K(\pi, 1)$ is any map, and that F is the fixed point set of the action. Then there is a characteristic class $c(\nu)^2$ of the equivariant normal bundle to F, so that*

$$\text{sign}(M) = \langle c(\nu) \cup L(F), [F] \rangle$$

(so that, if $F \neq \varnothing$, $\text{sign}(M) = 0$).

Theorem 5.6 (Weinberger, 1985b, 1987) *Suppose that $\mathbb{Z}/n$-action acts homologically trivially on M,[3] that $f: M \to K(\pi, 1)$ is any map, and that F is*

---

[1] This formula makes sense and is true even for topological actions – at least if the fixed set is an ANR. That it makes sense is due to the fact (see Borel, 1960) that the fixed set of a circle action is automatically a rational homology manifold.

[2] Here $c(\nu)$ is the average over the generators of $\mathbb{Z}/n$ of the characteristic classes arising in the formula for tr_g G-signature in Atiyah and Singer (1968b).

[3] This means that the $\mathbb{Z}/n$-action lifts to the universal cover, commutes with the action of the covering translates, and acts trivially on the rational homology there.

the fixed-point set of the action. Then for the characteristic class $c(v)$ of the equivariant normal bundle to F mentioned above, the formula for the higher signature of M:

$$\text{Sign}_\alpha(M) = \langle f * (\alpha) \cup c(v) \cup L(F), [F] \rangle$$

is true iff the Novikov conjecture is true for the group π.

We shall discuss more the interaction between the Novikov conjecture and group actions below in this and the succeeding chapters – because it turns out to be actually a somewhat different problem, and it takes some work[4] to find an equivariant version that is provably exactly equivalent to the original problem! (The first things one thinks of seem to be of the same depth as the Novikov conjecture – i.e. proofs of the Novikov conjecture usually affirm these as well – but not quite provably equivalent to it.)

Let me mention one last example of an equivariant problem that we will see works out rather differently:

Theorem 5.7 *Define an action of G on X to be* pseudo-trivial *if $(X \times EG)/G \cong X \times BG$. If $G = S^1$ or $\mathbb{Z}/p\mathbb{Z}$ acts pseudo-trivially on a (noncompact) manifold (or manifold with boundary) homotopy equivalent to a closed manifold M, then, if the fixed set X^G is a compact manifold, we have*

$$\text{sign}(M) = \text{sign}(X^G).$$

The proof of this follows from Smith theory[5]. The map $X^G \to M$ can be seen to be a rational homology equivalence, and *a fortiori* preserves signatures. Later, we will discuss what happens in the non-simply connected case is a provocative problem.

As we move from topology to differential geometry and beyond, the problems we study do not seem to have direct implications for the original Novikov conjecture. They are in the *spirit* of the problem; they are analogues and can be studied simultaneously and profitably.

The most prominent example is the question of "which manifolds have metrics of positive scalar curvature?" Recall that if M is a Riemannian manifold, then the scalar curvature is a function on M that measures infinitesimally the

---

[4] This is all meant philosophically. Conceivably all the currently unknown versions of the Novikov conjecture are true, and then they will be equivalent to each other . . . However, we will see that working on the equivariant version very quickly leads one to introducing coefficients and other refinements and extensions of the original problem.

[5] One can improve this to where G is a p-group or an extension of a torus by a p-group. But for non-p-groups or Lie groups with nonabelian identity components, the relationship between M and M^G is much more tenuous, even for pseudo-trivial actions. Indeed, in the noncompact case, one can always arrange for M^G to be empty (as the reader should be able to prove after reading §7.1). In the compact case, achieving this also requires a condition on $\chi(M)$.

extent to which the Riemannian volumes of balls of radius r deviate from the Euclidean volumes of balls of the same radius:

$$\text{Vol}(B(r)) = \omega_n r^n - [K(p)/6(n+1)]r^{n+2} + O(r^{n+3})$$

so positive scalar curvature means that balls are infinitesimally smaller than they "should be." The Gauss–Bonnet theorem implies that the only connected oriented surface with positive scalar curvature is the sphere.

Using the index theorem for the Dirac operator and a Bochner-type formula that Lichnerowicz discovered, Atiyah, Lichnerowicz, and Singer gave the first obstructions to any manifold of dimension greater than 2 having positive scalar curvature in the following theorem:

Theorem 5.8 (Atiyah and Singer, 1968b)   *If M is a compact spin manifold with a Riemannian metric of positive scalar curvature, then $\langle A(M),[M]\rangle = 0$.*

This suggests, according to the same pattern:

Conjecture 5.9 (Gromov–Lawson–Rosenberg; see Gromov and Lawson (1980a, 1980b))   *If M is a compact spin manifold[6] with positive scalar curvature, and $f: M \to K(\pi,1)$ is any map, then*

$$f * (\alpha)\langle \cup A(M),[M]\rangle = 0,$$

for all $\alpha \in H^*(K(\pi,1)\mathbb{Q})$.

In particular, no closed (spin) $K(\pi,1)$-manifold should admit a metric of positive scalar curvature. The special case of tori of dimension ≤ 7 was established by Schoen and Yau (1979a). Gromov and Lawson (1980a,b) observed that, for all dimensions, the torus cannot have a positive scalar curvature metric by combining the Atiyah–Lichnerowicz–Singer argument with the argument of Lusztig's thesis. Further, they proved the non-existence for closed non-positively curved manifolds with residually finite fundamental group, and in Gromov and Lawson (1983) they removed, by developing enough index theory on the universal cover, the residual finiteness.[7]

Rosenberg (1983, 1986a,b) directly connected this problem to the work of Kasparov (and Mischenko and Fomenko) on the Novikov conjecture, greatly clarifying the situation and showing that more than analogies were involved here – this chapter owes a great debt to him.

The third important operator studied by Atiyah and Singer is the $\bar{\partial}$ operator on a complex manifold, whose study leads to the Hirzebruch Riemann–Roch

---

[6] It turns out to be reasonable to only ask for a spin structure on the universal cover of M.
[7] Surely this resonates with earlier discussions.

theorem. It, too, gives rise to a characteristic class statement in the simply connected case that one can try to generalize.

If M is a complex manifold and $E \downarrow M$ is a holomorphic vector bundle, then the Hirzebruch Riemann–Roch theorem calculates

$$\sum (-1)^i \dim H^i(M; E) = \langle \mathrm{ch}(E) \cup \mathrm{Td}(M), [M] \rangle.$$

(Here $\mathrm{ch}(E)$ is the Chern character of E, and $\mathrm{Td}(M)$ is a graded characteristic class in the Chern classes of M.) The arithmetic genus is the alternating sum of the dimensions of the space of holomorphic k-forms.

Theorem 5.10 (Corollary to Hirzebruch Riemann–Roch) *If M and M' are birational smooth algebraic varieties, then*

$$\langle \mathrm{Td}(M), [M] \rangle = \langle \mathrm{Td}(M'), [M'] \rangle.$$

A birational equivalence is an almost-everywhere-defined isomorphism that is locally a quotient of polynomials. In fact, it is automatically defined in the complement of a complex codimension-2 subvariety[8] (of domain and range); this implies, in light of Hartog's theorem, that holomorphic functions on the complement extend over the subvariety – and that even the individual (holomorphic) cohomology groups are isomorphic.

A consequence of the fact that the singularities of a birational map being complex codimension-2 is that, if M and M' are smooth birational varieties, then they have the same fundamental groups. This led Rosenberg (2008) to conjecture the following theorem:

Theorem 5.11 (Block and Weinberger, 2006; Borisov and Libgober, 2008; Brasselet *et al.*, 2010) *If M and M' are birational smooth varieties and $f : M \to K(\pi, 1)$ is continuous, then for any $\alpha \in H^*(K(\pi, 1); \mathbb{Q})$, we have*

$$\langle f(\alpha) \cup \mathrm{Td}(M), [M] \rangle = \langle f(\alpha) \cup \mathrm{Td}(M'), [M'] \rangle.$$

The goal of this chapter is to explain more about how to play the Novikov game, and to give some feeling for when the result of playing the game is a conjecture that tends to be a theorem (as in the examples of the Schultz conjecture, the signature localization theorem for S^1-actions, and the Rosenberg conjecture) and when the conjecture seems to be deeper than this – e.g. implying the Novikov conjecture, or at least only being currently provable for some class of fundamental groups. And then there are sometimes when you play and you lose: the new "Novikov conjecture" is just plain false.

In doing this, we will need to broaden our perspective from topology to

---

[8] In the smooth case.

index theory (as must surely be obvious) and develop the analogy between these fields. In doing this, it becomes possible to improve the various rational statements that we have been focusing on to more precise integral ones.

5.2 Anteing Up: Introduction to Index Theory

As might have been obvious in the examples of §5.1, almost all of the examples (except perhaps the one about pseudo-trivial actions) involve the Atiyah–Singer index theorem in some fashion. (That one involved Smith theory, although an index theorem would be involved to translate the posited equality to be one of characteristic numbers.)

The characteristic class (while perhaps rational) represents the index of an operator and our goal is to somehow boost the power of this result in the presence of a fundamental group.

Here we'll give a brief indication; more references can be found in §5.6.

Suppose that D is an *elliptic complex* on a manifold; that is, suppose that we have a sequence of vector bundles, $E_i, \downarrow M$, and $D_i \colon C^\infty(E_{i+1})$ are linear operators acting on the smooth sections of the E_i, given as differential operators (in local coordinates), so that $D_i D_{i-1} = 0$. Ellipticity is the condition that the Fourier transforms are exact away from the 0-section.

Concretely, let's consider the case of a single operator (i.e. a complex with just two bundles) on functions on the circle S^1. The operator $d/d\theta$ (acting on sections of the trivial line bundle) is elliptic; its Fourier transform is everywhere $\times \xi$, which is invertible (and hence gives an acyclic complex) when $\xi \neq 0$. Similarly the Laplacian on functions on the flat 2-torus, i.e., $\mathbb{T}^2$, given by $\partial^2/\partial X^2 + \partial^2/\partial y^2$ is elliptic, but the wave operator on functions given by $\partial^2/\partial X^2 - \partial^2/\partial y^2$ is not. (Its Fourier transform vanishes on the lines $\xi_x = \pm \xi_y$.)

By the "elliptic package," i.e. Sobolev space theory together with the theory of the Fredholm index, for any elliptic complex on a compact manifold, the cohomology groups

$$H^i(D) \cong \operatorname{Ker} D_i / \operatorname{Im} D_{i-1}$$

are all finite-dimensional. The individual groups can be quite subtle and depend on more information than just the symbol of the operators. For example, the Laplacian ∇ on $C^\infty(S^1)$ has $\operatorname{Ker} \cong \mathbb{C}$, but the perturbation by a small zeroth-order term $f \to \nabla f + \lambda f$ has no kernel unless λ is in the spectrum of ∇ (which then depends on the length of the circle!).

A similar but more topological example is the de Rham complex on a compact manifold M, but instead of considering real-valued forms, consider instead

ξ-valued forms, for a flat bundle ξ. The symbol will only see dim(ξ), but, say dim $H^0(M,\xi)$ is exactly the dimension of the part of ξ that has trivial monodromy.

The index theorem gives a topological calculation, though, of the index of D, ind(D) which is, by definition, $\sum(-1)^i \dim H^i(D)$. By taking an Euler characteristic, the subtleties of the individual cohomology groups are largely erased. The reader might enjoy seeing how this happens in these two examples. In the first case, one should use some Fourier series to see what's happening on both the kernels and cokernels, and in the second, the hint might be to consider why Euler characteristic is independent of the field used to define it (while the cohomology vector spaces have dimensions dependent on the characteristic of the field).

That the index is independent of "lower-order perturbations" is exactly the key property of the Fredholm index in functional analysis – invariance under compact perturbations.

The topological formula for the index involves the construction of a "symbol complex" (the analogue of the Fourier transform) over T^*M. It defines an element of $K^0(T^*M)$.[9] Noting that T^*M is an almost-complex manifold, and that the Bott–Thom isomorphism theorem (i.e. Bott periodicity for complex bundles, the form explained in the first chapter of Atiyah and Singer, 1968a) then says that almost-complex manifolds are oriented for K-theory, we have $K^0(T^*M) \cong K_{2m}(T^*M) \cong K_{2m}(M)$, where we are now dealing with the dual homology theory to K-theory.[10] The index is then given by pushing forward the symbol homology class under the map $M \to *$:

$$K_{2m}(M) \to K_{2m}(*) \cong \mathbb{Z},$$

$$[D] \to \text{ind}(D).$$

The equivariant index theorem holds for G a compact Lie group acting on the complex and is essentially exactly the same result! In that case the cohomology vector spaces are G-representations, and the relevant K-theory is equivariant K-theory, $K^G(M)$. The pushforward to a point is now the equivariant index, which is a (virtual) G-representation (G acts on both the kernel and cokernel).

[9] We follow the standard convention that K-groups of noncompact spaces are assumed to be with compact supports. (Despite this, we do not have a standard convention for ordinary homology or cohomology.)

[10] We have shifted our point of view away from the original (Atiyah and Singer, 1968a,b, 1971). They pushed forward a cohomology class to a point using a "wrong-way map" that was induced by Bott periodicity. Atiyah (1970) later gave a K-homology class associated to an elliptic complex. Brown, Douglas, and Fillmore and Kasparov later still gave a development of K-homology where elliptic operators on manifolds form its cycles. See Higson and Roe (2000), and references therein, for a very lucid account.

A similar situation arises when the (manifolds and the) elliptic complex is part of a family $(M_p D_p)$, with $p \in P$, a parameter space. In that case the cohomology groups $H^i(D_p)$ form a bundle over "most of" P. Atiyah and Singer show that it is possible to add a small perturbation d to the family so one has $D + d$, to repair this. In that case, the cohomology vector spaces all become vector bundles $H^i(D_p + d)$ over P, and the index is an element of $K^0(P)$. Atiyah and Singer (1971) explain how to compute this (and its Chern character).

These examples will interest us considerably in what follows (both have already been applied in earlier discussion – Lusztig[11] making use of the families index theorem to prove the Novikov conjecture for $\mathbb{Z}^n$, and the equivariant index theorem is relevant to the Atiyah–Hirzebruch and signature localization theorems mentioned in §5.1.) However, for now, we would like to focus on a formal algebraic aspect of all these theorems.

Notice the groups that the indices in these theorems take values in: for the index theorem it's in $\mathbb{Z}$, in the G-index theorem it is an element of $R(G)$, the representations of G, and in the families index theorem it is $K^0(p)$.

Indeed, even the $\mathbb{Z}$ in the index theorem is just $K^0(C)$, the Grothendieck group of finitely generated projective as $\mathbb{C}$-modules (i.e. finite-dimensional vector spaces). $R(G)$ is the same thing as projective modules over CG, if G is finite, and over C^*G (the C^* completion – to be discussed below) if G is compact. $K^0(p)$. can be thought of, thanks to a classical theorem of Swan, as the (Grothendieck group of finitely generated) projective modules over $C(P)$, the ring of continuous functions on P.

Thus we are led to thinking of the ordinary index theorem as a theorem about complex-valued elliptic operators, and the other index theorems as being about elliptic operators over other C^*-algebras, depending on the situation.

Commutative C^*-algebras correspond to functions on a space (the Gel'fand–Naimark theorem) – and one can[12] think of the general noncommutative case as some kind of index theorem for families "parameterized by a noncommutative space" – including situations where there is a group action as a very special case – and what then follows as a chapter in Connes, 1994).

Instead let's recall that, for elliptic operators P, there are parametrices, or "pseudo-inverses," Q, so that $PQ = I + \text{Compact}$, and similarly for QP. So, for operators over any algebra, we should think that we have a ring R of operators, and J an ideal, and then Fredholm (also known as elliptic) operators are those that are invertible modulo compacts, i.e. modulo J. In algebraic K-theory, there

---

[11] The need for perturbations for families of operators was already implicit in our discussion of Lusztig's method: even the cohomology of S^1 with coefficients in a flat bundle is not constant even for the Lusztig family of line bundles.

[12] If one chooses to.

is an exact sequence (see Milnor, 1971, pp. 33–34):

$$K_1(R) \to K_1(R/J) \to K_0(J) \to K_0(R).$$

The boundary is essentially the index map – one has $\partial[p] = \text{Ker}(P) - \text{cok}(P)$.

In the operator-theoretic setting of "A-algebras," the same is true, except that one has to add on a suitable "A-compact operator" to the elliptic operator to make the kernel and cokernel projective A-modules (e.g. vector bundles without singularities).

Critically important for our story is the following basic C^*-algebra associated to a representation. Suppose π is a group, and $\pi \to U(H)$ is a representation. We can therefore think of $C\pi$ as an algebra of operators on H. We then obtain a C^*-algebra by completing with respect to the operator norm.

There are two extreme cases of this construction that are of fundamental importance. The first is $H = L^2\pi$, and the completion is called $C_r^*\pi$, the reduced C^*-algebra of π. The other is when H is the sum of all irreducible unitary representations of π, and this yields $C_{\text{max}}^*\pi$, the maximal C^*-algebra of π.

The latter choice has better functorial properties: any group homomorphism gives a map between the associated algebras, but Property (T) shows that $C_{\text{max}}^*\pi$ can have rather large K-theory: $K_0(C_{\text{max}}^*SL_3(\mathbb{Z}))$ is an infinitely generated group with an infinitely generated subgroup generated by irreducible representations that factor through finite groups $SL_3(\mathbb{Z}/N)$.

Basically, if our interest is in the Novikov game, i.e. injectivity type statements,[13] this is not a problem and we can live cheerfully with $C_{\text{max}}^*\pi$, but when we begin playing the Borel game (which in this setting is called the Baum–Connes conjecture) and look for isomorphism theorems, $C_r^*\pi$, despite its functorial defects, will play the starring role. We will often be cavalier and just use $C^*\pi$ as notation when these details aren't important (or the completion is obvious).

With the above preliminaries, we have enough of a buy-in[14] to begin playing the Novikov game.

We do not look at every possible characteristic class formula, but rather ones associated to an elliptic operator.

Coupling this operator to an infinite-dimensional flat bundle $\times_\pi C^*\pi$, we try to get information from $K(C)^*\pi$ to improve the formula to one involving the fundamental group. Analogous to the assembly map in surgery,

$$H_*(K(\pi,1)L(e)) \to L_*(\mathbb{Z}\pi),$$

[13] See §4.7.
[14] For example, we have enough resources to be able to get a seat at the table.

there is a map (also called the assembly map, but mainly by reason of analogy[15] – it would be better called the index map[16])

$$K_*(K(\pi, 1)) \to K_*(C^*\pi).$$

Both of these maps "tend to be injective" at least rationally – as we have discussed in the case of L-theory and shall yet discuss more. These are Novikov-type statements and can be sometimes proved by expanding our point of view to consider L-groups of categories associated to and K-groups of algebras of operators associated to metric spaces (and other controlled settings).

Indeed, the usual Novikov conjecture can be studied from the operator-theoretic point of view by restricting attention to the signature operator. Thus, the operator-theoretic setting thus is a large extension of perspective.

Note that we can now try to extract some integral information, like the precise tangential information present in the Borel conjecture, and not just the rational that we have focused on in our discussion of Novikov. Note that the conclusion of the form of the Novikov conjecture that we have just proposed (and had proposed in Chapter 4 via surgery) contains some integral information. It asserts the vanishing of the pushforward of some operator under the natural map

$$K_*(M) \to K_*(K(\pi, 1)).$$

For example, it will include the statement that if we push forward the signature operator[17] (viewed as an element of $K_*(M)$ as above) into the K-homology of the fundamental group, we get an oriented homotopy invariant.

Even if we are interested in just the rational problem, working integrally is a good way to probe approaches, but for some applications, this information is absolutely necessary. The map $L \to K$ (for a point, for example) is split injective *away from the prime* 2, so away from 2 it is possible to deduce a topological injectivity from the analytic results. However, at 2, one cannot obtain strong L-theoretic results analytically (at least, not in any too direct a fashion).

---

[15] In a recent paper (Weinberger *et al.*, 2020) the analytic map is defined in a way that really looks like assembly.

[16] Or the higher index map, if one wants to emphasize how the higher cohomology of the group is implicated in this story.

[17] Here we are taking for granted the quite nontrivial point that the symbol class of the signature operator can be defined for topological manifolds. (See Rosenberg and Weinberger, 1990, for a discussion of this issue. In Weinberger *et al.*, 2020, a functorial map is built using just smooth manifolds, and a more substantial contribution from controlled topology.) In any case, this class can be defined in another way that we will explain in the next section.

Appendix: A Glimpse through the Looking Glass

... At a Parallel Universe whose Arrows are Reversed[18]

While we have focused on manifolds (and orbifolds), and in this setting there are very nice analogies between surgery theory and index theory, both subjects naturally encompass more territory where the analogies are not as evident (and one might imagine pessimistically that they don't extend or, if they do, they don't help[19]).

This mini-section is a brief about the noncommutative geometric perspective.

Many different kinds of things have L-theory: rings with (anti-)involution,[20] pairs (i.e. relative L-groups), additive categories, stratified spaces, etc. Each of these opens up new ranges of application.

On the index theory side, the object one takes the K-theory of is always a C^*-algebra. Rather than generalize the setting of K-theory, which was the surgical route, one instead creates innovative constructions of these algebras in various geometric situations, as is explained lovingly and inspiringly in Connes (1994) and Connes and Marcolli (2008).

More generally, in noncommutative geometry[21] one tries to take seriously noncommutative algebras as analogues of spaces, and then one mimics important geometric constructions in this setting.

The basic example from which everything generalizes is that of X, a locally compact Hausdorff space, to which one associates $C^*(X)$, the ring of continuous complex-valued functions on X, with respect to complex conjugation and the uniform norm as the norm.[22] This is a *contravariant* functor that defines an equivalence of categories. K-theory enters by Swan's theorem that projective modules over $C^*(X)$ are the same thing as vector bundles over X. On the other hand, K-homology is associated to extensions[23] (by the algebra of compact

---

[18] That humans are remarkably symmetric makes it the case that what we see through the looking glass is the same kind of object as we are. Of course, the looking glass of C^*-algebras assigns an infinite-dimensional algebra to a beautiful finite-dimensional space. It might be fun to create a mixture of a mirror with night goggles, so that one gets a similar feeling of strangeness on looking through the mirror.

[19] Even for positive-dimensional group actions on manifolds, the two theories have very different flavors. I am not a pessimist, however.

[20] I'm old fashioned: I stick in the "anti-" but many writers don't bother and use the word involution for the same concept.

[21] We ignore the algebraic geometric side of this philosophy, just as in the commutative world we went continuous from differentiable, not analytic.

[22] This itself is in analogy to the very beginnings of algebraic geometry (initially, at least, over an algebraically closed field) where one associates to each affine variety the coordinate ring of polynomial functions on the variety.

[23] See Higson and Roe, 2000 for an excellent exposition of the Brown–Douglas theory and the beginnings of Kasparov's KK-theory.

operators on a separable Hilbert space) and (therefore) to generalized elliptic operators over X.

Having K-theory, one wants then to generalize the Chern character $K(X) \to \oplus H^{2i}(X, \mathbb{Q})$ to get (computable) invariants of elements[24] of these groups.

Cyclic homology was introduced by Connes (1985) to be the target of such a generalized Chern character from K-theory to something more immediately computable and definable algebraically without the commutativity: it is closely related to de Rham cohomology in the commutative case (see Loday, 1976, 1998, for an excellent treatment; needless to say, having computable invariants for K-theory is important in many situations where the ring[25] whose K-theory is taken is not a C^*-algebra.) It is thus an excellent example of the noncommutative philosophy: "commutative" invariants (i.e. geometric invariants of spaces) that can be interpreted noncommutatively are much more powerful and natural.[26]

In this very short appendix, I will describe just a few of the noncommutative algebras that arise in geometric situations.

Example 5.12 $(A = B)$ Suppose we start with two points, A and B. We form $X = \{A, B\}$ and $C^*(X) = \mathbb{C}^2$, where addition and multiplication is coordinate-wise. A function on X requires two values, one for each of A and B, and there is no communication between them.

Suppose we now set $A = B$; then the functions need to assign the same values to A and B and we obtain the algebra $\mathbb{C}$.

But suppose we just had an equivalence relation $A \sim B$: then A would communicate with B and vice versa. It would not be crazy to consider associated to this system $M_2(\mathbb{C})$, the 2×2 matrices that have off-diagonal entries, that reflect the communication between A and B.

We can consider this as being governed by the groupoid[27] associated to the equivalence relation. Then we get $C(A \cup B)$ acted on by the bounded operators, one for each arrow in this category. (Assuming boundedness[28] will make the convolution product of the operators defined in this way to be defined.)

Interestingly $\mathbb{C}$ and $M_2(\mathbb{C})$ are closely related algebras: they are Morita

---

[24] This is very relevant to the Novikov conjecture, which is, after all, a lower bound on K-theory – so a suitable Chern character, rich enough to detect the image of the assembly map, would be a dream come true.

[25] Randy McCarthy's thesis (see McCarthy, 1994) generalized the definition of cyclic homology and the trace map to the setting of exact categories.

[26] This is also true of working with stratified spaces. Frequently, arguments that must work in full generality are constrained, and therefore easier to find, than ones that just apply to very particular classes.

[27] Recall that a groupoid is a category in which every arrow is invertible.

[28] The boundedness is automatic in this setting, but it seemed like a good idea to say it explicitly anyway.

equivalent – they have equivalent categories of projective modules. So in this case $A \sim B$ and $A = B$ have very similar effects. But for more complicated equivalence relations, it is indeed important to remember that equivalent does not mean equal, and that the noncommutative perspective includes the price of seeing that points are equivalent in a quotient space and keep track of these "transaction costs" when one does further analysis.[29]

The Morita equivalence of various ways of producing quotient objects (and, in particular, when there's a reasonable commutative choice) is common in tame situations.

Example 5.13 (Group actions)   If G is a finite group acting nicely on a space X, one can form X/G. The continuous functions on this are $C(X)^G$. But the G-space X has much more information than X/G.

Even from the theory of vector bundles, one knows that it's much more exciting to consider equivariant bundles on X as opposed to bundles over X/G. So, what kind of space is this?

So for X that is a point, we want a vector space with a self-identification associated to each self-identification of X given by $g \in G$. This means that we need modules over $\mathbb{C}[G]$ which are representations of G. In general, we should have modules over the semidirect product $C(X) \rtimes G$. For G a compact Lie group, one needs more general convolution algebra.

Of course the K-theory of this convolution algebra is exactly the Grothendieck group of equivariant vector bundles over X. A similar construction can also be made for locally compact groups acting properly on X.

When the action on X is not proper, then things become less clear geometrically. However, one can still form convolution algebras and study their properties. This could well lead you to the Baum–Connes conjecture (with coefficients) if you were bold enough.

Example 5.14 (Tilings, bounded geometry, etc.)   Suppose[30] that every day

---

[29] Mathematicians are trained from early years to take equivalence classes, and form quotient objects. Try explaining this to a non-mathematician! It is not at all easy to do. (Maybe you remember wondering whether real numbers were really equivalence classes of sequences of rational numbers or whether they were "really" numbers.)

   Logicians know well the importance of distinguishing between = and $\sim$, and in model theory have special rules for the interpretation of =. Because (as Bill Clinton famously explained) it is not always clear what it is.

   Respecting that one should have to pay for every use of an equivalence relation is a good idea (and is behind things like Dehn functions in geometric group theory). The C^*-algebra approach to quotients does this naturally.

[30] This example is the result of many conversations with Jean Bellissard and Semail Ulgen-Yildirim and inspired by the work of Bellissard (1995) and Abert *et al.* (2017), among others.

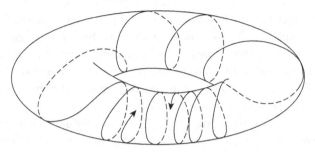

Figure 5.1

were exactly like the previous and the next. So 9:00 a.m. today would be like 9:00 a.m. any other day. Then my experience of time would be indistinguishable from a circle. It would be a matter of debate for the metaphysicians (who would come to either no conclusion or the very same conclusion every day) whether time was "really" a line – with certain regularities holding – or whether it was a circle.

I think it could be an amusing project to write a (series of) novel(s) that would have such a periodic structure. But it would be a contrived project in that even if the last page of the last volume were identical to the first page of the first, in the absence of determinism one would not expect "the continuation" to be the same as the first time around.[31] (Very interesting are the random samples from a periodic distribution.)

Returning to the real world, we can consider the experience of a creature with bounded memory and resources on various spaces. Suppose that there were one glitch in the periodic "time-line" universe. Then for all of early eternity till that glitch, time would be a circle, but from the glitch on, depending on how good the creature's memory was, it would seem like there was non-periodicity – after all, there was some time from "the beginning" or "the change" but that it would asymptote also to a circle. (Figure 5.1 describes this – assuming the stable states at $\pm\infty$ are the time-reversed for entertainment. Without the identification, there would be a similar picture taking place on a subset of the cylinder.)

This topological space is the space where experience takes place. Geometric processes that weight local geometry and involve large scales in increasingly damped fashions should extend to this space that compactifies the time-line.

Now let's think still about a time-bound creature, but time, while a line, is

---

[31] It might be more interesting to double the number of volumes and have a second series not identical to the first. I'm not sure what I want the last page of the last volume of the second series to be like.

tiled by tiles A and B, with some rules about how the A and B are put down. If all the A were in positive time and all B in negative, then one would get the two-circle space mentioned above. But if instead we can put our tiles down in a random fashion then time would formally be $\mathbb{R}$: there is no periodicity, but if our memories get weak, each time will have been anticipated infinitely often, and indeed history will yet repeat itself infinitely often. We can embed this $\mathbb{R}$ in $\mathbb{R} \times \{A, B\}^{\mathbb{Z}}$/the disagonal action of $\mathbb{Z}$ where the product is bi-infinite. A real number is mapped to r (this being the label of the tile of which it's a part), with boundary points resulting in the identification on the right. The space most appropriate for modeling the experience of our creature will be the same as this limit space, i.e. the closure of the $\mathbb{R}$-orbit in this space. If the placement of tiles were truly random, the closure would be the whole space. (Almost all $\mathbb{R}$-orbits in this space are dense.)

I like this example a lot. It is easy to build into topology a theory that takes into account only balls of some, perhaps unspecified, size. However, allowing far-away points to have influence that is decaying seems much better suited to analysis.

This is an example of a foliated space, which is a slight generalization of the notion of a foliation. And, there is a C^*-algebra associated to such spaces which, in the case of a fiber bundle foliated by fibers, would be Morita-equivalent to the continuous functions on the quotient.

This can be done for tilings of $\mathbb{R}^n$ rather than just on $\mathbb{R}^1$.

A lot of the information of a tiling on $\mathbb{R}^n$ can be described by giving the centers of gravity of the tiles.[32] (In higher dimensions, we can get interesting examples without the expedient of labeling as we did in one dimension.) We are interested in tilings where these point sets are:

(1) C-dense for some C (every point is within C of some center);
(2) sparse, i.e. no two points are within δ of each other; and
(3) repetitive, so the pattern of every ball of radius R repeats in a $C(R)$ dense way everywhere.

From conditions (1) and (2), one can see that the set of patterns has an embedding in a nice locally compact space. This space of tilings has an $\mathbb{R}^n$-action on it (actually the whole Euclidean group acts on it, and when we take that into account with condition (3) one gets a broader notion). Condition (3) is of a different nature, and guarantees that the set of patterns arising forms a minimal dynamical system with respect to this action.

---

[32] If one is just interested in convex tiles, then one can use a slightly different point set S: a set of points such that the tile containing p is defined as the set of points closer to p than to any of the other points of S.

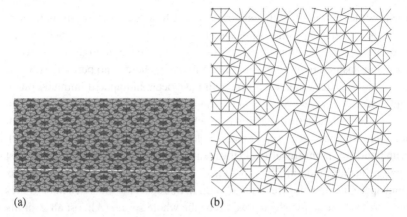

(a)   (b)

Figure 5.2 (a) The Penrose tiling. (b) A pinwheel tiling. (From Sadun, 1998. Reprinted by permission of Springer Nature.)

Two examples of aperiodic tilings are shown in Figure 5.2. The pinwheel tiling does not have recurrence entirely on the basis of translations, but that is irrelevant to the story we are discussing, which allows all isometries.

The closure of this orbit is called the *hull* of the tiling. It consists of the pointed Gromov–Hausdorff limits of centered balls in the tiling.[33] It consists of the tilings that cannot be locally distinguished from the tilings that occur within the original one, but is usually much larger. If the tiling is aperiodic, then it has countably many "subtilings" (just recenterings, essentially) but there are always uncountably many possible limits.

For aperiodic tilings, the hull usually has the local structure of the product of a manifold with a Cantor set, but if one relaxes condition (3) then very different kinds of structures can occur. (The reader might enjoy considering the hulls of graphite versus diamond.)

This can also be done with respect to any manifold with bounded geometry. These limits can be thought of as doppelgangers, whose properties restrict the original manifolds. They are like the way a novelist can use pieces of our personalities to create characters that resemble us yet are more extreme than we are – in order to shed light on what we are like. More subtle than just the

---

[33] Following an unpublished note with Bellissard and Ulgen-Yildirim, it can be defined as the closure of the natural map of the manifold into the inverse limit of the pointed Gromov–Hausdorff space of centered Riemannian balls of various radii (under the natural map that sends a centered ball of radius R to one of radius r if $R > r$). Each point in M is mapped to the consistent set of balls in M centered at that point. Generically this map is an embedding, and the closure is a foliated space, but for special manifolds that have a lot of symmetry, this mapping can have some strange properties.

individual limit points within this construction are the properties that describe the size of the hull.

Periodicity, like regular covers, gives rise to compact hulls with the manifold not embedded in the hull. The difference between crystals, quasicrystals, and glass is apparent in the structure of the hull. One is led very naturally to fascinating problems like when such constructions have transverse measures (e.g. when one has a reasonable notion of "typical behavior" on a manifold), etc.

Example 5.15 Connes introduced C^*-algebras associated to foliations (as in the previous examples); these are variations on the ones associated to groupoids – and they are all essentially convolution algebras acting on functions (or half-densities[34]) on the total space. If the foliation were a fibration, this algebra is Morita-equivalent to the continuous functions on the quotient, but interesting foliated spaces can frequently have all leaves dense (like the foliation of a torus by irrational lines or planes, and the example of aperiodic tilings above).

There are many more examples that come up from physics (e.g. the standard model) or number theory (e.g. spaces of $\mathbb{Q}$-lattices, etc.). I won't discuss these, but see Connes and Marcolli (2008), and references therein.

When one has a C^*-algebra/noncommutative space, one begins using it to do mathematics. At first, there are questions about which algebraic properties of the algebra hold, or how are they reflected in the geometry of the situation. For example, a trace on the C^*-algebra of a foliation corresponds to a transverse measure.

After such a dictionary is established, it becomes possible to implicate various functors such K-homology and K-theory and develop and apply appropriate index theorems in this setting. Foliations (say, of compact manifolds) always have an implicit dynamics as noncompact leaves recur, and such results can have profound implications.

Cyclic homology was invented to be a noncommutative analogue of de Rham cohomology. It's close, but doesn't quite agree with this. Nevertheless, it provides an important invariant of algebras and a place to map K-theory to.

In dimension 0, it captures the idea of a trace, and these are associated to the simplest index theorems. Higher homology reflect higher index theorems (with the important, seemingly technical, issue that one frequently has to go to dense subalgebras[35] to get nontrivial higher homology).

Let's color in the outlines of the picture we tried to draw. Here's how the dream would go in the special case of $R = \mathbb{C}\pi$), a group ring: we are interested

---

[34] Half-densities are useful for creating appropriate L^2-spaces.

[35] Such algebras describe types of smoothness – and, once one says this, the idea of passing to such algebras is not at all shocking if we are to use these for doing analysis.

in invariants on projective modules over R (i.e. $K_0(R)$) generalizing the notion of dimension. Consider the formula

$$\dim P = \operatorname{rank} P,$$

which almost looks too tautologous for words! This formula exploits our habit of writing the same letter P for a projective module and a projection $P^2 = P$ whose image is that module. For projections we have

$$\operatorname{rank} P = \operatorname{trace} P.$$

The key defining property of a trace is that for all matrices $\operatorname{trace}(AB) = \operatorname{trace}(BA)$ (from which follows the key property that it is an invariant of an automorphism, not merely of a matrix). This suggests considering

$$R/[R, R] = R/\{ab - ba\}$$

as the best target for a trace that we could possibly hope for (and why hope for any less?).

Here we are considering a quotient abelian group in this formula. (The set of elements $ab - ba$ is not closed under addition, and we consider the subgroup generated by such commutators.) In the case of a group ring, the right-hand side breaks up into pieces, $\bigoplus \mathbb{C}(g)$, where (g) is the equivalence class of g under conjugation. A little thought shows that taking the "ordinary" trace of a matrix and collecting all the coefficients of the elements within a conjugacy class actually is a well-defined operation.

This quotient is exactly $HC_0(\mathbb{C}\pi)$, and this algebraic trace is called the Hattori–Stallings trace. The trace map from K_0 is quite nontrivial, as the case of π finite readily indicates. Indeed, it shows that for any non-torsion-free group, the modules induced from finite subgroups can be non-free projective modules for the group. (The converse, of course, is part of the Borel package to which we have alluded several times.)

It also clearly has connections to Nielsen fixed-point theory (where we generalize the Lefshetz number just by changing the meaning of the word trace).

In the K_0 setting it is an interesting and natural question whether the coefficient of a conjugacy class of infinite order can ever be nonzero. (That this is impossible is sometimes called the Bass conjecture: see Bass, 1976.) This is known in many cases – and it seems worth noting here that the theory of cyclic homology and the higher trace maps has been effectively deployed in this direction (see Eckmann, 1986)).

In passing to the completion there is very serious (analytic) trouble. If an element g has infinitely many conjugates, perhaps an element in $\mathbb{C}^* \pi$ might want to give that trace an infinite value.[36] That there are homomorphisms from $K(\mathbb{C}^*\pi)$ to $\mathbb{C}$ corresponding to conjugacy classes of finite order does not seem any less deep than the Novikov conjecture.[37]

Cyclic homology actually closely resembles the homology of $E S^1 \times_{S^1} \Lambda X$, where Λ denotes the free loop space, and S^1 acts on this space by rotation of loops (see, e.g., Burghelea, 1985; Goodwillie, 1985). The case relevant to us is $B\pi$: $\Lambda B\pi$ has components corresponding to the conjugacy classes of elements of π – the HC_0 that we saw earlier. The component of loops freely homotopic to a given g is itself aspherical, and is a $K(Z(g), 1)$, where $Z(g)$ is the centralizer of g.

When we take the Borel construction on the action of S^1 by loop rotation, we get different behavior when g is finite order and when it's infinite. In the finite case we get a $K(Z(g), 1) \times \mathbb{CP}^\infty$, but in the infinite case, it is $K(Z(g)/\langle g \rangle, 1)$. Rationally, the Baum–Connes conjecture asserts that $K(\mathbb{C} * *\pi)$ is isomorphic to $\bigoplus K(C(g))$ where the sum is taken over g with finite order.[38]

The general theory provides for maps $\mathrm{HC}_n \to \mathrm{HC}_{n-2}$ dual to Bott periodicity, which here is dual to the cup product with the Euler class of the circle bundle, and the trace with target HC_{n-2k} factors through this. This is used in Eckmann's work on the Bass conjecture, and is also completely reasonable in the C^*-setting where K-theory has a Bott periodicity isomorphism. Of course, the $K(Z(g), 1) \times \mathbb{CP}^\infty$ results in certain homology groups being counted many times, but, inverting the periodicity, they each are rationally counted once.

5.3 Playing the Game: What Happens in Particular Cases?

In §5.2 we explained that the formal framework of index theory is rather similar to that of surgery theory,[39] and, as a result, the Novikov phenomenon applies more broadly to other operators.

This point of view is fine as a starting point, but it is way too formal to be a

---

[36] On the other hand, if g has only finitely many conjugates, then there is no trouble defining a trace associated to (g) and this can be exploited for geometric gain. This, for example, arises for groups with a finite normal subgroup whose quotient is torsion free.

[37] More precisely, it actually does seem less deep, but I have never found an *a priori* argument that doesn't involve Novikov technology. Weinberger and Yu (2015) is a failed struggle with this problem.

[38] This is compatible with what one would expect from the topological conjectures if one replaced $\mathbb{Q}\pi$ or $\mathbb{R}\pi$ with $\mathbb{C}\pi$ (with the involution being complex conjugation on $\mathbb{C}$).

[39] Indeed, the signature operator gives rise to a *functorial* mapping of surgery theory into operator K-theory, as mentioned above (see Higson and Roe, 2005a,b,c). It is important, though, to note that the infinite loop space structures on the two theories are different at the prime 2 (see Rosenberg and Weinberger, 2006) because of the nature of the boundary map in the exact sequences of pairs.

stopping point – it oversimplifies, missing the exquisite texture of the subject and the true benefit of unification.

Some methods develop naturally within the context of one problem, and the informal parallels between subjects leads to a search for cognate results for other problems – ultimately leading to multiple analogous theorems. In some sense, every time we play the Novikov game, we get a new test, a new area that is suggestive of techniques internal to it – that afterwards we can hope will shed light on the original problem, or, if not to it, perhaps to some other analogues, and it also leaves us with the puzzle of understanding why we didn't succeed in this export.

Surely the richest two cases were the original Novikov problem, which we have already discussed and shall have to review in light of the index-theoretic perspective, followed by the problem of positive scalar curvature.[40] We shall first discuss this latter problem before returning to the first and then to the general discussion.

The first results on the positive scalar curvature problem, after the Atiyah–Lichnerowicz–Singer vanishing result, was the proof by Schoen and Yau (1979a) of the nonexistence of positive scalar curvature metrics on the torus and related manifolds (e.g. those that resemble Haken manifolds or have maps of nonzero degree to them) and have dimension ≤ 7.[41] This method has the feel of the original methods on the Novikov conjecture using codimension-1 splitting, and it has consequences that we do not yet know how to approach by Dirac operator techniques (coupled to the fundamental group).[42] The geometric nature of this method makes it possible to describe it before any discussion of applications of the index theorem.

The very elegant idea in Schoen and Yau (1979a), in its embryonic form, is this. Let M be a manifold with positive scalar curvature and a nonzero degree map to $\mathbb{T}^n$. Then one finds in M an area-minimizing minimal hypersurface dual to any class in H^1. This hypersurface is smooth if M has dimension ≤ 7 (and this is where the dimension hypothesis enters) and has a map of nonzero degree to $\mathbb{T}^{n-1}$.

---

[40] Needless to say, for many people the positive scalar curvature problem is the more interesting one, because, for example, it has important connections to general relativity (the positive mass conjecture).

[41] See also Lohkamp (2006) for an announcement of a method for getting around the fact that minimal surfaces can develop singularities in dimension greater than 7, and a more recent paper of Schoen and Yau (2017) that gives a different approach.

[42] Results using the Dirac operator require some spin structure. Indeed, for non-spin simply connected manifold of dimension greater than 4, Gromov and Lawson (1980a,b) have shown that positive scalar curvature metrics always exist! On the other hand, the Schoen–Yau result shows that if V is any manifold of dim < 8, then $V \# \mathbb{T}$ never has positive scalar curvature. This manifold's universal cover is, of course, not spin.

A calculation then shows that the induced metric on a minimal hypersurface in a positive scalar curvature manifold is naturally conformally equivalent to a positive scalar curvature metric. (The conformal factor is a power of the eigenfunction associated to the first eigenvalue (necessarily positive) of $\Delta - (n-3)K/4(n-2)$, where k is scalar curvature on the hypersurface.)

One repeats this argument till one gets down to dimension 2, and the result follows from the Gauss–Bonnet theorem.

Let us now turn to the Dirac operator method. We shall not review the definition of the Dirac operator, leaving the reader to standard references for it and its theory (e.g. Atiyah *et al.*, 1964; Atiyah and Singer, 1968a,b, 1971; Lawson and Michelsohn, 1989; Roe, 1998; Berline *et al.*, 2004; Higson and Roe (20XX)).

Lichnerowicz showed the following "Bochner type" formula relating the Dirac operator on a manifold and the Laplacian on forms: $\mathbf{D}^2 = \nabla^*\nabla + k/4$ (see e.g. Atiyah and Singer, 1968b; Roe, 1998). This implies, assuming that the scalar curvature is everywhere positive,[43] that $\mathbf{D}$ and $\mathbf{D}^*$ can have no kernel (as the Laplacian is semidefinite). In other words "M can have no harmonic spinors," and, in particular, $\text{ind}(\mathbf{D}) = 0$. The index theorem then gives (see, e.g., Atiyah and Singer, 1968b for the calculations of the symbol of the Dirac operator, and how the index theorem works out in this case) that

$$\langle A(M), [M] \rangle = \text{ind}(\mathbf{D}) = 0.$$

In every dimension that is a multiple of 4, there is a spin manifold whose A-genus is nonzero, and these give examples of simply connected manifolds with no positive scalar curvature metrics.

Actually, one can do somewhat better than this (without using the fundamental group at all). The Dirac operator naturally has a *real* structure, allowing more subtle *real* index theorems to be applied. The index then takes values in $KO_i(*) = \mathbb{Z}/2, \mathbb{Z}/2, 0, \mathbb{Z}, 0, 0, 0, \mathbb{Z}$ (depending on i mod 8) – thus providing a refinement at the prime 2. Hitchin (1974) showed that, under the assumptions of spin and positive scalar curvature, for an i-manifold

$$\text{ind}(\mathbf{D}) = 0 \in KO_i(*).$$

Thus spin manifolds of dimensions 1, 2 mod 8 there is an extra mod 2 obstruction to having positive scalar curvature. There are even examples of manifolds homeomorphic to the sphere that do not have positive scalar curvature metrics.

---

[43] Kazdan and Warner (1975) showed that if M has a metric that is nonnegative everywhere, and positive somewhere, then *any* smooth function on M is the scalar curvature function of some metric. Therefore, we can weaken the curvature conditions in this argument.

Remarkably, for spin manifolds of dimension greater than 4,[44] Stoltz (1992) has shown that every simply connected manifold with vanishing ind(**D**) (in $KO_i(*)$) actually has a positive scalar curvature metric.[45]

Lusztig's proof of the Novikov conjecture (§4.4) applies: when we couple the Dirac operator to a bundle ξ, the formula $\mathbf{D} * \mathbf{D} = \Delta * \Delta + K/4$ gets another term coming from the curvature of the bundle, but if the bundle is flat, then the same argument gives the vanishing of ind($\mathbf{D}_\xi$). As we vary ξ over a parameter space, especially over the dual torus $\mathbb{T} = (\mathrm{Hom}\,(\pi_1(M)\colon S^1)$, we get a zero-dimensional trivial bundle as the index $\in K^0(\mathbb{T})$.

This then gives the result that, if M is a spin manifold with positive scalar curvature, then for any $f\colon M \to \mathbb{T}$, and $\alpha \in H^*(\mathbb{T})$,

$$\langle f * (\alpha) \cup A(M), [M] \rangle = 0.$$

Gromov and Lawson (1980a,b) suggested a beautiful alternative, using families.

If we use only a single (finite-dimensional) flat bundle, then we gain nothing from the index theorem; we still have ind($\mathbf{D}_\xi$) = ind(**D**) = 0, but the index theorem just has on the topological side

$$\langle \mathrm{ch}(\mathbf{D}_\xi) \cup A(M), [M] \rangle.$$

But $\mathrm{ch}(\mathbf{D}_\xi) = \dim\xi$ has no positive-dimensional component.[46] However, if we allow ξ to be a bundle with *very small* curvature (in comparison to inf $k/4$ that we assume is $> \varepsilon > 0$), then conceivably $\mathrm{ch}(\mathbf{D}_\xi) = \dim\xi \neq 0$, but the new curvature term in the Lichnerowicz–Bochner formula is still positive enough to give the vanishing of the index.

So, how do we get bundles with *arbitrarily small curvature* and nontrivial Chern class c_n so we can implement this idea?

This is impossible on a single compact manifold (because the cohomology classes represented by Chern classes are *integral*, so, if they are sufficiently small,[47] they will integrate to 0 on all cycles). However, we can find (sometimes) a sequence of bundles ξ_i on covers M_i of M with nontrivial c_n, and whose curvatures tend to 0.

For example, suppose $M = K \backslash G / \Gamma$, where Γ is a uniform lattice, and suppose (without loss of generality for our current purposes) that $\dim(G/K) = 2n$ is

---

[44] In dimension 4, the Seiberg–Witten invariants give additional obstructions to the existence of positive scalar curvature metrics (see, e.g., Morgan, 1996).

[45] Gromov and Lawson had earlier used cobordism methods to show that *every* simply connected non-spin closed n-manifold ($n > 4$) has a positive scalar curvature metric, and Stoltz extended their cobordism arguments using very clever algebraic topological arguments.

[46] Chern classes are integral, so sufficiently small curvature implies that they vanish rationally. (Note that it's only the real Chern classes that have a description in terms of curvature, so even zero curvature is compatible with a torsion Chern class.)

[47] And of a fixed dimension.

even. Then, by the residual finiteness of Γ we can find finite normal covers $K \backslash G / \Gamma_i$ whose injectivity radii R_i are arbitrarily large. We can use the logarithm map, i.e. the inverse of the exponential map, followed by the pinch map that wraps the Euclidean ball of radius $R < R_i$ (for i large) onto the standard round sphere S^{2n} of curvature equal to $+1$. Let $L_i : K \backslash G / \Gamma_i \to S^{2n}$ be this logarithm map, associated to some base point where there is an embedded geodesic R-ball. As $i \to \infty$, the Lipschitz constant of $L \to 0$ (because of the rescaling by R; the logarithm map itself is 1-Lipschitz for non-positively curved manifolds).

Note that by Bott periodicity there is a bundle ξ over S^{2n} with $C_n(\xi) \neq 0$ (indeed, the "Bott element" has $c_n = (n-1)!$). Let $\xi_i \downarrow K \backslash G / \Gamma_i$ be $L_i * (\xi)$. It is an almost-flat family yet $c_n(\xi_i) = \langle c_n(\xi), [S^{2n}] \rangle [K \backslash G / \Gamma_i]$, a nontrivial multiple of the fundamental class. Given a manifold M with fundamental group Γ we can use a map inducing the isomorphism of fundamental groups $M \to K \backslash G / \Gamma$ to pull back the almost-flat bundle and see that at least the higher A-genus associated to the fundamental class of Γ obstructs positive scalar curvature.

If you are unhappy with the ξ_i being over different manifolds (although this is completely irrelevant to the application of the argument!), we can push them forward with respect to the covering map $K \backslash G / \Gamma_i \to K \backslash G / \Gamma$. These pushforward bundles will still be almost-flat – in the sense of having decreasing curvature approaching 0 – and have nontrivial Chern class, but increasing dimension.

See Hanke and Schick (2006) and Hanke *et al.* (2008) for papers that explain how this argument fits into the assembly map perspective.

In Gromov and Lawson (1983), the linearity condition on the fundamental group of the nonpositive curvature manifold were removed. Above, we used this condition to produce the sequence of finite covers with arbitrarily large injectivity radius – unfortunately, as far as we know, there might be a nonpositively curved manifold whose fundamental group is simple![48]

However, there's always the universal cover, which has infinite injectivity radius – or alternatively $\mathbb{R}^{2n}$ has a bundle with compact support, which has connections on it which are trivial outside a ball, and have arbitrarily small curvature (simply rescaling of the Bott element on $\mathbb{C}^n$) – that can be used if only you are bold enough to give up on compactness in index theory.[49] Gromov and Lawson develop the relevant index theory and, using this one almost-flat bundle with compact support, proved the result without any residual finiteness hypothesis.

---

[48] No such Riemannian example is known, but Burger and Mozes (2001) have given simple finitely presented groups that act properly discontinuously on a product of trees.

[49] This is a different type of noncompact index theory than the L^2 index theorem of Atiyah for infinite regular covers that we had discussed in Chapter 3.

These almost-flat ideas were later turned around and applied to define the notion of the signature of a manifold with coefficients in an almost-flat bundle (or family) (Connes *et al.*, 1990). Then the relevant index theorem applies to give the Novikov conjecture in the nicest possible way: it gives a simple-to-understand homotopy invariant that expresses the reason that the higher-signature characteristic classes are homotopy invariant.

The reader must have noticed that the discussion above sufficed to explain the higher A-genus (and signature) associated to the fundamental class – but not the other cohomology classes. In Gromov and Lawson (1983) and Connes *et al.* (1993), various approaches to the other cohomology classes are given, via families (very similar to the "descent argument" given in §4.9).

Rosenberg (1983, 1986a,b) modified this argument by using infinite-dimensional flat bundles in place of finite-dimensional almost-flat bundles – this is essentially the work described in §5.2. (He also explained in Section 2 of Rosenberg (1991) the relevant aspects of the *real C^*-theory* and its K-theory, to get the correct obstructions for the prime 2 to include the Hitchin obstructions.)

In the infinite-dimensional setting there is relatively little difference between a single operator or a family. Asserting that one can fill $K(K(\Gamma, 1))$ by Chern classes of flat $C^*\Gamma$ bundles is essentially a form of the Novikov conjecture.

The next step in continuing our development of the parallelism between surgery and operator theory is to develop analogues of the topological theories of noncompact manifolds, and, perhaps most importantly, the analogue of bounded control. We will see that the replacement for $L^{Bdd}(\Gamma)$ is the K-theory of a C^*-algebra (the "Roe algebra" of the discrete metric space Γ, often denoted by $|\Gamma|$). I refer the reader to Roe (1993, 1996) and Higson and Roe (2000) for detailed discussions.

This Roe algebra is the algebra of "bounded propagation speed operators" on $|\Gamma|$. The idea is this (and is entirely analogous to the algebraic description of $L^{Bdd}(\Gamma)$ that we did not give!): Imagine that at each point of Γ we attach a Hilbert space, and we only allow operators that map the Hilbert space at p to (the sum of) ones that are at most some distance away (where the distance bound is independent of p).

Except that in order to make a C^*-algebra it is necessary to take the closure of such operators, and this allows some amount of infinite speed, but "very little that is going very fast." An example is the heat flow $e^{-t\nabla}$, where ∇ is the Laplacian, for $t > 0$, which is bounded propagation speed in the most naive sense: the combinatorial model of ∇ propagates at a scale of 1 unit, and when we truncate the exponential after finitely many terms we get a norm converging sequence of bounded propagation speed operators, but whose speed keeps growing.

The algebra is denoted by $C^*|\Gamma|$ or for general metric spaces[50] by $C^*(X)$.

If X is a complete manifold, then the "geometric operators" on X have bounded propagation speed (see Cheeger *et al.*, 1982). Elliptic operators give rise to elements of $K_x^{lf}(X)$ and there is an "index map" (that is, a cousin of the "assembly map" in bounded surgery theory) that assigns an index to each elliptic operator over X an element of the group $K(C^*(X))$:

$$K_x^{lf}(X) \to K_x(C^*(X)).$$

This index contains, for example, for spin manifolds, an obstruction to the existence of a positive scalar curvature metric on X so that the map $(X, g) \to X$ is Lipschitz (in the large) (so the propagation speed still is finite from the $|X|$ point of view).

Note that if X is uniformly contractible, then one can conjecture that the index map is an isomorphism, and that, even without this,

$$KX_x^{lf}(X) \to K_x(C^*(X))$$

is. (The manifold in Dranishnikov *et al.* (2003) gives a counterexample to the first statement, but it makes use of the map from $KX_x^{lf}(X)$ and remarkably, even though X is uniformly contractible, the left-hand sides of these seemingly identical constructions do not coincide.) Guoliang Yu (1998) gave a very elegant example showing that this latter map is not even injective. Let X be the disjoint union of rescaled copies of the sphere:

$$X = \bigcup \sqrt{n} S^{2n}.$$

This is a spin manifold with positive scalar curvature bounded away from 0. The rescalings mean that, at any scale, only finitely many of the spheres can be ignored. We have an isomorphism

$$KX_x^{lf}(X) \approx \prod(\mathbb{Z}) / \bigoplus \mathbb{Z}$$

with the Dirac operator representing the element $(1, 1, 1, \ldots)$ which is nontrivial, and hence an element of the kernel of the assembly map.

This is a very striking example[51] but (like the Dranishnikov *et al.*, 2003, example) it involves unbounded geometry. I'd be curious to know what $K(C^*(X))$ is in this example.

Despite these counterexamples, the reverse is true: there are many situations (such as complete non-positively curved manifolds) where one can show that

---

[50] This only depends on the coarse quasi-isometry type of X, or even just of the bounded category of X (in the sense of §4.8).

[51] And it would be very interesting to know if it gives an example for noninjectivity of the coarse assembly map in bounded L-theory.

the coarse index map is an isomorphism, and then the method of descent applies to give the Novikov conjecture for the group Γ (from the metric space $|\Gamma|$). See §5.6 for more details and further extensions that take into account the fundamental group of the noncompact manifold.

Now let us turn to the problems of group actions, where there are different phenomena in the case of the circle and the case of finite groups, and then to the birational invariance of higher Todd genus, where the result is actually true unconditionally (and integrally!). In all of these cases[52] it is not too hard to promote the simply connected argument to a proof that's conditional on the injectivity of the index homomorphism $K(B\Gamma) \to K(C^*\Gamma)$ (also known as the strong Novikov conjecture[53]). However, what we would like to understand is "why" these are now known to be true unconditionally. The mechanism is actually different in the two cases, but the "reason" seems to be the same.

Let us start with the results about S^1-actions, and, for simplicity, let's assume that the action is "semi-free," namely that every orbit is either trivial (corresponding to fixed points) or free,[54] leaving the reader to refer to the original papers for the general case (which is philosophically the same, but does show some different aspects in detail).

Let's start with the case of the higher A-genus:

Proposition 5.16 *Suppose S^1 acts on M semi-freely with nonempty fixed point set* F; *then the map $M \to M/S^1$ is split injective on the fundamental group.*

We suppose that codimension F is at least 4: this can be achieved by taking the product with $\mathbb{C}^2$, giving the latter the obvious circle action thinking of S^1 as unit complex numbers. In that case, the map is actually an isomorphism on the fundamental group by a simple application of Van Kampen's theorem (and general position: in that removing F will not change the fundamental group).

Now, recall from §4.5 that to prove the vanishing of the image of $A(M) \cap [M]$ in $\bigoplus H_{m-4i}(K(\pi, 1); \mathbb{Q})$, it just suffices to show that, for all "subcycles" X of $K(\pi, 1)$ that have a trivial normal bundle neighborhood, the transverse inverse image $f^{-1}(X)$ has vanishing A-genus. But this is true, since by first taking the

---

[52] Except for the case of invariance of higher signatures for fixed sets of pseudo-trivial actions.

[53] So-called because it implies the homotopy invariance of higher signatures – i.e. the original Novikov conjecture. It also implies, away from 2, integral refinements that we will discuss in the next section.

[54] Actually, the free case is even easier: M bounds a $\mathcal{D}^2$ bundle over M/S^1 and then we can use cobordism invariance of characteristic classes to see that, for any $\alpha \in H*(M/S^1)$, the higher signature of M and the higher A-genus of M vanish. Alternatively, dually, and somewhat more precisely, the pushforwards of [Sign] and [$\mathcal{D}$] in $K_m(M/S^1)$ vanish. (They are the boundary of the natural classes in $K^{lf}_{m+1}(E)$, where E is the total space of the vector bundle whose unit sphere bundle is the circle bundle defined by $M \to M/S^1$.)

transverse inverse image of X in M/S^1 and then taking its inverse image in M, we obtain an inverse image for X that is still spin (it has a trivial normal bundle in a spin manifold) and has a nontrivial S^1-action. Thus, the ordinary Atiyah–Hirzebruch theorem gives the conclusion.

The result about higher-signature localization for S^1-actions follows from similar reasoning. There is a cobordism (see below) from M to a union of $\mathbb{CP}^{c/2-1}$ bundles over the components of F, where c is the codimension of the component. We need the following lemma that tells us about the L-classes of this total space in terms of the L-class of F.

Lemma 5.17 *If* $\pi: \mathrm{E} \to \mathrm{B}$ *is a (block) bundle with (homotopy) fiber (a homotopy)* $\mathbb{CP}^k$, *whose monodromy is trivial on* $H^2(\mathbb{CP}^k)$, *then* $\pi_*\big(L(\mathrm{E}) \cap [\mathrm{E}]\big) = \mathrm{sign}\,(\mathbb{CP}^k) L(M) \cap [M]$.

This boils down (by the same reasoning as before) to the fact that for all such bundles $\mathrm{sign}\,(\mathrm{E}) = \mathrm{sign}\,(\mathbb{CP}^k)\mathrm{sign}\,(M)$. This is a theorem of Chern, Hirzebruch, and Serre (1957).[55]

The cobordism is explicit: it is $M \times [0, 1]/\sim$ where we identify points on $M \times 1$ that are on the same orbit if that orbit does not touch a tubular neighborhood of F. The structure comes from the equivariant tubular neighborhood theorem (in the smooth case, and the proof of the existence of block bundles in the PL case[56]). Explicitly, this proves:

Theorem 5.18 *If* S^1 *acts on a manifold* M *with nonempty fixed set* F, *then the higher signatures of* M *are those of* F, *i.e. for all* $\alpha \in H^*(\mathrm{B}\pi)$, *one has*

$$\langle f^*(\alpha) \cup L(M), [M]\rangle = \langle f^*i^*(\alpha) \cup L(\mathrm{F}), [\mathrm{F}]\rangle.$$

For $\mathbb{Z}_p$-actions, the cobordism argument fails (the inverse image has a $\mathbb{Z}_p$-action, but its homological properties are not restricted). We will give one argument now about the connection to the Novikov conjecture – another one can be made based on the ideas from Chapter 6 when we study the equivariant Novikov (and Borel) conjectures more systematically. The current argument is based on preliminary remarks about rational homology manifolds.

The reasoning given in §4.5 for the definability of L-classes for PL-manifolds actually produces homology L-classes for oriented rational homology mani-

---

[55] Their theorem assumes the monodromy of the bundle is trivial – for the $\mathbb{CP}^k$ case one could have monodromy of order 2 (i.e. inducing complex conjugation on the fiber). However, in that case, the 2-fold cover of this fibration has trivial monodromy, and signature is multiplicative for all finite sheeted covers (as a consequence of the Hirzebruch signature theorem) and the result follows again.

[56] The topological case requires some form of Smith theory to make these arguments (that are essentially locally homological and sheaf-theoretic, rather than completely geometric).

folds[57] (that agrees with the Poincaré dual of the L-cohomology class): all that one needs to produce L-classes is a cobordism-invariant definition of signature – and one has this using rational cohomology.

Moreover, this characteristic class can be pushed into group homology to give a "higher signature." It turns out (see immediately below) that if the Novikov conjecture is true for manifolds, then it is true for rational homology manifolds in the sense that the higher signature in the homology of $K(\pi, 1)$ will be preserved by maps $f: X \to Y$ that are orientation-preserving and induce $\mathbb{Q}$-homology isomorphisms on the regular covers induced from the map $Y \to K(\pi, 1)$.

The most straightforward argument for this uses Ranicki's algebraic theory of surgery (Ranicki, 1980a,b) (mentioned earlier in §4.7) or its predecessor (Mischenko, 1976). Ranicki views the L-groups of surgery as cobordism groups of certain chain complexes with duality over $\mathbb{Z}\pi$. Inverting 2,[58] we can view $X \cup X \to X$ as a surgery problem over the ring $\mathbb{Q}\pi$ (since we are only assuming a $\mathbb{Q}$-homology manifold) and thus gives an element $\sigma^*(X^n) \in L_n(\mathbb{Q}\pi)$. This element is homotopy invariant since it is defined using chain complexes (the chain complex of a homotopy equivalence gives a cobordism in the appropriate sense). Under the assembly map,

$$H_n(K(\pi, 1); L(\mathbb{Q})) \to L_n(\mathbb{Q}\pi),$$

the homology L-class gets mapped to $\sigma^*(X^n)$.[59] It is a general and remarkable algebraic result of Ranicki (1979a) that, for any π, the map $L(\mathbb{Z}\pi) \to L(\mathbb{Q}\pi)$ is an isomorphism away from the prime 2 – indeed, has kernel and cokernel annihilated by multiplication by 8 – so the injectivity of this assembly map – away from 2 – is equivalent to the injectivity of the usual one and thus the

---

[57] In the PL case: for the topological situation, this can be done using controlled topology (see Cappell *et al.*, 1991 – as a topological definition of L-homology classes that works for manifolds implies Novikov's theorem on Pontrjagin classes, such a definition cannot be too trivial.

[58] In the coefficient ring, since this is a degree-2 map, so we need to multiply the Poincaré duality isomorphism by 2 in the range – which is OK if 2 is inverted in the coefficient ring. Ranicki actually does something different and better. He defines a slightly different cobordism group of chain complexes with self-duality for rings with anti-involution, and then observes that the chain complex of X is naturally an element of this different group. Wall's L-groups, the surgery L-groups, are denoted using subscripts, and Ranicki denotes these modified groups with a superscript: $L_s(\mathbb{Z}\pi)$, for instance (despite their remaining covariantly functorial) and christens them *symmetric L-groups* (and refers to Wall's as *quadratic L-groups*). In any case these only differ at the prime 2. This is not merely an academic issue though: alive versus dead, yes versus no, being and nothingness, are all mod 2 issues.

[59] This, while useful, is just unraveling all the definitions of the objects and morphisms involved. Of course, deducing the Hirzebruch formula from this approach then involves identifying two different homology L-classes!

Novikov conjecture in this setting follows from (is equivalent to) the usual one.[60]

Now, if $G \times M \to M$ is an orientation preserving action of a finite group, then M/G is a $\mathbb{Q}$-homology manifold, and sign (M/G) is just the signature of the G-invariant part of the cohomology of M. In terms of the G-signature of M, which is a representation, we are looking at the multiplicity of its trivial component – which can be computed using character theory as

$$\text{Sign}\,(M/G) = 1/|G| \sum \chi_g (G\text{-signature})$$

and the right-hand side can be computed by characteristic classes of the fixed sets M^g and their equivariant tubular neighborhoods. If $G = \mathbb{Z}_p$ the formula is

$$\text{Sign}\, M/\mathbb{Z}_p = \frac{1}{p}\text{sign}\,(M) + \frac{p-1}{p}\langle v^8(\xi) \cup L(F), [F]\rangle,$$

where $v(\xi)$ is $p/(p-1)$ times[61] the sum of the local contributions in the G-signature formula from all of the generators of $\mathbb{Z}_p$ of the characteristic class of equivariant normal bundle to F from Atiyah and Singer (1968b). If the action is homologically trivial, then Sign $(M/\mathbb{Z}_p) = \text{sign}\,(M)$, so we get

$$\text{Sign}\,(M) = \langle v(\xi) \cup L(F), [F]\rangle.$$

If one combines the various formulas, one obtains[62] that, after applying the assembly map,

$$\sigma_*(M) = \sigma_*(M/\mathbb{Z}_p) = A_* f_* \big(v(\xi) \cup L(F) \cap [F]\big) \in \bigoplus H_{*-4i}\big(K(/\pi, 1); \mathbb{Q}\big),$$

so that assuming the Novikov conjecture, one gets the localization formula. Conversely, if M is a manifold with $\chi(M) = 0$,[63] so that $f_*(L(M) \cap [M])$ lies in the kernel of the assembly map,[64] then the surgery problem $M \to M \times K(\mathbb{Z}_p, 1)$ in $L_m(\mathbb{Q}[\pi \times \mathbb{Z}_p])$ will have vanishing obstruction[65] (after tensoring with $\mathbb{Q}$).

---

[60] Of course, this leaves some room for differences in the integral theory.

[61] We put this factor in to make later formulas more pleasant.

[62] Somewhat profligately, since it is possible to only invert 2 and p in this formula.

[63] If the Euler characteristic is nonzero, then it is impossible to have a finite complex with a free homologically trivial action, by the Lefshetz fixed-point theorem. It occurs formally in our setting, in ensuring that the infinite complex $M \times K(\mathbb{Z}_p, 1)$ has rational chain complex chain equivalent to a finite complex. (The element of $K_0 \times (\mathbb{Q}[\pi \times \mathbb{Z}_p])$ represented by the complex $C_*\big(M \times K(\mathbb{Z}_p, 1)\big)$ is $\chi(M)[\mathbb{Q}\pi]$ where $\mathbb{Q}\pi$ is clearly a nontrivial projective module over $\mathbb{Q}[\pi \times \mathbb{Z}_p]$.

[64] Note that the map from bordism to group homology with coefficients in the (symmetric) L-spectrum, given by $[?] \to f_*(L(?) \cap [?])$ is onto; this is certainly elementary, and all we need, if one tensors with $\mathbb{Q}$.

[65] Surgery with coefficients in a subring of $\mathbb{Q}$ measures the obstruction of a degree-1 normal map being cobordant to one that is a local homology equivalence (in the universal cover) (see, e.g., Taylor and Williams, 1979b).

The result of the surgery will be a manifold with free homologically-trivial $\mathbb{Z}_p$-action cobordant to a multiple of M, and hence with nontrivial higher signature. It will be a counterexample to the localization principle. To summarize, we have explained:

Theorem 5.19 *If $\mathbb{Z}_p$ acts on a manifold, trivially on π_1 and on twisted homology, then one gets a localization formula*

$$f_*(L(M) \cap [M]) = f_*(\nu(\xi) \cup L(F) \cap [F]) \in \bigoplus H_{*-4i}(K(\pi, 1); \mathbb{Q})$$

iff the Novikov conjecture is true for the group π.

Why is there no localization principle for (untwisted) higher signatures for pseudo-trivial actions? We note that the whole problem is anomalous from the point of view of the game: although there is an index equality in the simply connected case, it isn't based (solely) on index theory – but rather Smith theory, a homological result, played a key role.

That doesn't mean that one can't play the game – only that we don't see how to win. The actual failure is based on two principles.

The first is that Smith theory, i.e. the homologically trivial theory of group actions, essentially gives p-adic information for p-groups, rational information for tori, but almost nothing at all for non-p-groups or nonabelian compact Lie groups.

In any case, for p-groups, the only information one can hope for is $\mathbb{F}_p[\pi]$ information, with $\mathbb{F}_p$ some finite field with p elements, and, with some effort, for any manifold M' that is $\mathbb{F}_p[\pi_1(M)]$-homology equivalent to M, one can construct a quasi-trivial $\mathbb{Z}_p$-action on the product of M with a disk, with M' being the fixed-point set.[66]

The second is the very general[67] homological surgery theory of Cappell

---

[66] We have ignored some algebraic K-theory problems that actually arise even if $\pi = \mathbb{Z}$ (as we will show in forthcoming work with Cappell and Yan). As usual, such can be gotten rid of by the violent act of crossing with a circle.

In the smooth category, there are additional bundle-theoretic considerations even to obtaining actions in a neighborhood of M', since a real vector bundle can only admit a $\mathbb{Z}_p$ action if it has a complex structure. For the PL and topological situations, there are results that put a $\mathbb{Z}_p$ action on many neighborhoods (see Cappell and Weinberger, 1991a) and this is being tacitly invoked here. Extending the action outside a neighborhood in the semi-free case (even in low codimension) assuming suitable K-theory conditions is the main result of Assadi and Vogel (1987).

[67] Although, over the past decade there have been a number of occasions when I had wished for a yet more general theory: the Cappell–Shaneson theory is very well adapted to the problems for which it was invented, codimension-2 embedding theory, a.k.a. knot and link theory, but it does not describe the obstructions to doing surgery to obtain a map that is a homology equivalence so that the map is a homology equivalence with coefficients in rather general bundles, or to handle general Serre classes that are not associated to localizations.

Even $\mathbb{Z} \to \mathbb{R}$ is not "officially" part of their theory, although of course a map of $\mathbb{Z}$ chain and Shaneson (1974). They define obstruction groups to performing surgery in this setting, and the relevant group is $\Gamma_m(\mathbb{Z}\pi \to \mathbb{F}_p[\pi])$. The even-dimensional groups are quite hard to get one's hands on, and there are interesting phenomena to be unraveled, but the odd-dimensional groups are typically very small: there is a natural map

$$\Gamma_m(\mathbb{Z}\pi \to \mathbb{F}_p[\pi]) \to L_m(\mathbb{F}_p[\pi])$$

that is automatically one-to-one for m odd. Since $L_m(\mathbb{F}_p[\pi])$ is a module (by tensoring) over the Witt group of nonsingular quadratic forms $W(\mathbb{F}_p)$, and $W(\mathbb{F}_p)$ is always of exponent at most 4 (Milnor and Husemoller, 1973), these Γ groups are exponent 4. Consequently, we can easily produce manifolds in odd dimensions that are $\mathbb{F}_p[\pi]$ homotopy-equivalent to M, whose higher signatures deviate almost at will from those of M.[68]

Then it is necessary to produce group actions with these manifolds as fixed sets, and for this there is well-developed machinery (Assadi and Browder, 1985; Weinberger, 1985a, 1986; Jones, 1986; Assadi and Vogel, 1987; Cappell and Weinberger, 1991a).

We close this section with a *very* brief discussion of the argument given in Block and Weinberger (2006) for Rosenberg's algebraic–geometric Novikov conjecture on birational invariance of higher Todd genera. It follows the pattern we saw above for the S^1 localization formula (or vanishing of higher A-genus).

If V and V' are birational smooth varieties, then according to Abramovich *et al.* (2002)[69] one can move from V to V' by a sequence of blowings up and down. Thus one needs only to check that if

$$\pi : V' \to V$$

is a blowup, then $\pi_*(O_V) = O_v$, and then rely on the topological Riemann–Roch theorem of Baum, Fulton, and MacPherson (1975) to map further (and give commutativity of the diagram) $K(V) \to K(K(\pi, 1))$. This is a local result (and is essentially Hartog's argument given for the birational invariance of the Todd class).

---

complexes is an $\mathbb{R}$ isomorphism iff it is a $\mathbb{Q}$ isomorphism, so one can apply their theory by replacing $\mathbb{R}$ by $\mathbb{Q}$.

[68] In even dimensions some higher signatures survive the map $L_m(\mathbb{Z}\pi) \to \Gamma m(\mathbb{Z}\pi \to \mathbb{F}_p[\pi])$ while many don't. For instance, for free abelian groups, only the ordinary signature survives, but for surface groups of genus greater than 2, the higher signature associated to the fundamental class also survives (as a consequence of the Atiyah–Kodaira fiber bundle).

[69] This result is a kind of Hironaka theorem for maps.

5.4 The Moral

What have we learned from playing several rounds of "the Novikov game"? I think there are two lessons:

There really is a gap in level of depth between the problems that seem to be conjectural and the ones that we know how to prove. The latter tend to be essentially local statements, and the difficulty (nowadays) is to prove theorems from essentially *global* hypotheses.

Novikov's theorem is (as we have seen) the statement that L-classes can be preserved by *hereditary* homotopy equivalences (i.e., CE-maps). The whole problem with the Novikov *conjecture* is determining for which cycles homotopy equivalences "descend" or can be "inherited" and after how much work. Also, the birational invariance is of the same sort: the key to the proof is that a map that is birational is birational on all of its Zariski-open subsets.[70]

The group actions overwhelmingly ratify this perspective. Being a circle action is something that descends, by definition, to invariant submanifolds. However, being a homologically trivial group-action is a local condition.

The results about positive scalar curvature are apparently exceptional in this regard, but actually the point is that it is impossible to ever get a connection between the Dirac operator and its kernel from positive scalar curvature locally: this is a global phenomenon that requires completeness (and thus does *not* descend to open subsets). Indeed, on a manifold with "big" A-genus, nothing about the homological structure of this class is reflected in the structure of the negative scalar curvature set. The remarkable results of Kazdan and Warner (1975) imply that any closed manifold of dimension ≥ 3 has a metric whose scalar curvature is strictly positive outside of a ball – irrespective of fundamental group: the obstruction to positive scalar curvature doesn't "carry" to negative scalar curvature.[71]

On the other hand, the results about the Novikov conjecture all have global hypotheses. In order to play the game, we need an operator, and a hypothesis that combines well with flat bundles of arbitrary dimension (and that's what

---

[70] It is possible to write down modern proofs of these two theorems so that the diagrams look exactly the same, as can be specialized from the argument soon to follow. For the Novikov theorem, one thinks of the "canonical class" (that contains the signature operator, and the Poincaré dual of the L-class) as a self-dual sheaf that is preserved under hereditary homotopy equivalences. For the birational theorem, the canonical sheaf is a coherent algebraic sheaf that is preserved by birational equivalences.

[71] However, in some noncompact situations, something like this is true: see Roe (1988a,b) for a situation where a nonvanishing index on a noncompact manifold guarantees that the negative scalar curvature set is noncompact. This is also true for the results in Gromov and Lawson (1983) for the manifolds with bad ends. See also Chang and Weinberger (2010).

fails for the pseudo-trivial group-action situation: in passing to covers, we do not get any information that holds in characteristic 0 generally).

In light of this analysis, we can now formulate some additional theorems of Novikov type. That is, we need situations where our global conclusion is local from the point of view of some alternative space. For example:

Proposition 5.20 *If $f : M \to N$ is a Riemannian fiber bundle with spin structure so that the fibers $f^{-1}(n)$ have positive scalar curvature, then $f_*([D_m]) = 0 \in KO_*(N)$.*

Note that M does not immediately have positive scalar curvature in this situation: however, by rescaling the fibers to make them tiny, we can arrange for the vertical directions in this bundle to overwhelm the others and make the scalar curvature positive. Doing this indicates that the reason for the positive scalar curvature is *local from the point of view of the manifold N*, so the vanishing is to be expected, and it is not hard to prove.

Similarly we can generalize Novikov's theorem as follows:

Proposition 5.21 *If $f : M' \to M \to N$ is a homotopy equivalence over N, i.e. for all open subsets U of N, $f : f^{-1}\pi^{-1}U \to \pi^{-1}U$ is a proper homotopy equivalence, i.e. the map of pushforward sheaves $R\pi_* Rf_*(\mathbb{Q}) \to R\pi_*(\mathbb{Q})$ is a quasi-isomorphism of sheaves over N, then*

$$\pi_*[f_*([\text{sign}_M])] = \pi_*([\text{sign}_M]) \in K_*(N).$$

Note that Novikov's theorem is the special case of $\pi = $ identity $M = N$.[72]

Moreover, both of the statements above can be coupled to statements about the Novikov conjecture if the fibers are non-simply connected, but we know the Novikov conjecture for them, and we inflate the K-theory of N to include this additional information.

For example:

Proposition 5.22 *If $f : M' \to M \to N \times K(\pi, 1)$ is a map that is a homotopy equivalence (locally) over N (but not necessarily over $K(\pi, 1)$), then one has, assuming the Novikov conjecture for π, the equality*

$$\pi_*[f_*([\text{sign}_{M'}])] = \pi_*([\text{sign}_M]) \in K_*(N \times K(\pi, 1)) \otimes \mathbb{Q}.$$

(One could work integrally if one takes the Novikov conjecture for π to mean an integral statement.)

---

[72] Strictly speaking, Novikov's theorem is the rationalization of this statement. This integral statement about the signature operators is the main result of Pedersen *et al.* (1995). And the above refinement is neither simpler nor more difficult than this result (just as the rational version of this statement follows *mutatis mutandis* from Novikov's argument – or the one we gave in §4.5).

For N a point, this is the Novikov conjecture for π; for $\pi = e$, this is the generalized Novikov theorem.

These results can easily be understood from the point of view of controlled topology (see §4.8). We will explain this at the beginning of §5.5.

Another beautiful theorem that fits well into this philosophy is the following result of Borisov and Libgober (2008) that asserts that higher elliptic genera are invariants of K-equivalence. Recall that V and V' are K-equivalent if there is a (Z, ψ, ψ') with $V \leftarrow Z \rightarrow V'$ so that $\psi^* K v = \psi^* K_{V'}$ (where $K_?$ is the canonical divisor of '?'). (A motivating case is the Calabi–Yau case, where canonical divisors, by definition, vanish, so that one is asserting here the birational invariance of invariants of Calabi–Yau manifolds.)

Theorem 5.23 (Borisov and Libgober, 2008)   *For any fundamental group, all of the rational higher elliptic genera agree for any K-equivalent smooth varieties.*

The second answer is perhaps more pragmatic. Although the original Novikov conjecture was phrased in terms of rational invariants, we have already seen that, for torsion-free groups, one can conjecture an integral injectivity result – and this implies that many of the characteristic class formulas or restrictions that we have developed have, for torsion-free groups, integral refinements (with more refined definitions of the characteristic classes necessary: e.g. using K-theory and the cycle associated to the defining elliptic operator[73]).

We will discuss in Chapter 6 what to do for groups with torsion, but, for now, let us note that it is necessary to do something.

This is readily apparent in the case of cyclic groups $\mathbb{Z}_p$. If we do not invert p, then simple examples involving homotopy-equivalent linear lens spaces (of high dimension[74]) show that higher signatures are not always invariant, i.e. the pushforward of the signature class in the K-homology of $K(\mathbb{Z}_p, 1)$ Similarly, for the positive scalar curvature problem, this is even more obvious: lens spaces also have nontrivial Dirac classes, yet they all have positive sectional curvature.

Moreover, in these two problems at least, things seem to be rather deeper. We know as a consequence of functoriality that any homology class in (the homology of) $K(\pi, 1)$ comes a manifold homotopy equivalent to M. One can

---

[73] Or in topology using things related to the Sullivan orientation inverting 2 (Sullivan, 2005), and the Morgan–Sullivan class at 2 (Morgan and Sullivan, 1974). These are subsumed in the *controlled symmetric signature*, i.e., the *symmetric signature* in the sense of Ranicki (1980a,b), of M, thought of as a Poincaré complex controlled over itself (see, e.g., Cappell *et al.*, 1991), amplifying our earlier discussion in the chapter of signature-type invariants of $\mathbb{Q}$-homology manifolds.

[74] Three-dimensional lens spaces are all parallelizable. However, the K-homology of $K(\mathbb{Z}_p, 1)$ fills up using the differences of the signature operators of higher and higher-dimensional linear lens spaces.

show (using the Gromov–Lawson surgery theorem – see §5.6) that there is a similar statement possible for positive scalar curvature manifolds – there are no Dirac obstructions except for those detected in the $K(\pi, 1)$. However, in the local cases one can often refine the equalities to lie in more refined places than the group (K or L or Ell) homology. For example, note the refined Novikov and positive scalar curvature theorems discussed in this section that assert results in $K_*(N)$ for general (i.e. for not necessarily aspherical) N. (The result on Rosenberg's algebraic-geometric Novikov conjecture is also true integrally as follows from the argument of §5.3, showing how it fits into the realm of Novikov theorems rather than conjectures.)

Needless to say, until the Novikov conjecture is disproved, we do not know that there is a real difference between these classes of statements – and, moreover, the connection between statements about assembly maps and the geometric consequences are only tight in the topological case – conceivably the positive scalar curvature problem can work out differently than the higher-signature problem.[75]

5.5 Playing the Borel Game

It is time to take some stock again of where our journey has taken us so far.

Starting from the original Borel conjecture, we have seen how the geometry of lattices and ideas in geometric rigidity theory can lead to a great deal of information about the topological structure of these (and other much larger classes of aspherical) manifolds, if not their topological rigidity. We were inevitably led to consider the implications of functoriality (forced upon us by the π–π theorem of surgery).

In studying how the Borel conjecture restricts the variation of characteristic classes (and spectral geometry), we were led to the Novikov conjecture. This is a very broad phenomenon wherein the fundamental group of a manifold has strong implications for its global analysis, some of whose implications we studied in this chapter. The key to the breadth of what we've seen to this point always has involved elliptic operators and their properties – and was often the

---

[75] It is even conceivable that the Gromov–Lawson–Rosenberg conjecture is true, but the strong Novikov conjecture (i.e. the C^*-algebra version) fails and another mechanism is behind this truth. Nevertheless, the ordinary Novikov conjecture, not involving completions, has a definite chance of being true even if the strong Novikov conjecture is correct.

The main reason that one can imagine the first statement is that the work of Schoen, Yau, and Lohkamp gives methods *completely unrelated to Dirac operators* for the nonexistence of positive scalar curvature metrics on certain manifolds. The second statement can be suggested (to the authors of science fiction monographs) by some deviations between the topological analytic conjectures that will be discussed, for example, in Chapter 8.

consequence of the general injectivity of (related) assembly maps in K-theory or a Hermitian cousin, L-theory.

But it is natural to play the Borel game, as well, not just the Novikov. What can we say about the isomorphism statement regarding the assembly map, say, for torsion-free groups? Does this problem have relatives with a substantial family resemblance – that might themselves have beautiful implications?

It is also time to expand our perspective to situations that are not mediated by elliptic operators, topological K-theory, but exist within *algebraic K-theory*,[76] clearly an analogue (if only because of its name), but also directly connected to the difference between homotopy and homeomorphism. So, it could, in principle, also obstruct the Borel conjecture.

This section is just a first pass at this project. We will return to it in Chapter 6 when we deal more seriously with groups with torsion (after all, in this chapter, we have only dealt so far with products of torsion-free groups with finite ones).

5.5.1 Fibering and Controlled Surgery

Let us recall, temporarily ignoring the Whitehead group (see §4.1), the Farrell fibering theorem.[77] It describes the obstruction of fibering a manifold over a circle; given $f: M \to S^1$ one first considers whether the associated infinite cyclic cover is a finite complex up to homotopy type (or even finitely dominated). If so, then the mapping torus of the covering translate $T(\tau)$ is a finite complex, homotopy equivalent to M.

Indeed $T(\tau) \to S^1$ describes a *controlled Poincaré complex* over the circle. It is a Poincaré complex, and this is true for the inverse image of every open subset of the circle.[78] The homotopy equivalence $M \to T(\tau)$ is, among other things, a normal invariant for this Poincaré complex.

We can think of this situation in two different ways (that up to algebraic K-

---

[76] Actually, an analogue of the Novikov conjecture is known for the very large class of groups whose homology is finitely generated in every degree – according to a remarkable theorem of Bökstedt *et al.* (1993). (See also Dranishnikov *et al.* (2020) for a broader explanation of these ideas within the realm of algebraic K-theory.) It is conceivable that the correct Hermitian analogue of their technique could prove the Novikov conjecture for a similarly broad class of groups – although it is unlikely that the C^*-algebra version could ever succumb to such an approach.

[77] Taking the Whitehead group into account is more subtle than one might think. One quickly comes to the conclusion that Nil groups should be the source of non-approximate fibering, but Farrell *et al.* (2018) show that, in the presence of Klein bottles in the fundamental group of the base, there are a number of nil-type obstructions that all have to vanish.

[78] To be more precise, the inverse image of each open set is a proper Poincaré complex, satisfying the kind of Poincaré duality that open manifolds do – interchanging cohomology with support – having compact projection to the circle with ordinary homology. One could also define an *approximate* Poincaré complex, in an ε–δ fashion where deviations of duality at one scale are trivial in a somewhat larger one

theoretic obstructions are equivalent). First of all, we have a Poincaré complex blocked over the circle S^1. That is, we have a Poincaré complex for each vertex (in a triangulation) and a Poincaré cobordism between these over each edge. And associated to this we have a blocked surgery obstruction, which will be a map $[S^1 : \mathbf{L}_{m-1}(\pi)]$ (where π is the fundamental group of the fiber, and $\mathbf{L}$ indicates the space which encapsulates the obstruction to blocked surgery).

Alternatively, we can try to do controlled surgery, which is to build a map $M \to T(\tau) \to S^1$, so that over each open subset of S^1 the map restricts to a proper homotopy equivalence.

Both alternatives are a little weaker than fibering: the s-cobordism theorem (or, alternatively, obstructions that naturally give an element in $H^1(S^1; \mathrm{Wh}(\pi))$ is used to straighten the h-cobordisms in the first theorem to be a fibration. For the second, there are issues related to $K_0(\mathbb{Z}\pi)$ as well – there is no guaranteed way to find the fiber over a point from the fiber over open intervals,

$$[S^1 : \mathbf{L}_{m-1}(\pi)] \cong \mathbf{L}_m(\Gamma),$$

where $\Gamma = \pi_1(M)$. Needless to say, the π on the left-hand side really means $\pi_1(F_\theta)$, the fundamental group of the fiber over a point θ in the circle. That means the left-hand side should be thought of as sections of a fibration rather than as a function space in general. (The monodromy of the bundle over the circle with fiber $\mathbf{L}_{m-1}(\pi)$ is induced by the covering translate on the infinite cyclic cover.)

In the controlled situation, cohomology is the wrong variance (block bundles, and their obstructions, pull back): we push forward a controlled surgery problem to obtain a problem with somewhat looser control. This leads to the conclusion – and it is one that we had earlier seen in some situations using the α-approximation theorem of Chapman and Ferry – that controlled surgery theory should be a homology theory (again twisted if $\Gamma \neq \mathbb{Z} \times \pi$),

$$\mathbf{L}^{\mathrm{controlled}}(T(\tau) \to S^1) \cong H_1(S^1 : \mathbf{L}_{m-1}(\pi)) \cong \mathbf{L}_m(\Gamma),$$

where the first statement is a "general" calculation (and would be correct were S^1 replaced by some other space X) and the second statement is a consequence of the fibration theorem.

Note that in the Borel conjecture we had the assembly map

$$H_m(M^m : L(e)) \to \mathbf{L}_m(\Gamma),$$

being an isomorphism when M is a $K(\Gamma, 1)$-manifold.[79] But surely it is now irresistible to suggest that $H_m(M^m : L(e)) \to \mathbf{L}_m(\pi_1(E))$ is an isomorphism

---

[79] This statement does not require M to be closed if we use compact supports, as suggested at the end of Chapter 3 and §4.7.

where $E \to M$ is a fibration, and where π is the fundamental group of the fiber (and thus the left-hand side should be interpreted in a (co-)sheaf-theoretic way).

These statements are conjecturally the case, and are included in what I call the *Borel package*, a collection of statements yet more general. They have an interpretation in terms of some kind of fibering of manifolds over aspherical ones.

Note, of course, that the Borel conjecture itself is the statement that if M is aspherical and M' is homotopy equivalent to it, we can find a homotopy of this homotopy equivalence $M' \to M$ to one that is a fibration over every open subset (which is, therefore, a controlled homotopy equivalence, which is the same a CE map – when we are mapping between manifolds of the same dimension – which in turn is a limit of homeomorphism, by the theorem of Siebenmann or Edwards).

Needless to say, also, that this is compatible with the Borel conjecture if the group π satisfies the Borel conjecture. One nice feature of this viewpoint is that it tautologously builds in a closure of the Borel under short exact sequences: $1 \to \pi \to \Gamma \to \pi' \to 1$ (i.e., the result for Γ follows from that for π and π').

The Borel package itself needs at least two further amplifications. The first is a modification or expansion to include algebraic K-theory; we give the modification immediately and the expansion later in this section. Whitehead groups are not always trivial; the product of h-cobordant manifolds with the circle are Cat-isomorphic,[80] so there can never be a uniqueness of fibering without taking algebraic K-theory into account. Moreover, there is also an algebraic K_0 condition to being able to compactify the infinite cyclic cover, which would surely be possible if the manifold fibered over the circle. However, if the statements we had written were correct "on the nose," then, for example, blocked surgery theory would indeed give existence and uniqueness of fibering.

However, it is pretty close. It turns out that all algebraic K-theoretic obstructions die after crossing with a torus, and that, by using tori, one can make the solutions essentially unique. (The uniqueness is typically another algebraic K-theory obstruction). As a result, one way to get around the algebraic K-theory issue is to "stabilize." We can just cross all the groups involved with $\mathbb{Z}^\infty$ and then the arguments would work as described above. This is a little awkward, and it is best to replace $L_k(\pi)$ by $\lim L_{k+d}^{\mathrm{Bdd}}(\pi \times \mathbb{R}^d \downarrow \mathbb{R}^d)$, where we map $L_{k+d}^{\mathrm{Bdd}}(\pi \times \mathbb{R}^d \downarrow \mathbb{R}^d) \to L_{k+d+1}^{\mathrm{Bdd}}(\pi \times \mathbb{R}^{d+1} \downarrow \mathbb{R}^{d+1})$ by crossing with $\mathbb{R}$.

This limit is referred to as $L_k^{-\infty}(\pi)$. The map $L_k(\pi) \to L_k^{-\infty}(\pi)$ is an isomor-

---

[80] Recall that the obstruction to an h-cobordism being a product is the Whitehead torsion; torsions are multiplied by Euler characteristic in products (see e.g. Milnor, 1966; Cohen, 1973).

phism away from the prime 2. (I don't know any example where the kernel and cokernel don't have reasonably small exponent,[81] but I can't imagine a proof of such a statement either given our current state of knowledge.)

The second comment and amplification is that L-groups have a completely algebraic definition, and all of the constructions, while we have made or explained them geometrically, can also be algebraicized. Note that $\mathbb{Z}[A \times B] \cong \mathbb{Z}[A][B]$ (and similarly with a twisted group ring for semidirect products, and a more complicated but obvious enough expression for the situation where one has an extension that is not split). It suggests therefore that, even for a nontrivial family of rings R over $K(\Gamma, 1)$, there should be an isomorphism

$$H_m(K(\Gamma, 1), L^{-\infty}(R)) \to L^{-\infty}(\text{``} R\Gamma \text{''}),$$

where "$R\Gamma$" is the group ring (twisted, when the family demands it). Frequently, geometric techniques for the Borel conjecture will initially apply directly to the assembly map where $R = \mathbb{Z}$, and then have an extension to fibered situations, allowing R to be a group ring – but, with some algebraic variation of the method, one can get this whole package.[82]

The above statement is equivalent to the statement that the forget-control map from "controlled L-theory (in the $-\infty$ sense) with coefficients in R" to $L^{-\infty}$ ("$R\Gamma$") is an isomorphism.

This package can have useful applications geometrically that go beyond the Borel and Novikov conjectures themselves. We mention three examples that have bearing on issues that we've already discussed.

The first is the proof of the combined Novikov conjecture/Novikov theorem made in §5.4. If we have a controlled homotopy equivalence as in the hypothesis of that proposition, then we would get equivalence of the signature classes in

$$H_m\big(N, L^{-\infty}(\mathbb{Q}, \pi)\big).$$

However, if we know that the map $H_*(K(\pi, 1)L(\mathbb{Q})) \to L^{-\infty}(\mathbb{Q}\pi)$ is (rationally) split injective, then generalities about homology theories gives injectivity of composition,

$$H_*(N \times K(\pi, 1), L(\mathbb{Q})) \to H_m(N, L^{-\infty}(\mathbb{Q}\pi)),$$

and, therefore, controlled homotopy invariance in the domain of this map (rationally, if that's our assumption on π).[83]

---

[81] It's not hard, though, to give examples where these are infinitely generated.

[82] See Weinberger (1985b, 1987) and Bartels and Reich (2007).

[83] This argument is only being asserted for untwisted fundamental group situations. Frequently proofs of the Novikov conjecture are natural enough to accommodate twistings, but this is a stronger hypothesis.

As a second application, we observe that the proper Borel conjecture for the Q-rank-2 case follows from the Borel package of the underlying lattice. In this case the Borel–Serre boundary is aspherical, as we have already noticed, and proper rigidity of the original manifold follows from the rigidity of the compactified manifold.[84] This, in turn, by the exact sequence of a pair in surgery and group homology, reduces to the isomorphism statement for the lattice and the boundary separately. The first is the ordinary Borel conjecture, but the second is the twisted one in a situation where we have a (non-split) group extension (associated to the Borel–Serre boundary), and the relevant ring is the group ring $\mathbb{Z}[F_\infty]$, which is not associated to a finitely presented group (and so would require more effort to deal with geometrically, since the relevant fiber could not be a compact manifold).

Finally, in our discussion of the higher-signature localization formula for homologically trivial group actions on non-simply connected manifolds, we gave a particularly symmetric expression of the formula that ended up being equivalent to the Novikov conjecture. The characteristic class on the right-hand side of the formula was the average of classes introduced by Atiyah and Singer, as one goes over the generators of the cyclic group. However, the reasoning suggesting the formula suggests that one can use any generator to get a characteristic class formula: and all generators should give the same result – i.e., included in such a formula would also be a vanishing theorem for certain higher characteristic classes.

This is indeed feasible, except that the argument that one would naturally give would be phrased in the ring $L(\mathbb{Q}[\xi][\Gamma])$, where ξ is a primitive root of unity.[85] I do not see any way to deduce from a statement about the $\mathbb{Q}[\Gamma]$ the full necessary statement about $\mathbb{Q}[\xi][\Gamma]$.[86] However, the "Novikov package," which is also available in as wide a generality as the Novikov conjecture (at this point in time) would give this.

5.5.2 The C^*-algebra Setting (the Baum–Connes Conjecture, First Meeting)

In the C^*-algebra setting we also have an assembly map (interpreted as an index map)

$$K(K(\Gamma, 1)) \to K(C^*\Gamma),$$

---

[84] Modulo issues about vanishing of Whitehead groups that follow by the K-analogue of the L-argument we are now giving (and appropriate work on the K-theoretic Borel package for the lattice).

[85] $\mathbb{Q}[\xi]$ arises naturally as a piece of $\mathbb{Q}[\mathbb{Z}_n]$.

[86] This would not be an issue in the C^*-algebra framework – the abstract algebra $C^*(\mathbb{Z}_n \times \Gamma)$ is obviously a finite product of m copies of $C^*(\Gamma)$.

with the Novikov conjecture being a statement about (rational) injectivity. The Baum–Connes conjecture is the isomorphism statement that goes along with this injectivity statement. Thus, the BC conjecture would assert (in its strong package form) that an assembly map involving Γ–C^*-algebras going to a cross-product algebra should always be an isomorphism for Γ torsion-free (again, leaving the discussion of groups with torsion to Chapter 6).

Recall that $C^*\Gamma$ is a completion of the group ring $\mathbb{C}\Gamma$ which we think of as an algebra of operators either on $L^2\Gamma$, in which case we get $C^*_{\text{red}}\Gamma$ (the reduced C^*-algebra), or acting on arbitrary unitary representations, which then produce $C^*_{\text{max}}\Gamma$. To have a mathematical statement, surely it is necessary to specify which completion should be used. (Note that there is a map $C^*_{\text{max}}\Gamma \to C^*_{\text{red}}\Gamma$ so injectivity for the reduced assembly map implies injectivity for the max.)

The problem is this. Given a homomorphism of groups $\Gamma \to \Delta$, there is a map between their Eilenberg–Mac Lane spaces, so there is functoriality of the left-hand side, but there is no induced map $C^*_{\text{red}}\Gamma \to C^*_{\text{red}}\Delta$ (except for the situation where the kernel of the map is amenable).

Using $C^*_{\text{max}}\Gamma$, there is an induced map, so both parts of the picture do have the same functoriality. However, there is no chance that this map can be an isomorphism in general: If Γ has Property (T), then the trivial representation is a projective module over $C^*_{\text{max}}\Gamma$, and it lies in the cokernel of the assembly map. (In fact, for a group like $\text{SL}_3(\mathbb{Z})$ – or a torsion-free congruence subgroup thereof – there are infinitely many $\mathbb{Z}$ summands in $K_0(C^*_{\text{max}}\Gamma)$ coming from the infinitely many irreducible representations coming from finite quotients that are all isolated in the Fell topology. The domain of the assembly map is a finitely generated abelian group.)

So we have a dilemma for those who would make conjectures: To be true, one must work with $C^*_{\text{red}}\Gamma$, else Property (T) immediately explodes the conjecture, yet doing so posits a highly non-obvious functoriality for the K-groups that does not appear to make any sense at the level of the algebras themselves.

The latter is what Baum and Connes (2000) boldly did in an influential paper (that appeared many years after its initial circulation!).[87] It was a major advance[88] when V. Lafforgue (2002) gave an example of a group that has Property (T) and satisfies the conjecture. Subsequently, building on these techniques, the Baum–Connes conjecture was verified for all hyperbolic groups by Lafforgue (2002) and Mineyev and Yu (2002) – with Lafforgue (2012) subsequently giving a proof in this situation with coefficients, as well (see Puschnigg, 2012).

---

[87] I have to admit to having been offended by this reckless behavior – as a penance for my timid skepticism.

[88] Thereby causing me to doubt my skepticism.

However, we now know that the Baum–Connes conjecture with coefficients is false in general (we will discuss this further in Chapter 8). It remains an extremely important insight – injectivity is known for a very large class of groups, as we shall see – and understanding the extent of its full validity is a major problem, e.g. for lattices or linear groups.

5.5.3 Algebraic K-Theory

So, finally, let's turn to the long overdue issue of algebraic K-theory and how it connects to this story. This subject fills bookshelves in a library: we shall devote only a few pages to this.

Classical algebraic K-theory centered around two functors of rings (that were linked) $K_0(R)$ and $K_1(R)$. These have important applications in topology and are analogues of (say, complex) vector bundles over X and ΣX if R is the space of continuous functions on a compact Hausdorff space X (a C^*-algebra). (A bundle over ΣX can be viewed as two trivial bundles over each of the two cones, that are "clutched" or identified over X: this is a family of changes of bases – i.e. a map $X \to \mathrm{GL}_k(C)$, i.e. an element of $\mathrm{GL}_k(C(X))$.)

Even earlier, these functors, in the case of number rings, had important arithmetic interpretations, and consequently served as a bridge between topology and arithmetic.

Subsequently, the functors and their range of topological applications grew to include $K_i(R)$ both for i negative and for $i > 1$, and also deep connections to algebraic geometry and arithmetic developed. The negative groups having direct meaning using controlled (or bounded) algebra, the positive groups being related to the homeomorphism and diffeomorphism groups of manifolds.

The case $i = 0$, i.e., $K_0(R)$, is the Grothendieck group of finitely generated projective R-modules. It is thus the the group that contains the most general possible dimension for finitely generated projective modules.

It arises frequently in geometric topology as the Euler characteristic of a chain complex that has finiteness properties. Note that a chain summand of a chain complex of finitely generated free modules is such a chain complex; homological vanishing theorems can often detect that a chain complex is chain-equivalent to one of this form.

Thus $K_0(\mathbb{Z}\pi)/K_0(\mathbb{Z})$ contains an obstruction to a finitely dominated cell complex (e.g., a cell complex that's a retract of a finite complex) to being homotopy-equivalent to a finite complex, and indeed this is the whole obstruction according to the finiteness theory of Wall (1965). It also measures (according to Siebenmann, 1965) the obstruction to putting a boundary on a noncompact manifold that is "tame at ∞."

Perhaps even simpler, consider this: if X is a finite complex with a PL G-action (for G finite), then the cellular chain complex is projective over $\mathbb{Q}G$. (All finitely generated modules over $\mathbb{Q}G$ are projective.) As Euler characteristic is the same on the chain and homology level (when the latter is projective), we can identify this invariant on homology, and then, via characters, with the invariant at the chain level. In other words, we see that

$$\operatorname{Tr}_g \chi(G, X) = \chi(X^g),$$

and the Lefshetz fixed-point theorem is thus encoded in this functor (i.e. the equivariant χ is a multiple of the regular representation – the image $K_0(\mathbb{Q})$, which is equivalent to the vanishing of the character of the representation on all nontrivial elements).

Whitehead actually defined $K_1(\mathbb{Z}\pi)$ earlier. We can think of it as the Grothendieck group of automorphisms of finitely generated projective modules. By adding on a complementary projective module with the identity automorphism (a complement to P is a finitely generated module Q so that $P \oplus Q$ is free), one can think of this as made from automorphisms of free modules, i.e. elements of $\mathrm{GL}_n(R)$ which are allowed to be stabilized.

This is the same as the Grothendieck group of finitely generated free based[89] acyclic chain complexes.

We can therefore think of K_1 as the universal target for determinants of invertible matrices (over the ring).

Thus for any finite-dimensional orthogonal representation of π, there is a Norm map that assigns to an invertible matrix over $\mathbb{Z}\pi$ the determinant of the associated matrix with real entries. Thus for $\pi = z_p$, with p an odd prime, the trivial representation gives nothing, but the remaining $(p-1)/2$ representations all give interesting invariants: however, the products of all of these determinants must be ± 1 (because it is the norm of a unit of an algebraic integer). That this is the complete dependency is the content of the Dirichlet unit theorem.

Just as $K_0(\mathbb{Z}\pi)$ measures existence of finite complexes within a homotopy type, $K_1(\mathbb{Z}\pi)$ measures the uniqueness of the finite complex. Given two finite homotopy-equivalent complexes, the mapping cone of the homotopy equivalence is almost a based acyclic $\mathbb{Z}\pi$-complex – the chain complex under discussion uses cells of the universal cover, but each cell has two orientations, and there is no canonical lift of a cell to the universal cover, so we also have an indeterminacy by multiplying by elements of π. So we get in this situation a "torsion" which is an element of

$$\mathrm{Wh}(\pi) \equiv K_1(\mathbb{Z}\pi)/(\pm\pi).$$

---

[89] A based chain complex is a chain complex where each chain module is given a specified basis.

The equivalence relation this puts on finite complexes is called *simple homotopy equivalence*: it is the equivalence relation generated by viewing any finite L as equivalent to $L \cup e$, where e is a cell and the attachment is along a face in its boundary (i.e. elementary expansion). See Figure 5.3 for a schematic elementary expansion.

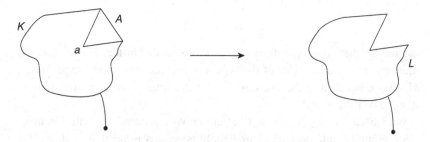

Figure 5.3 A schematic elementary expansion. Reproduced from Cohen (1973) with permission of Springer.

The s-cobordism theorem connects this to manifold theory. A manifold with boundary that deform-retracts to each of its two boundary components is an h-cobordism. It is a product – if the dimension is at least six – iff one of (and therefore both[90] of) the boundary inclusion(s) is (are) a simple homotopy equivalence.

If W is an h-cobordism, then $W - \partial_- W \cong \partial_+ W \times [0, 1)$, so $\mathrm{Wh}(\pi)$ can be thought of as measuring the uniqueness of the solution to the problem of putting a boundary on an open manifold.

There is an important relationship (Bass *et al.*, 1964) between K_1 and K_0 called the fundamental theorem of K-theory. It asserts that

$$K_1(R[t, t^{-1}]) \cong K_1(R) \oplus K_0(R) \oplus \mathrm{Nil}(R) \oplus \mathrm{Nil}(R),$$

where $\mathrm{Nil}(R)$ is the Grothendieck group of nilpotent automorphisms of free modules. It frequently vanishes (e.g. when R is a regular ring) but is nontrivial, and indeed infinitely generated, when $R = \mathbb{Z}[\mathbb{Z}_p \times \mathbb{Z}_p]$.

The way to get a map from the right-hand side to the left is like this. On K_1 it's obvious. From K_0 consider assigning to P a finitely generated projective module with a complement Q, the isomorphism $P \oplus Q$ to itself, sending (p, q) to (tp, q), and to a nilpotent automorphism A of R^k, $I + tA$ or $I + t^{-1}A$ (hence two copies of the Nil terms).

---

[90] In general there is a formula, called the *Milnor duality formula* (see Milnor, 1966), relating $\tau(W, \partial_+ W)$ to $\tau(W, \partial_- W)$. It depends on the dimension, the orientation character, and the involution on $\mathrm{Wh}(\pi)$ induced by $g \to g^{-1}$.

Since algebraic K-theory forms a spectrum, we can write this isomorphism as

$$K_1(R[\mathbb{Z}]) \cong H_1(S^1; K(R)) \oplus \text{Nil}(R) \oplus \text{Nil}(R).$$

Loday (1976, 1998) defined an assembly map in algebraic K-theory, and thus the fundamental theorem of algebraic K-theory[91] can then be interpreted as the statement that, for $\pi = \mathbb{Z}$, the assembly map is always split injective (a Novikov conjecture) and is an isomorphism if R is a regular ring.

There are transfer maps associated to the self-covers $S^1 \to S^1$. On the $K_1(R)$ summand, this map is multiplication by the index of the cover. On the $K_0(R)$ factor, this map is the identity (i.e. $K_0(R)$ is the transfer-invariant part of $K_1(R)[\mathbb{Z}]$)), and on the Nil terms, the transfer is nilpotent, i.e. each element dies on passing to sufficiently high covers.[92]

Considering $K_0(R[\mathbb{Z}])$ and insisting that the fundamental theorem holds gives rise to a definition of $K_i(R)$.[93] We can go further, with $\mathbb{Z}$ replaced by $\mathbb{Z}^d$ to get negative K-groups. These have interpretations in terms of controlled topology. The controlled Whitehead group of $\mathbb{Z}\pi$ over $\mathbb{R}^d$ is $K_{i-d}(\mathbb{Z}\pi)$ – and it obstructs controlled h-cobordisms from being products (or controlled simple homotopy equivalent complexes from having controlled homeomorphic thickenings).

Higher algebraic K-groups, when introduced by Quillen, were also discovered to satisfy a fundamental theorem. Thus we can hope for a statement like:

Conjecture 5.24 *The assembly map*

$$H(K(\pi, 1); K(R)) \to K(R[\pi])$$

is always split injective for π torsion-free (we assume that the nonconnective spectrum K is used; the extension to the general case will be given in Chapter 6) and is an isomorphism if, in addition, R is regular.

Of which the case of $\pi = \mathbb{Z}^d$ would be the theorem.

By the way, when we actually apply K-theory to topology as in the examples above, we use the reduced class group, i.e. we mod out by the image of $H_0(K(\pi, 1); K(\mathbb{Z}))$, and the Whitehead group, where we mod out by $H_1(K(\pi, 1); K(\mathbb{Z}))$, which is $H_1(\pi) \times \{\pm 1\}$. As a result, we are often interested in the cofiber of the assembly map – Whitehead theory more than the K-groups.

Further, note that the Borel conjecture actually implies (exercise, using the h-cobordism theorem) the vanishing of $\text{Wh}(\pi)$ at least when $K(\pi, 1)$ is a finite

---

[91] Farrell and Hsiang (1970) gave the generalization to twisted polynomial extensions.

[92] However, there are always elements that live arbitrarily long.

[93] Ferry (1981) gives a very nice geometric approach to the Wall finiteness theory via the Whitehead simple homotopy theory and this perspective on K_0.

complex, which is the above conjecture for $R = \mathbb{Z}$ and for homotopy groups in dimension ≤ 1. In particular, the above conjecture would imply that $\mathrm{Wh}(\pi) = 0$ for π torsion-free, and that, when this holds, the Borel conjecture for π really boils down to the isomorphism of the L-theory assembly map.[94]

Needless to say, it is important to understand what happens when R is not regular. In that case, let me mention a beautiful special case that shows what one can hope for.

Theorem 5.25 (Farrell and Jones, 1993a; Bartels *et al.*, 2008) *If π is a torsion-free hyperbolic group, then the above conjecture is true. Moreover, in general, there is an isomorphism*

$$H_i\big(K(\pi, 1); K(R)\big) \oplus \bigoplus \mathrm{Nil}_i(R) \to K_i(R[\pi]),$$

where the sum is over conjugacy classes of nontrivial elements of π that are not proper powers.

The split injectivity result is not that difficult: it follows from the principle of descent, just like other Novikov conjecture results that we've discussed (see Ferry and Weinberger, 1991, 1995; Carlsson and Pederson, 1995), together with a trick (transferring to the infinite cover corresponding to the various $\mathbb{Z}$s in π and thinking of this in a suitably controlled at ∞ way) for detecting the Nil terms.

The surjectivity statement is much deeper, and we have not yet seen any mechanism (other than codimension-1 splitting methods[95]) that can yield it.

In Chapter 8, I will explain where these summands come from, at least in the original situation of closed hyperbolic manifolds, where Farrell and Jones proved this using dynamical properties of geodesic flow. The Farrell–Jones conjectures describe in both K-theory and L-theory a more comprehensive picture of what happens that goes beyond the cases predicted by the Borel conjecture.

The higher K-groups have a close topological cousin invented by Waldhausen, called A-theory. Waldhausen's $A\big(K(\pi, 1)\big)$ is a kind of group completion of $\mathrm{BGL}\big(\Omega^\infty\Sigma^\infty K(\pi, 1)^+\big)$ (which surely looks close to $\mathrm{BGL}(\mathbb{Z}\pi)$), and the assembly map for them enters into an understanding of the higher homotopy of diffeomorphism and homeomorphism groups. We cannot do justice to this here, but instead refer the reader to Cohen (1987); Waldhausen (1987); Weiss

---

[94] One can think of $\mathrm{Wh}(\pi)$ as being an analogue in K-theory of $S(B\pi)$ in surgery. (Early on, historically, because the difference between K_1 and Wh is so small, this point was obscured, and people would think of $\mathrm{Wh}(\pi)$ and $L(\pi)$ as being analogues.)

[95] Waldhausen (1978) developed such methods and, for example, proved that Whitehead groups vanish for fundamental groups of Haken manifolds.

and Williams (2001) and Rognes and Waldhausen (2013) and just discuss a little piece of the story that directly bears on the Borel philosophy.

We discussed in §1.2 the notion of pseudo-isotopy, and observed that part of the Borel conjecture should be the statement that homotopic homeomorphisms are pseudo-isotopic.

A pseudo-isotopy is essentially a homeomorphism of $M \times [0, 1]$ and we can ask whether it is isotopic to an isotopy, i.e. a level-preserving homeomorphism of $M \times [0, 1]$.

This is kind of like uniqueness of the product structure in the s-cobordism theorem, and so should involve K_2. This is the case. It is a beautiful theorem of Cerf (1970) that for simply connected manifolds pseudo-isotopies are always isotopic to isotopies (in high enough dimensions), but Hatcher (1973) showed that it is *never* true in the non-simply connected case. This starts already on $S^1 \times \mathcal{D}^n$ (rel ∂) and gives rise to homeomorphisms pseudo-isotopic to the identity but not isotopic on all aspherical manifolds. The space $A(S^1)$ is quite complicated, and the fundamental theorem is already not unobstructed in this case: the cofiber of the assembly map is an analogue of Nil; and $\Omega^\infty S^\infty$ is not a regular ring.

5.6 Notes

The notes in this chapter really divide up by problem more than by section.

Regarding the index theorem and K-theory, which plays a critical role in this chapter, good references, from various points of view, are the original papers of Atiyah and Singer (1968a,b, 1971), and the more recent Lawson and Michelsohn (1989), Roe (1998), Berline *et al.* (2004), Higson and Roe (2010), and Bleecker and Booss-Bavnbek (2013). The basic relevant functional analysis and understanding of elliptic operators can be found in many more places, such as Zimmer (1984), Evans (2010), and Taylor (2011a,b,c).

Topological K-theory, and the K-theory of C^*-algebras cannot be separated from index theory. For example, for compact Lie groups, equivariant Bott periodicity still only has the analytic proof given by Atiyah (1968), as far as I know. For the thrilling initial chapters of this story, nothing beats Atiyah's collected works.

Good sources for K-theory of C^*-algebras are Wegge-Olsen (1993), Blackadar (1998), and Higson and Roe (2000). At some point, you will surely want to look at Connes' (1994) masterpiece for a view of the world, indeed of the universe, centered at this mathematics. I hope my brief appendix is not useless in stimulating an interest in doing this sooner rather than later.

We started the chapter by mentioning the celebrated theorem of Atiyah and Hirzebruch on circle actions on spin manifolds. This has had a celebrated extension, conjectured by Witten by arguing heuristically about the equivariant Dirac operator on ΛM (the free loop space, thought of as an S^1 space by rotating loops) and proved by Bott and Taubes (1989), with another proof by Liu (see Liu, 1995; Liu and Ma, 2000) that gives a K-theoretic refinement.

Stoltz (1996) has observed that Witten's heuristic can be developed to give a vanishing theorem for the so-called Witten genus, for manifolds of positive Ricci curvature that have $W_1 = W_2 = 0$ and $p_1/2 = 0$ ("string manifolds"). Unfortunately, this has never been proved.

It is not hard to see that all of this work has the expected non-simply connected generalization, by the method of Browder and Hsiang.

The twisted higher-signature localization theorem for $\mathbb{Z}_n$-actions discussed is equivalent to the Novikov conjecture for any fixed n. However, for other choices of characteristic class $c(v)$, this seems to be related to the Novikov conjecture with coefficients in a ring other than $\mathbb{Z}$ or $\mathbb{Q}$. It is for this reason that I was led to introduce this problem in Weinberger (1988a). The "simplest" formula is the Galois invariant one (i.e., invariant under change of generator of $\mathbb{Z}_n$). Rosenberg and Weinberger (1988) is an attempt to understand these equivariant geometric and topological phenomena in a coherent way.

Weinberger (1988a) also gives other formulations of the Borel conjecture with suitable coefficients in terms of being able to solve transversality problems in the setting of "homologically trivial group actions." Thus, if one has a free homologically trivial $\mathbb{Z}_p$-action on a manifold with fundamental group π satisfying the Borel conjecture, then, with a suitable dimension restriction, one can arrange for an equivariant map from $M \to K(\pi, 1)$ to have the transverse inverse image of any cycle (with manifold normal bundle) to have inverse image homologically trivial.

For smooth SU(2)-actions (or in general any smooth nonabelian connected group actions) there is a connection between the equivariant index theory and the positive scalar curvature problem: Lawson and Yau (1974) showed that any manifold with effective SU(2)-action has an invariant metric of positive scalar curvature. As a consequence we get a vanishing result of the (equivariant) Dirac operator for such manifolds. When combined with Hitchin's theorem, we discover that certain exotic spheres, for example, of dimensions 1, 2mod 8, do not have any (positive-dimensional) nonabelian group actions. For SU(2)-actions on non-simply connected manifolds, we then get a vanishing result in

$KO(K(\pi, 1))$ for all fundamental groups.[96] (Note that this integral statement is true even for groups with torsion!)

It is interesting to note that some of these exotic spheres do possess S^1-actions (by work of Schultz, 1975). Thus the Atiyah–Hirzebruch vanishing phenomenon is indeed quite subtle – the vanishing is not due to the vanishing of the KO-theoretic index of the Dirac operator.[97]

Understanding positive scalar curvature metrics has many parallels to surgery theory, but also some essential differences. As mentioned in the text, the most striking difference is the mysterious role of the spin condition: for simply connected manifolds of dimension greater than 4, every non-spin manifold has positive scalar curvature (Gromov and Lawson, 1980a), but in the spin case, according to Stoltz (1992), the necessary and sufficient condition is the triviality of the index of the Dirac operator in $KO_n(*)$ (i.e. the Atiyah–Lichnerowicz–Singer and Hitchin conditions).

Besides the appearance of indices of Dirac operators that are analogues of the indices of signature operators (or symmetric signatures of Poincaré complexes), a key role is played by the surgery theorem of Gromov and Lawson (1980a) (see also Gajer, 1987; Rosenberg and Stoltz, 2001) – it gives rise to the analogue in the positive scalar curvature problem of the π–π theorem (that has suitable relative versions, as well): a spin manifold with boundary $(M, \partial M)$ of dimensions ≥ 6, that satisfies the π–π condition (namely, $\pi_1 \partial M \to \pi_1 M$ is an isomorphism) always has a positive scalar curvature metric that is a product in a neighborhood of ∂M. Having a positive scalar curvature metric is thus a spin cobordism invariant (with respect to the fundamental group of the relevant manifold). This is often referred to as the "surgery theorem" since it is proved by showing that it is possible to do surgery on spheres of codimension ≥ 3 and maintain positive scalar curvature.[98]

In this analogy, the concordance classes of positive scalar curvature metrics is closer to the surgery group than to the structure set. Thus, in the π–π setting,

---

[96] Note that since SU(2) is simply connected, the SU(2)-action lifts to an action on the universal cover of M. The group of all lifts of this action is then $\pi \times SU(2)$ (since all automorphisms of SU(2) are inner). Thus, we can build an equivariant map $M \to K(\pi, 1)$ where we give the latter the trivial action. This implies the vanishing of the Dirac class.

[97] In this way, playing the Novikov game for the Atiyah–Hirzebruch theorem is a more bold departure than playing it for the positive scalar curvature problem (or for birational invariance).

[98] This theorem is behind the positive results of Gromov and Lawson and of Stolz mentioned above. In the non-spin case, Gromov and Lawson use surgery to reduce the problem to special generators of oriented bordism. Stoltz uses spin cobordism, which is not fully analyzed, but shows that there are enough classes that are total spaces of $\mathbb{HP}^2$ bundles (with its usual isometry group as structure group) to produce, by scaling the fibers to be very small, positive scalar curvature metrics on the kernel of $\mathrm{ind}D : \Omega^{\mathrm{spin}} \to KO(*)$.

In the missing case of dimension 4, where surgery methods fail, positive scalar curvature has additional obstructions that come from Seiberg–Witten theory.

there is a unique concordance class of metrics. It is an important open problem whether there is a unique isotopy class.

The analogue of the "main result" of Chapter 3 – the problem of existence and nonexistence of complete positive scalar curvature metrics on arithmetic manifolds in the noncompact case – was settled much earlier in Block and Weinberger (1999). The low $\mathbb{Q}$-rank case (rank ≤ 2), where one is looking for obstructions, is settled using Novikov conjecture technology (*not* Borel conjecture technology, as is necessary for the rigidity statement, see §5.5.1). We made use of a souped-up version of an index theorem of Roe (1988c): this index theorem is a special case of the index theorem for bounded propagation speed operators on a metric space, adapted to the situation where the "corona" (i.e., the space at ∞) is disconnected. The higher-rank case follows from the surgery theorem.

For more on this theme as it refers to closed manifolds, I highly recommend Stoltz (1995) and Rosenberg and Stoltz (2001).

The situation for noncompact manifolds is much stickier. As indicated throughout our text, the parallels continue into the noncompact setting. One key difference is caused by the problem that C_r^* is not functorial, so we are forced to make use of $C_{\max}^*$ – where one is surely dealing with an algebra that is "further away" from geometry than feels reasonable. "Surely" the extra elements of $K(C_{\max}^*\pi)$ coming from, say, Property (T), should not arise, e.g. from relative indices associated to a pair of positive scalar curvature metrics on M with fundamental group π? In any case, for noncompact manifolds that are tame at ∞, one defines an index that lies in a relative group, $K(C_{\max}^*\pi, C_{\max}^*\pi')$. The assembly map

$$KO(K(\pi, 1), K(\pi', 1)) \to K(C_{\max}^*\pi, C_{\max}^*\pi')$$

seems to have a tendency to be injective (rationally, or for torsion-free groups), although there is no legitimate 5-lemma reason to believe that this should be true.

Even in the absence of tameness, one can define an algebra that gives a *prima facie* place for index-theoretic obstructions (Chang et al., 2020). This includes, $\lim^1$ type obstructions to the existence of positive scalar curvature metrics, and other "phantom" phenomena in the theory.

Another problem in global analysis that has been connected to the Novikov conjecture is the "zero in the spectrum" problem.

Conjecture 5.26 (Gromov) *If M is a compact aspherical manifold, then the Δ on forms on the universal cover of M always has 0 in its spectrum. Indeed, it should be non-zero in the "middle dimension."*

Middle dimension means dimension k if $\dim M = 2k + 1$. At the moment the only evidence for this is of the following form:

(1) It is observed to be true for $K \setminus G$, so it is true for the classical aspherical manifolds.

(2) It follows from the Novikov conjecture, for otherwise the index of the signature operator in $K(C * \pi)$ would vanish, but the "1" in dimension 0 of the L-class should give rise to a nonzero image in $K(C^*\pi) \otimes \mathbb{Q}$ if the Novikov conjecture were true.

(3) The chain complex of M (thought of as $R\pi$-modules) is not chain-equivalent to one with a zero morphism in the middle (this would contradict the cohomological dimension of $\pi_1 M > K$). We are alright, therefore, for arbitrary group rings. The issue is entirely one caused by completion.

It is also likely true for all uniformly contractible Riemannian manifolds (with bounded geometry[99]). In any case, neither of these problems is known to imply anything about the original Novikov conjecture, but both of them can be studied jointly with the Novikov conjecture.

As mentioned in the text, the analytic version of the Novikov conjecture can be proved, just like we did in Chapter 4 in L-theory, by a principle of descent. The analogue of the bounded category is the Roe algebra.

Other related ideas are the Dirac–dual Dirac argument (see Kasparov, 1988, for the exemplar of this), and the use of almost-flat bundles (as in the text), which are not completely unrelated. There are three ways of getting around the basic fact that the (rational) Chern classes of a finite-dimensional bundle on a compact polyhedron are trivial. The first is the use of families, as in the Lusztig method. The second is to use families of almost-flat bundles with increasing dimensional fiber, following Gromov and Lawson (1980a,b) and Connes *et al.* (1993), as we explained in the text. This naturally could lead one towards using infinite-dimensional fibers – which is essentially the problem of understanding $K(C * \pi)$!

The third method for producing almost-flat bundles on finite-dimensional spaces, which allows one to keep the same ground manifold and not increase fiber dimension, is to use compact support. (This is like the use of the Bott element on $\mathbb{R}^n$ with compact support in Gromov and Lawson, 1983.)

This is quite similar in spirit, if not completely in detail, to the use of

---

[99] A good test of the depth of this question is whether one can construct a complete uniformly contractible manifold with $0 \notin \operatorname{spec}(\Delta)$ even with bounded geometry. Currently one doesn't know any example, even, of a complete contractible manifold without 0 in its spectrum – or of such a manifold that is homotopy-equivalent to a finite complex (although surely these must exist!).

the K-theory of Higson corona to obtain useful indices in Roe (1993). The Higson compactification of a locally compact metric space is an analogue of the Stone–Čech compactification, but one does not require that *all* bounded continuous functions extend – rather only those whose variation decays at infinity (diam $f(B(R,p)) \rightarrow 0$ as $p \rightarrow 0$ for any fixed radius R).

Any reasonable compactification, i.e. one where restrictions to the interior have decaying variation (such as the ideal boundary of G/K or the Gromov–Tits compactification of a word hyperbolic group), admits a map from the Higson compactification, so objects on any of these can effectively be pulled back to the Higson compactification. In any case, bundles on the Higson corona (i.e. the ideal points of the Higson compactification) can be paired with bounded propagation speed operators to give useful obstruction indices. It is as if there were a Lipschitz map to the cone on the Higson corona, and a rescaling construction would produce tiny curvature (although this is not literally the case).

Needless to say, all of these techniques can be viewed as the simply connected versions of a more general phenomenon. Thus one can study, on a non-simply connected manifold, the bounded propagation speed operators taking values in $C^*\pi$ and get more subtle and useful information, just as we can do in the situation of bounded L-theory. The small-scale version of this is precisely what we discussed in controlled K- and L-theories in giving, for example, a Novikov theorem for situations where we have controlled homotopy equivalences.

We used this added flexibility in proving that there are no complete positive scalar curvature metrics on $\mathbb{Q}$-rank-2 lattices. Stanley Chang (2001) proved by this method (i.e. marrying the Roe algebra to a fundamental group) that for *no* $K \setminus G/\Gamma$ is there a coarse quasi-isometric metric of positive scalar curvature and thus the metrics of Block and Weinberger when $\mathbb{Q}$-rank > 2 must be quite distorted.

In §5.3, the method of Thom, Milnor, Rochlin, and Schwartz gives rational Pontrjagin classes for PL homology manifolds. Sullivan (2005) gave a refinement which gives (anachronistically describes) a class

$$\sigma^*(X) \in H_x(X; L^*(Q)).$$

(Note that since 2 is inverted in the coefficient ring Q, we have no issues regarding the difference between quadratic and symmetric L-theory.) This class, when we invert 2, then lies in $KO_x(X) \otimes \mathbb{Z}[1/2]$ (see Taylor and Williams, 1979a, for an explanation of the work of Sullivan on the relation between L-spectra and K-theory away from 2, and the structure of L-spectra at 2 to see what we're throwing away by making this discussion somewhat crude). This is essentially the class of the signature operator on X.[100]

This class assembles to $\sigma^*(X) \in L^*(\mathbb{Q}\pi)$ just like in the case of manifolds. This can be viewed as forgetting control, or can be viewed along assembly lines (for the PL case, see e.g. Siegel, 1983, and Weinberger, 1987, for how such arguments go). It is important to note that there is an issue for the $\mathbb{Q}$ situation that we don't have for $\mathbb{Z}$ – namely, that homotopy equivalences can have degrees other than ± 1. Thus, the assembled characteristic class cannot be expected to be an oriented homotopy invariant in this setting – integrally. However, since the quadratic form $(d) \oplus (-1)$ is torsion (of exponent at worst 4) in $L^\circ(\mathbb{Q})$ (which equals Witt($\mathbb{Q}$)), this only affects the prime 2 – so, assuming that the assembly map with coefficients is injective, we get $\mathbb{Q}$-homotopy equivalence of higher signatures, with respect to orientation preserving maps of arbitrary (positive) degree – if the usual assembly map is an injection (away from the prime 2). Note that we are making use of Ranicki's localization result (Ranicki, 1979a) that tells us that the integral and $\mathbb{Q}$ assembly issues are equivalent (for all π) away from the prime 2.

The integral statement, allowing for 2, must take into account the degree of the map. Also, as far as I can tell, the $L^*(\mathbb{Q})$ injectivity statement is not equivalent to the $L^*(\mathbb{Z})$ – although both are part of the "Novikov package."

If one moves from the PL setting, then assuming that X is an ANR, controlled methods – e.g. following Yamasaki (1987) and Cappell *et al.* (1991) – allow the same statements to be made for topological $\mathbb{Q}$-homology manifolds.

Turning to §5.5, first of all let me call attention to Rosenberg (1996), a book that is a very useful introduction to many of the ideas of K-theory of all flavors, with many hands-on examples. There are still a few things that need to be discussed in view of our (belated) discussion of torsions and algebraic K-theory.

The first is that we have ignored all along "decorations" in surgery theory, and we now have the ingredients to set this straight. If X is a finite complex which satisfies Poincaré duality, then there are two natural questions to ask: (1) Is X homotopy-equivalent to a closed manifold? (2) Is X simple homotopy-equivalent to a closed manifold? The second takes advantage of the finite complex structure that X has – and is *not* a homotopy-invariant question.

However, (2) is not a reasonable question without some additional condition.

---

[100] This is literally the case for X smooth; for X a PL or Lipschitz manifold, this makes sense by the work of Teleman (1980). Recently Albin *et al.* (2018) have taken off on the seminal work of Cheeger on the L^2-cohomology of stratified spaces and the duality induced by $*$ (a variant of intersection homology) and have used microlocal analysis to give a suitable signature class on Witt and "Cheeger spaces."

If M is a manifold, then the Poincaré duality isomorphism $C_*(M) \to C^{n-*}$ is actually a *simple* chain equivalence. If we change basis on $C_*(X)$ via A and dually to $C^{n-*}(X)$, then the torsion of the equivalence is change by $[A] + (-1)^n[A^*]$; here $*$ is induced by $g \to w(g)g^{-1}$ (where w is, as usual, the orientation character). Note, by the way, that the self-duality of the cap product tells us that the isomorphism $C_*(M) \to C^{n-*}$ is "self-dual."

Thus, there is a "simplicity obstruction" lying in

$$\{\tau \in \mathrm{Wh}(\pi)|\tau = (-1)^n\tau^*\}/\{\sigma + (-1)^n\sigma^*\};$$

this is the Tate cohomology of the involution $*$ on $\mathrm{Wh}(\pi)$, i.e. $H^n(\mathbb{Z}_2; \mathrm{Wh}(\pi))$.

This is the obstruction in the category of finite complexes; that is, can we take a given X with a chain-level Poincaré duality map that is not a simple equivalence homotopy-equivalent to one where the duality map is a simple isomorphism? If this obstruction is non-zero, then there's no chance of X being homotopy-equivalent to a manifold!

In the relative situation this is very simple to appreciate. For the mapping cylinder of a homotopy equivalence between closed manifolds, the τ of the duality map for the relative Poincaré chain complex is essentially the torsion of the homotopy equivalence.

However, more fundamentally, this discussion suggests that, for question (2) above, we only ask it for X a *simple Poincaré complex*, i.e. one for which the duality map is a simple equivalence.

Surgery theory, as we have discussed it, makes sense in both settings and gives slightly different obstruction groups. If X is a Poincaré complex, or (Y, X) is a Poincaré pair, we can ask if X or (Y, X) is homotopy-equivalent to a manifold (pair) – and the obstruction is finding a degree-1 normal map with vanishing surgery obstruction that lies in $L_n^h(\pi_1(X))$ or $L_n^h(\pi_1(Y), \pi_1(X))$.

If X is a simple Poincaré complex, or (Y, X) is a simple Poincaré pair (which implies that X is a simple Poincaré complex), then we can ask if X is simple homotopy equivalent to a manifold or (Y, X) to a manifold pair $(M, \partial M)$ – and the obstruction lies in $L_n^s(\pi_1(X))$ or $L_n^s(\pi_1(Y), \pi_1(X))$.

These groups have a π-π theorem, and fit into the obvious exact sequences, and further satisfy a Rothenberg sequence (Shaneson, 1969):

$$\cdots \to H^{n+1}(\mathbb{Z}_2; \mathrm{Wh}(\pi(X))) \to L_n^s(\pi_1(X))$$

$$\to L_n^h(\pi_1(X)) \to H^n(\mathbb{Z}_2; \mathrm{Wh}(\pi(X))) \to \cdots.$$

Thus the L-groups only differ at the prime 2, and only if the Whitehead group is nontrivial. Thus, conjecturally for torsion-free groups, for example, these groups are isomorphic. However, in general they are different – even for cyclic groups.

Even for manifolds, the choice of decoration makes important sense: $S^h(M)$ measures how unique the manifold homotopy equivalent to M is in the "h-sense," i.e. up to h-cobordism. The version $S^s(M)$ measures the manifolds simple homotopy equivalent to M up to s-cobordism, i.e. up to homeomorphism.

Note that, as a formal consequence of the Rothenberg sequence and a diagram chase, we get an exact sequence

$$\cdots \to H^{n+1}\Big(\mathbb{Z}_2; \mathrm{Wh}\big(\pi_1(M)\big)\Big) \to S^s(M) \to S^h(M)$$
$$\to H^n\Big(\mathbb{Z}_2; \mathrm{Wh}\big(\pi_1(M)\big)\Big) \to \cdots.$$

However, it is not so hard to understand it directly. The map $S^h(M) \to H^n(\mathbb{Z}_2; \mathrm{Wh}(\pi_1(M)))$ measures whether a homotopy equivalence can be h-corded to a simple homotopy equivalence. The map

$$H^{n+1}(\mathbb{Z}_2; \mathrm{Wh}\big(\pi_1(M)\big)) \to S^s(M)$$

also comes out of the s-cobordism theorem. If I take an h-cobordism from M, then the torsion of the homotopy equivalence $M' \to M$ is $\sigma - (-1)^n \sigma^*$. If this vanishes, then I get a new simple homotopy equivalence.

If I take an h-cobordism from M to M' and "turn it upside down" to get one from M' to M, the torsion is changed by $\tau \to (-1)^{n+1}\tau^*$. I can glue these together to get a nontrivial s-cobordism for M to itself. These torsions – the obviously self-dual ones – never change the structure!

Finally, with these concepts, we can properly describe what happens for the product formula:

$$L_n^S(\pi \times \mathbb{Z}) \cong L_n^s(\pi) \times L_{n-1}^h(\pi).$$

For $L_n^h(\pi \times \mathbb{Z})$ we would need to introduce a new group $L_{n-1}^p(\pi)$ to obtain a formula: the p indicating the use of projective modules in the definitions rather than free modules in the quadratic forms used to define the L-groups. And so on. We will be forced to descend into negative K-theory to give a comprehensive approach.

For example, for fibrations over the circle with nontrivial monodromy α, we would be led to "intermediate L-groups" between L^s and L^h. Instead of allowing arbitrary torsions of homotopy equivalences in L^h, we should use the theory associated to allowing torsions that are elements of $\mathrm{Ker}(1-\alpha_*) \subset \mathrm{Wh}(\pi)$.

We have simplified, and will continue to simplify, our discussion by working with the $L^{-\infty}$-theory – which has the interpretation as having to do with being able to obtain a homotopy equivalence after crossing with some (unspecified-dimensional) torus. This will, by a sequence of Rothenberg sequences, only affect the prime 2.

Our discussion of Waldhausen's work in §5.5 was highly inadequate. We will leave the question of whether a pseudo-isotopy is pseudo-isotopic to an isotopy, just saying that it is analogous to the question of whether an h-cobordism is h-cobordant to a product. The analysis requires consideration of an involution on pseudo-isotopy theory and its action on the homotopy types of the topological groups, Homeo(M) and Diff(M). Waldhausen's work gives a kind of description of these in a stable range that grows linearly with dim(M). (Unstably, we know very little about these groups: one could hope – although I believe that this is dubious – that the components of Homeo(M) are $\mathbb{Q}$-acyclic for closed aspherical manifolds with centerless fundamental group.[101])

More importantly, but this is a direction that has not yet been well integrated into the Novikov/Borel philosophy, $A(X)$ is a deformation or extension of $K(R)$ and allows the modification of problems involving $\mathbb{Z}$ to ones involving $\Omega^\infty S^\infty$, which has a lot of internal structure. In this analogy, one obtains that the analogue of the result mentioned about $K(R\pi)$ for π the fundamental group of a hyperbolic manifold is that for such a manifold:

$$\text{Wh}(M) \cong \prod \text{Wh}(S^1),$$

where here Wh is the cofiber of the A-theory assembly map, and where the product is taken over primitive closed geodesics.

Similarly, the basic trace

$$K_0(R) \to R/[R, R]$$

that assigns to a projection the trace (i.e. the sum of its diagonal elements thought of as lying in R – as an additive group, modulo the additive subgroup generated by elements of the form $[r, s] = \{rs - sr\}$), of any matrix representing it, has a two-stage generalization. The first is Connes's trace map[102]

$$K_n(R) \to \text{HC}_n(R)$$

from K-theory to cyclic homology. The second leaves the world of rings, and

---

[101] And, if there's center, rationally equivalent to a torus. Later we will see that there is not, in general, a homomorphism $\mathbb{T} \to$ Homeo(M) inducing such a putative rational homotopy equivalence.

[102] Connes's interest in the trace was to deal with K-theory of C^*-algebras and then prove the operator-theoretic Novikov conjecture. Needless to say, executing this involves analytic difficulties in addition to the algebraic ones – however, in several important examples, this has been achieved – and the issues involved are in any case central for proving isomorphism conjectures (see Chapter 8.)

moves into stable homotopy theory (i.e. of spectra) and is an analogue of this, called the cyclotomic trace, developed by Bökstedt and Madsen, and used in Bökstedt *et al.* (1993) to detect $K(\mathbb{Z})$ rationally[103] and therefore a proof for all π with finitely generated homology of the algebraic K-theory Novikov conjecture for the ring $\mathbb{Z}$. Hesselholt and Madsen (2003) have applied this to give a great deal of information on the algebraic K-theory of, for example, the ring of integers in a local number field.

Alas, these methods do not prove an integral version, and are highly sensitive to the ring $\mathbb{Z}$ – it is not at all routine to replace $\mathbb{Z}$ by another ring of integers. Currently, such a modification would require deep number-theoretic conjectures[104] so that, for example, certain p-adic L-functions would be guaranteed to have nonvanishing properties.

We refer the reader to the book by Dundas *et al.* (2013) that explains trace technology, and the Goodwillie calculus that shows that the trace is not just accidentally successful in these problems: the trace gives a calculation of relative K-theory $K(R, S)$ if the map $R \to S$ is "1-connected," so the trace is an effective linearization of K-theory.

I believe that the ideas of Waldhausen's K-theory, concordance theory, and traces are related to the embedding theory calculations in Chapter 6 that give rise to a class of counterexamples to the equivariant Borel conjecture, and that there should be some unification of all these – but, at the moment, this is too vague.

---

[103] Recall that this space was rationally analyzed by Borel by relating the cohomology of lattices to the Lie algebra cohomology of the Lie group containing them.

[104] See Lück *et al.* (2017) on this. We will discuss this paper somewhat in Chapter 6 – to avoid misconception, I note that some of its implications are indeed unconditional.

6

Equivariant Borel Conjecture

6.1 Motivation

Remember our founding myth, whereby we pretended that Borel obtained his topological rigidity conjecture starting from Mostow rigidity:

Theorem 6.1 *Suppose that M and M' are closed irreducible Riemannian manifolds covered by G/K for some semisimple Lie group G that is not $\mathrm{PSL}^2(\mathbb{R})$. Then, any isomorphism $\phi\colon \pi_1 M \to \pi_1 M'$ is induced by a unique isometry $\phi\colon M \to M'$.*

We have concentrated so far on the existence of this isometry and its topological analogues, but now let's consider the implications of uniqueness.

It is worth noting that the uniqueness is not replaced by "a contractible space of choices" even in the case that M is locally symmetric but not semisimple. For instance, when $M \cong M'$ are isometric flat tori, ϕ is an arbitrary translational isometry. Thus the space of equivalences *is* a torus. This actually is more reasonable, because as Borel had noted:

Proposition 6.2 *If M is an aspherical complex, then the identity component of $\mathrm{Aut}(M)$ (the space of self-homotopy equivalences of M) is aspherical with abelian fundamental group $\cong Z(\pi)$ (the center of the fundamental group).*

This has two consequences: (1) rigidity will be somewhat stronger in situations that avoid center (or even normal abelian subgroups); and (2) one should not, in any case, want more topological rigidity than occurs homotopy-theoretically.

Alas, we have seen that the most obvious topological variant, the contractibility of the space of homeomorphisms, say, if the fundamental group is centerless, is unfortunately rarely true in high dimensions. On the other hand, we have also seen that the "cubist variant" of contractibility – namely, uniqueness up to pseudo-isotopy (and higher "block" analogues of this statement) – are well-founded conjectures (e.g. are consequences of the Borel conjecture itself, albeit for other groups).

Thus, we concentrate on problems and statements that are visible at the level of individual manifolds, rather than families – the essential difference between the block and fibered worlds. This is critical. I can't overemphasize this difference between the smooth category and the PL and topological categories.

In the smooth category, understanding *objects* more complicated than manifolds requires some understanding of families. For example, submanifolds have tubular neighborhoods that are unstable vector bundles. Maps are often thought of as "singular fibrations" like Lefshetz pencils, and one considers "Whitney stratified spaces" and so on.

In the PL situation, it is more natural to break things up over simplices in the base (i.e. not over individual points). Over vertices, one has a fiber F, and over edges, one has something isomorphic to $F \times I = [0, 1]$, but not with any particular projection map to I, and more generally over a simplex Δ one has a space isomorphic to $\Delta \times F$ (compatibly with the face relations of the simplex, but not with respect to anything going on over points – see Figure 6.1). It results in a "cubist" decomposition of the space. (See Figure 6.2 for an example of how a typically smooth object becomes polyhedral in a cubist perspective.) Analyzing such an object is never more complex than analyzing a manifold with boundary – since those are all that occur inductively. Spaces of such block bundles are effectively understood using blocked surgery.

We will be most interested in the topological category, where there are obstructions in algebraic K-theory to this structure. What exists is an even more smeared out structure, where *no point or edge* (or lower-dimensional sub-object) is given a particular pre-image. This is the content of the "teardrop neighborhood theorem" (Hughes *et al.*, 2000) and we will discuss it Chapter 7. As in Chapter 5, we will start by ignoring the constraints of K-theory to form intuitions, and then following it up by discussing the inevitable changes that K-theory necessitates.

To return to our story, one discovers, as a consequence of Mostow rigidity that:

Corollary 6.3 *If $M = K\backslash G/\Gamma$ with G semisimple, M irreducible and not a hyperbolic surface, then* $\mathrm{Out}(\Gamma)$*, the outer automorphism group of Γ, is finite and isomorphic to the isometry group of M. Furthermore, the action is unique up to conjugacy by isometry.*

In the excluded case of surfaces, one does not have the finiteness, but one still has the theorem (the "Nielsen realization conjecture") of Kerckhoff, and proved

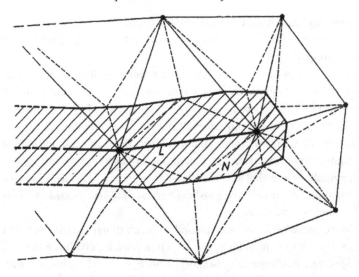

Figure 6.1 Regular neighborhood of a PL-manifold with boundary is an example of a block bundle – note the absence of fibers over many points of the submanifold. (Reproduced from Rourke and Sanderson (1982) with permission of Springer.)

Figure 6.2 PL wave. (Courtesy Esther Segal-Weinberger.)

several times since,[1] that finite subgroups of Out(Γ) act on M. A consequence of the proof, by hyperbolic geometry, is that the action is unique up to topological

---

[1] See §6.11 at the end of the chapter.

conjugacy – from the moduli space point of view, these actions are the fixed set of an action on Teichmüller space, and this fixed point set is contractible. Thus, our interest in this and the next chapters is called to:

Problem 6.4 (Nielsen realization problem) If M is an aspherical manifold with centerless fundamental group, can one realize finite subgroups of $\text{Out}(\pi_1 M)$ by group actions?

Problem 6.5 If G acts on M and M' is a compact group action on an aspherical manifold, and $f: M' \to M$ is an equivariant homotopy equivalence, then is f equivariantly homotopic to an equivariant homeomorphism?

We shall see that the answer to both problems is negative, but we shall also see that there are many interesting problems raised by their study. We shall study Problem 6.5 first and then return to Problem 6.4 in Chapter 7.[2]

Remarkably enough, the Novikov version of this conjecture does not yet have any known counterexamples despite the counterexamples to the rigidity statements. (Moreover, these Novikov statements have additional interesting applications to closed manifolds, even without group actions.)

6.2 Trifles

This section is devoted to several examples of group actions that show different kinds of phenomena that are present for different kinds of actions. We will ultimately focus on the topological category, and, for the sake of rigidity, carefully confine our attention to the type of group actions we allow: by the end of this section we will see that, if we are not somewhat picky, many lattices have infinitely many (or even uncountably many) co-compact properly discontinuous (C^0) actions on Euclidean space.

The picture of a smooth (compact) group action is kind of simple: the manifold decomposes according to orbit types. Each orbit type defines a stratum. They are essentially principal bundles over their quotient spaces, which are manifolds. These strata have neighborhoods that are equivariant vector bundles. They are put together in a reasonably comprehensible way.

This is proved using the elementary Riemannian geometry of any invariant metric (and such a metric can be obtained by averaging any particular metric over the group) and playing around with the exponential map (see Bredon, 1972). It has many straightforward consequences: e.g., the quotient of the set

---

[2] Actually, these two questions aren't the usual two sides of a coin that we usually look for in existence and uniqueness: the actions demanded in Nielsen would not – in a relative version – be enough to give us an "equivariant h-cobordism."

of free points for a group action on a compact manifold is a (noncompact, if the action is not free) manifold that has a canonical compactification as a manifold with boundary. The boundary is essentially the set of points of distance ε from the singular points for some ε smaller than the normal injectivity radius of the fixed set.

Equivariant vector bundles are studied via detailed understanding of the topology of various Lie groups. Unfortunately, this does not reduce to K-theory because the bundles involved will be unstable, even if the codimension(s) of the fixed set (and other strata) is(are) very high. For example, if F is the fixed set, the equivariant normal bundle decomposes canonically into bundles associated to the irreducible representations of G. Obviously, the sub-bundle corresponding to the trivial representation is trivial, but some other representation might occur with a very low multiplicity, forcing an unstable bundle as part of the data.

This picture is essentially correct in the PL situation, except that the fixed set need not be a manifold: if we insist that it *is*, by fiat, then the rest follows, except that, instead of vector bundles, there are block bundles. The proof of this is even simpler: one writes down formulas for the neighborhoods, just like in ordinary regular neighborhood theory (see Rourke and Sanderson, 1968a,b,c). As we mentioned in §6.1, the liberation of the "structure group" from a compact Lie group to the complicated space of "block automorphisms of the fiber" is actually a blessing when it comes to rigidity, which will become clearer as we proceed.

In any case, it still is the case that the quotient of the set of free points for a group action on a compact manifold is a (noncompact, if the action is not free) manifold that has a canonical compactification as a manifold with boundary.

In the topological case, this is not true.

Our first task is to give a bunch of examples of what group actions on some simple manifolds look like and how the different categories compare to each other. For instance, although there are only finitely many smooth structures on a compact topological manifold (except in dimension 4), this is not at all the case equivariantly. We will see very crude and also some subtle differences between these categories.

In the beginning there were linear actions. The orthogonal group acts on Euclidean space, preserving unit spheres. Every subgroup therefore acts linearly on the sphere, and the most obvious thing to do is to try to compare arbitrary actions to linear ones.

And, indeed, this works quite well[3] (smoothly and PL) in dimension ≤ 3.

Let us consider, as a starting place, $\mathbb{Z}_k$-actions and S^1-actions on the disk

---

[3] See §6.11.

that are semi-free and equivariantly contractible. Semi-free means that there are only two possible isotropy groups – the whole group and the trivial group. Equivariantly contractible is equivalent to asserting that the map to a point is an equivariant homotopy equivalence, which is equivalent to asserting that, restricted to all fixed sets (including the trivial and whole group), the map is a homotopy equivalence, which, in our situation, just means that we are assuming *a priori* that the fixed set, henceforth denoted F, is contractible.

This *is* a nontrivial restriction. Smith theory would tell us that F is $\mathbb{Z}/p$-acyclic (or $\mathbb{Z}$-acyclic for the case of S^1-actions). In a moment we will see that any $\mathbb{Z}$-acyclic manifold V^n is the fixed set of a smooth S^1-action on a disk $\mathcal{D}^{n+2}$ unless $n = 3$.

But let's start with the slightly simpler case where we start with a manifold V (of dimension at least 5) that is contractible (so the action will be of the desired sort). In that case, $V \times [0,1]$ is homeomorphic to the disk. (This is a straightforward application of the h-cobordism theorem.[4]) Thus *a fortiori* $V \times \mathcal{D}^i$ is a disk, possessing an action of O(i) on it with fixed set V. Restricting to a semi-free action on $\mathcal{D}^i$ gives us an appropriate semi-free action on $\mathcal{D}^{n+i}$ with fixed set V.

(Note that if we restrict this action to the boundary, we get action on S^{n+i-1} with fixed set ∂V; by a theorem of Kervaire (1969), the boundaries of contractible manifolds in dimension ≥ 4 are exactly the integral homology spheres in the PL and topological categories – in the smooth category, there is a unique differential structure on the sphere that one must connect sum with to get it to be a boundary. In short, *every homology sphere of dimension $d > 3$ is the fixed set of a semi-free S^1-action on the sphere S^{d+2}.*)

The question of which integral homology sphere – or mod p homology spheres – are fixed sets of *smooth* semi-free S^1- or $\mathbb{Z}_p$-actions is more subtle and studied by Schultz in a remarkable series of papers[5] (see, e.g., Schultz, 1985, 1987).

Example 6.6 (A PL action whose fixed set is not a manifold)   Take a homology sphere that bounds a contractible manifold. If we consider the action on the sphere with that as the fixed set, then we can cone (suspend) the action. It gives a PL-action on the disk (sphere) whose fixed set is a polyhedron with a (two) singular point(s) (but one can shrink an arc connecting them to a point, to get a new action on the sphere with one singular point). This action on the disk is equivariantly contractible.

Remark 6.7   Some information is given by Smith theory. In this semi-free

---

[4] It's a contractible manifold with simply connected boundary, which must be a disk.

[5] As well as what happens for dimension 3.

case, one knows that the fixed set, F, is a homology manifold (with coefficients in $\mathbb{Z}$ for the circle, and $\mathbb{Z}/|G|$ for a finite group G). In the case of S^1-actions, the converse holds and any acyclic PL homology manifold is the fixed set of a semi-free PL-action.[6] The proof of this is an induction in the spirit of Cohen (1970).

Now let us address the uniqueness question. How many actions are there with a given fixed set? The following is a slight variant of an old theorem of Rothenberg and Sondow (1979).

Theorem 6.8   *If the codimension of* F $> \dim(G) + 2$, *then the smooth semi-free G-actions on a disk with the contractible fixed set* F *and given local representation (which we assume is a free representation of dimension equal to the codimension of* F) *are a one-to-one correspondence with* $\mathrm{Wh}(\pi_0(G))$.

(The normal representation is determined by differentiating the action of G at a fixed point.) Near the fixed set, bundle theory and the tubular neighborhood tell us that the action is a product action. Then the rest of the proof is an application of the h-cobordism theorem.

Remark 6.9   The condition on codimension is important, so that we can use homological methods to control the homotopy theory – in other words, we want to be able to conclude that complements are simply connected.

It is not very hard to construct "exotic" actions, even smoothly, with codimension-2 fixed-point sets, on the sphere or the disk once dim > 3. These are called "counterexamples to the Smith conjecture"(see Giffen, 1966). Here's a sketch of a construction in dimension 5 and higher based on the Poincaré conjecture,[7] and making use of a nontrivial knot K, so that π_1 of the knot complement is $\mathbb{Z}$.

Consider a free action of $\mathbb{Z}_p$ on the sphere with an invariant codimension-2 subsphere S (which might even be assumed unknotted for simplicity). It is easy enough to see that $S\#K\#K\#K\cdots K$ (with p copies) is invariant under the action as well. Now do "surgery on this action," i.e. remove this invariant knot, and glue in $S \times \mathcal{D}^2$ with the action that is trivial on the S direction and semi-free on the $\mathcal{D}^2$ direction. (The reader can check that this is possible.) This gives a $\mathbb{Z}_p$-action on the sphere (here we use the Poincaré conjecture) whose fixed set is connected of p copies of K, and thus is an example.[8]

---

[6] However, not necessarily in *all* even codimensions because of the contribution of Rochlin's theorem (as in Schultz, 1987), but in codimensions that are a multiple of 4, this is OK.

[7] Giffen worked in the smooth category, and gave some examples in dimension 4 because he was also able to avoid use of the Poincaré conjecture in that elegant paper.

[8] It is a theorem of Levine (1965) that a knot in a high-dimensional sphere is trivial iff its complement has the homotopy type of a circle; as a consequence, one can't unknot a knot by taking its connected sum with another knot.

In the PL category, we don't necessarily have a bundle structure. Nevertheless, for locally linear actions, the theorem is correct – because local linearity implies enough homogeneity to give ("simple") block structures, which are determined by maps into classifying spaces.

At this point there is a subtlety related to torsions (the "simple" in the previous paragraph) and therefore ultimately to decorations in L-theory (see §5.5) as we now explain:

Example 6.10 (PL neighborhoods) A basic fact about the smooth category is that the neighborhood theory near the fixed points is bundle theory, and therefore homotopical in nature: the germ neighborhoods of $F \times [0, 1]$ are just those of the product of F with $[0, 1]$).

However, suppose F is a point, and we start with a semi-free linear action of $\mathbb{Z}_p$ on a disk $\mathcal{D}^n$ with 0 as the isolated fixed point. The quotient of the boundary of an invariant neighborhood of 0 is a lens space. Now suppose that $p \geq 5$, so $\mathrm{Wh}(\mathbb{Z}_p)$ is nontrivial. Erect an h-cobordism on this lens space. We can take universal covers, and cone the two boundaries separately to obtain a nonlinear action on $\mathcal{S}^n$ that has fixed set $\mathcal{S}^0 = 0 \cup \infty$ (with obvious conventions). Near 0 the action is linear, but near ∞ it is not: the Whitehead torsion of the homotopy equivalence from this quotient to the linear lens space is nontrivial. (Exercise, but see §5.5.3 if you need a hint.)

Now consider the cone on this action.

We obtain a PL $\mathbb{Z}_p$-action, whose fixed set is an interval I. However, the germ neighborhood is not trivial. For then the "normal representations" at the two fixed points would have to be the same.

Moral: In the PL category one has to do one of two things. Either assume a local model: this is perhaps not so unreasonable if one recalls that assuming the fixed set is a manifold is an assumption – not guaranteed, as Example 6.6 shows.

Or, alternatively, one can work up to concordance: view two neighborhoods of F as equivalent if there is a neighborhood of $F \times [0, 1]$ which restricts to each on the boundaries. (Better, we should allow F to change, and allow h-cobordisms into the equivalence relation on the blocks over F. A neighborhood of F will be equivalent to a neighborhood of F' if they are h-cobordant, and there is a neighborhood of the h-cobordism that restricts to each.) Both of these theories give rise to block bundle theories and have classifying spaces. The relation between these theories is established in Cappell and Weinberger (1991a) and is determined by "Rothenberg classes" that lie in the cohomology of F.

In any case, the smooth (Rothenberg–Sondow) examples we discussed do not become PL equivalent.

Example 6.11   The neighborhoods of 0 and ∞ in the previous example are topologically equivalent.[9] Thus, in the topological category there cannot be uniqueness of "closed regular neighborhoods of fixed points" as there is in the PL category.

This is a simple consequence of the h-cobordism theorem. Since L and L' are h-cobordant, they are diffeomorphic after crossing with S^1 (as torsions multiply by $\chi(S^1) = 0$.) Passing to infinite cyclic covers gives a diffeomorphism $L \times (-\infty, \infty) \to L' \times (-\infty, \infty)$. Taking covers and extending to a point at $-\infty$ gives an equivariant homeomorphism between the two open neighborhoods.

Example 6.12   With a little more care one can see that all the smooth actions with fixed set F and given normal representation are topologically equivalent. This is surely plausible, as we've seen that torsion does not obstruct, and the torsion is all there is in the smooth category.

The result actually follows from the following beautiful fact, due to Stallings, whose proof goes back to Euler and Eilenberg, and then to Mazur and Stallings (Stallings, 1965b)[10] (and oft exploited since).

Proposition 6.13 (Stallings)   *If $(W, \partial_+, \partial_-)$ is an h-cobordism (of dimension greater than 4), then $W - \partial_- \cong \partial_+ \times [0, \infty)$.*

Let V be the h-cobordism with $\tau(V, \partial_-) = -\tau(W, \partial_+)$. We have $W \cup V \cong \partial_+ \times [0, 1]$, by the h-cobordism theorem, and similarly, $V \cup W \cong \partial_- \times [0, 1]$. Then

$$
\begin{aligned}
\partial_+ \times [0, \infty) &\cong (W \cup V) \cup (W \cup V) \cup \cdots \\
&\cong W \cup (V \cup W) \cup (V \cup W) \cup \cdots \\
&\cong W \cup \partial_- \times [0, 1] \\
&\cong W - \partial_-.
\end{aligned}
$$

Thus, we've now seen infinitely many PL-inequivalent topologically-equivalent smooth actions for reasons that can be attributed to K-theory.

Topological actions that cannot be triangulated exist for many reasons, of various degrees of subtlety.

---

[9] This example is a variation on the trick used by Milnor in his disproof of the general *hauptvermutung*.

[10] Our scholarship is inadequate to the task of justifying this folklore description of the history of the series $1 - 1 + 1 - 1 + 1 - \cdots$.

Example 6.14 We can take an infinite connect sum of S^1-actions on S^n with fixed set Σ (a non-simply connected homology sphere). This will give an action on S^n with fixed set the one-point compactification of the infinite connected sum $\Sigma \# \Sigma \# \Sigma \# \cdots$. This fixed set is not even an ANR – its fundamental group is uncountable!

Example 6.15 Bing (1959) gave an example of a non-manifold X whose product with $\mathbb{R}$ is $\mathbb{R}^4$, and so that $X^+ \times \mathbb{R} \cong S^3 \times \mathbb{R}$. Now we know that such examples abound (i.e that there are uncountably many different such spaces in every dimension[11]). In any case, $X^+ \times S^1$ is a manifold whose quotient under a circle action is a non-manifold.

Bing also gave uncountably many $\mathbb{Z}_k$ actions on $\mathbb{R}^3$ whose fixed sets are different non-locally flat $\mathbb{R}$s.

In this chapter we will often assume that the fixed sets (and quotient spaces of the free parts) of our topological actions are ANRs or even compact topological manifolds.

Example 6.16 For this example, we will make use of Siebenmann's proper h-cobordism theorem Siebenmann (1970b). For W a paracompact manifold, Siebenmann defined $\mathrm{Wh}^P(W)$, which classifies proper h-cobordisms with one boundary component W. This group can frequently be computed, if W is "not too wild." For our purposes, we just note for L a compact manifold

$$\mathrm{Wh}^P(L \times \mathbb{R} \cong K_0(\pi_1(L)).$$

If one takes a prime with nontrivial class group, we can start with a linear action, and erect an h-cobordism with given element of K_0 as its torsion, and, with a one-point compactification, obtain an action with fixed set an interval, but that does not have *any* invariant closed tubular neighborhoods.

(The reader with some number theory experience can use the analogue of the Milnor duality theorem in this setting to give examples with fixed set a circle by arranging for the "other end" to be trivial and then gluing the ends together.)

These kinds of examples occur very naturally when one studies possible fixed sets of group actions. Certain lens spaces (with odd-order fundamental groups) occur as fixed sets of, for example, Q_8 actions only if one allows actions where

---

[11] These can be obtained by shrinking quite general decompositions to a point, and ultimately using Edwards's theorem. We refer the reader to Daverman (2007) for a discussion of this beautiful area of topology.

there are no invariant closed tubular neighborhoods.[12] (See Figure 6.3 for a picture decribing how such a construction works.)

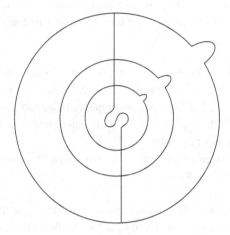

Figure 6.3 The actions are created on an ascending union of mod 2 homology balls on which PL-actions can be constructed, the mod 2 homology being present to get around a finiteness obstruction. The action on the sphere is the one-point compactification of the union.

It is a remarkable fact, discovered by Quinn (1979, 1982a,b,c, 1986, 1987b), that actions which have locally flat fixed point sets (and manifold quotients of pure strata[13]) – actions that have come to be called "tame" – have some topological homogeneity: if one has an arc entirely within a pure stratum of this action (i.e. where the isotropy group does not change along the curve) then there is an equivariant isotopy covering this. This means that if an action is locally linear at one point, it will be locally linear over the whole stratum. As a result, the seeming singularities that arise at one-point compactifications and related constructions often are not there at all topologically.

In some sense the differences between Top and PL (when one has local triangulability of fixed sets and quotients) can be attributed to K-theory (and the Kirby–Seibenmann obstruction). Of course, the non-ANR situation is a serious problem in general for the topological category. The smooth category differs from these categories for other local reasons as well.

---

[12] A three-dimensional lens space with odd-order fundamental group $\mathbb{Z}_n$ is a PL fixed set iff n is $\pm 1 \bmod 8$. This is due to the Swan homomorphism that associates to n the projective module given by the kernel of the reduction of the augmentation map $\mathbb{Z}Q_8 \to \mathbb{Z} \to \mathbb{Z}_n$ (the one-dimensional free module $\mathbb{Z}Q_8$). (For any G, this defines – nontrivially – a homomorphism $\mathbb{Z}/|G|^* \to K_0(\mathbb{Z}Q)$.) See Example 6.18 below and Example 6.28 in §6.6.

[13] We obviously only need to add this as a hypothesis when a positive-dimensional group acts (and then we do because of Bing-type examples as above).

Example 6.17 Let's think about $\mathbb{Z}_p$-actions. In the smooth case, the fixed set F will be smooth and the neighborhood of F has the structure of an $\mathbb{R}[\mathbb{Z}_p]$-module (with no trivial piece). So, for $p = 2$, the neighborhood is essentially an arbitrary vector bundle.

In the PL (and Top is essentially equivalent in this case) case, we can analyze the situation using "blocked surgery." For the orientation reversing situation, $BPL_{2k-1}(\mathbb{Z}_2) \cong BAut(\mathbb{RP}^{2k-1}) \otimes \mathbb{Z}[1/2]$ (as the L-spaces $L_{2k}(\mathbb{Z}_2, -)$ are 2-torsion). Thus, the map $BO(2k - 1) \to BPL_{2k}(\mathbb{Z}_2)$ loses most of the rational Pontrjagin classes that are present in the smooth case!

On the other hand, for even codimensions, when the action is orientation-preserving, the PL space is much richer because $L_{2k}(\mathbb{Z}_2, +)$ is rationally a product of BO × BO or its second loop space (depending on k).

However, the reader should not jump to any conclusions. The stabilization map $BPL_{2k}(\mathbb{Z}_2) \to BPL_{2k+2}(\mathbb{Z}_2)$ (and these are approximately both isomorphic to $BO \otimes \mathbb{Q}$ aside from a few homotopy groups coming from the BAut) factors through $BPL_{2k+1}(\mathbb{Z}_2)$ which is essentially trivial.

Consequently, although the maps $BO(n) \to BO(n + 1)$ become highly connected with n, the equivariant PL versions *never* stabilize (even rationally).[14]

This is closely related to the failure of equivariant transversality in the PL and topological categories (see Madsen and Rothenberg, 1988a,b, 1989).

For p odd, there is another interesting difference between the categories. For the smooth category, one gets a decomposition of the vector bundle according to irreducible representations of $\mathbb{Z}_p$. In PL (and Top) there is no analogue of this.

In the smooth category, the structure of the neighborhoods is thus very dependent on what the local representation is: if, for example, the normal representation is a sum of n distinct irreps, then the bundle is equivalent to a sum of n-complex line bundles, while if it is untypical, the bundle is a $U(n)$ bundle. In the PL (and topological situations) the classifying spaces for the neighborhoods is pretty insensitive to the type of the local representation (e.g. to whether it is untypical or not).

Example 6.18 (Converses to Smith theory) The possible fixed sets of a group action are not arbitrary. Smith theory, and its generalizations, give connections between the group action, the homotopy type of the space acted upon, and the fixed set. In the extreme of a contractible space, the phenomena are rather stark and were pioneered by Lowell Jones (1971) and Oliver (1976b), respectively.

---

[14] Nor do the topological versions of these spaces for the very same reason (although the details necessary to rigorously verify this depends on Quinn's theory of controlled ends – or something equivalent).

For G a p-group, the fixed set must be mod p-acyclic. This condition is essentially necessary and sufficient (although for complicated G, in the PL case, there is a $K_0(\mathbb{Z}G)$ obstruction – see Assadi, 1982). To be really concrete, if p is prime, then a manifold M is the fixed set of a semi-free PL locally linear $\mathbb{Z}/p^n$-action on a high-enough-dimensional sphere iff M is a mod p homology sphere of the same parity of dimension as that of the sphere.[15] In view of this, if the fixed set is not unique in its homotopy type, we can easily get equivariantly homotopy-equivalent actions on the sphere by realizing these different manifolds as fixed sets.

For G a non-p-group, the homotopy types of possible fixed sets are determined by a number n_G called the Oliver number. In this case F is a fixed set on a finite contractible complex (and hence homotopy-equivalent to the fixed set of some action on a disk) iff $\chi(\mathrm{F}) \equiv 1 \bmod (n_G)$. When $n_G = 1$ then every finite complex F is *actually* the fixed set of a G-action on some disk (such that F is embedded in the interior of the disk!).

Putting together these methods of construction with a few variations and the results we will explain later in this chapter, one obtains the following result (which answers a question of Borel).

Theorem 6.19 (Trichotomy; Cappell *et al.*, 2015) *Let G be a real Lie group, and suppose that the dimension d of G/K is at least 5[16] and suppose Γ is a uniform lattice in G. Then the number of properly discontinuous actions of Γ on $\mathbb{R}^d$ is either 1, $\aleph_0$, or c (the continuum). In the last case, there are (a continuum of) examples that are not locally rigid (e.g. arbitrarily C^0-close to the left action of Γ on G/K – indeed that are degenerations of this action).*

This trichotomy is determined by the nature of the singular set[17] of the isometric action of Γ on G/K. One has rigidity if the action is free (i.e. if Γ is torsion-free) and sometimes if the singular set is zero-dimensional,[18] but if the singular set is positive-dimensional then the number of actions is always c (and it's never uncountable unless the singular set is positive-dimensional).

---

[15] Indeed, if M embeds in the sphere and we are even codimension other than 2, it is the fixed set of PL locally linear action (see Weinberger, 1985b, 1987; Cappell and Weinberger, 1991a).

[16] The paper gives information in low dimensions as well.

[17] This is the set of points whose isotropy is nontrivial.

[18] $\aleph_0$ nonrigidity holds iff (the action has discrete singular set) and d is 2mod 4 (and greater than 2), and Γ contains an element of order 2. (As a comment whose significance will only become clear in the next section, Γ then automatically has at least two conjugacy classes of involutions, and, indeed, Γ contains an infinite dihedral group.)

Moral: For the equivariant version of the Borel conjecture, we shall assume that our actions are "tame": that the fixed sets are nice submanifolds and we shall also not allow codimension-1 and codimension-2 situations.[19]

In addition, we should assume that the G-action on M makes it into an "equivariant Eilenberg–Mac Lane space." This should have been obvious for reasons of functoriality (this is like the assumption of low $\mathbb{Q}$-rank in the proper Borel conjecture – we should not have made that mistake twice!). In any case, isometric actions on locally symmetric manifolds are Eilenberg–Mac Lane in the appropriate sense (as will be clear in a moment). This eliminates the converses to Smith theory examples (Example 6.18).

To make this condition clearer, recall that a $K(\pi, 1)$ is the terminal object in a category that includes some connected space, and only 1-equivalences (i.e. maps for which one can uniquely lift all maps of 1-complexes into the target). This same notion makes sense equivariantly. It boils down to – see tom Dieck (1979), Lück (1987), and May (appendix to Rosenberg and Weinberger, 1990) – all components of all fixed sets of all subgroups being aspherical.[20]

The smooth category, even more transparently than for the original Borel category, is not suitable for the equivariant version. The PL category also has no chance – there are too many K-theoretic obstructions. The K-theory that enables topological actions to exist that don't have closed equivariant neighborhoods haven't yet been implicated as an obstacle – but we shall have to study this more carefully. Depending on formulation, Nil is a problem or it is not. It will cause a perturbation in our understanding.

The topological category will be a reasonable one for studying the problem. Without assuming tameness, examples such as Examples 6.6 and 6.14 are unavoidable, and one cannot hope for equivariant homeomorphisms.

The need to avoid the low-codimension situation is because of the failures of the Smith conjecture. With codimension 2, one loses too much information on moving to closed strata from pure ones.

6.3 *h*-Cobordisms

In the case of closed manifolds, the Borel conjecture boils down to two statements: one about the vanishing of Whitehead groups, i.e. that h-cobordisms are products; and the second a statement about L-groups, that a certain assembly

---

[19] It is not impossible to incorporate codimension-1 and codimension-2 phenomena in an "isovariant Borel conjecture" – see Chapter 13 of Weinberger (1994). However, although the isovariant conjecture is "more true," the equivariant one is "more interesting."

[20] However, we will see that even this assumption does not save the day even for "equivariant Novikov conjectures" in the appendix to §6.7. Nevertheless, till then we will use the current guess as our guide till we are forced to abandon this as too naive.

map is an isomorphism[21] – which surgery then translates into the statement that homotopy-equivalent manifolds are h-cobordant, and therefore homeomorphic.

In the smooth and PL locally homogeneous categories, the h-cobordism theorem is quite straightforward:[22]

$$\mathrm{Wh}^G(M) \cong \bigoplus \mathrm{Wh}(\pi_1(M^H/(NH/H))) :$$

here NH/H is the normalizer of H divided by H and the sum is over conjugacy classes of subgroups; we use the convention that π_1 of a disconnected set is the sum of the π_1s of the components. Thus, on the right-hand side, we have a sum of the fundamental groups of all components of all strata. (Note that the group that acts on the stratum fixed by H is NH/H.)

The proof is a straightforward induction on the strata. Once one has a product structure on a stratum, the structure of neighborhoods (e.g. the tubular neighborhood theorem) extends it to a neighborhood, and then one uses the torsion on the complement and a relative form of the h-cobordism theorem to extend it to the outside.

Although the Whitehead group has a straightforward decomposition, the involution does *not* preserve the terms of the decomposition. It does at the level of an associated graded of a filtration, but not on the nose. To give an example, suppose that $G = \mathbb{Z}_2$ acts on a closed manifold W with codimension-1 fixed set. Then W/G is a manifold with boundary F:

$$\mathrm{Wh}^G(W) \cong \mathrm{Wh}(\pi_1(W/G)) \oplus \mathrm{Wh}(\pi_1(\mathrm{F})).$$

The involution, thought of as a matrix, has one non-diagonal term corresponding to $\pm$ (depending on conventions) the inclusion map $\pi_1(\mathrm{F}) \rightarrow \pi_1(W/G)$. So if $\pi_1(\mathrm{F}) \rightarrow \pi_1(W)$ is an isomorphism, the Tate cohomology $H^*(\mathbb{Z}_2; \mathrm{Wh}^G(M)) = 0$ – which would not be the case if the involution had actually preserved the pieces.

Note that these Whitehead groups can be quite large. If we consider a disk $\mathcal{D}$ (of dimension greater than 2) with a linear G-action (with no low-codimension situations), then[23]

$$\mathrm{Wh}^G(\mathcal{D}) \cong \bigoplus \mathrm{Wh}(NH/H),$$

---

[21] As noted at the end of Chapter 5, the statement about Whitehead groups can also be viewed as the bottom part of an isomorphism of assembly maps in algebraic K-theory.

[22] We do not allow any low-dimensional strata, or assume that the h-cobordism is assumed to be a product on those. Needless to say, when we work with h-cobordisms that are trivialized on a union of strata, we get the same answer, except that those strata do not come up in the right-hand side.

[23] Recall our convention that if we do not include the boundary in the notation, then we are working relative to the boundary. If we were not working relative to the boundary, then we would include the boundary into our notation, $\mathrm{Wh}^G(\mathcal{D}, \partial\mathcal{D})$.

where the sum is over conjugacy classes of isotropy subgroups (G always occurs as the isotropy of 0). This can be quite a large finitely generated group. (One can compute its rank using representation theory – it is the number of real irreducible representations that are not rational.)

Exercise 6.20 Use the PL Whitehead group to give equivariantly homotopy equivalent G-actions on an aspherical manifold that are not equivariantly PL homeomorphic.

Now, for a more typical situation, let's consider $\mathrm{Wh}^G(\mathcal{S}^1 \times \mathcal{D})$, where G acts trivially on the circle. In that case, we get the analogous decomposition:

$$\mathrm{Wh}^G(\mathcal{S}^1 \times \mathcal{D}) \cong \bigoplus \mathrm{Wh}(Z \times NH/H)$$
$$\cong \bigoplus \mathrm{Wh}(NH/H) \bigoplus K_0(NH/H) \bigoplus \mathrm{Nil}_\pm(NH/H)$$

by the Bass–Heller–Swan formula. For higher tori, we can iterate this formula. The Nil terms, when nonzero, give us infinitely generated torsion terms.[24]

However, as we saw in §6.2, in the topological case this formula is not quite right. It is not so hard to see that

$$\mathrm{Wh}^{G,\mathrm{top}}(\mathcal{D}) \cong 0$$

as a consequence of Siebenmann's thesis – which, while giving a condition for a manifold to be the interior of a manifold with boundary, gives, *inter alia*, a condition for a manifold V to be $\partial V \times [0, \infty)$ (rather analogous to the h-cobordism theorem – except that there is no K-theory obstruction[25]).

The situation is rather different for $\mathrm{Wh}^{G,\mathrm{top}}(\mathcal{S}^1 \times \mathcal{D})$. The $\bigoplus \mathrm{Wh}(NH/H)$ terms go away for the same reason as before. The $\bigoplus K_0(NH/H)$ also go away. We can explain this as follows.

The $K_0(NH/H)$ term corresponds to h-cobordisms that are isomorphic to their own 2-fold covers. So when doing a $1 - 1 + 1 - 1 + \cdots$ trick, one can have each term represent the 2-fold cover of its predecessor. When one does this, the homeomorphism produced in the limit will actually be convergent along the circle.

This leaves only the Nil terms; that is, the correct answer

$$\mathrm{Wh}^{G,\mathrm{top}}(\mathcal{S}^1 \times \mathcal{D}) \cong \bigoplus \mathrm{Nil}(NH/H).$$

Nil, which obstructs the fundamental theorem of algebraic K-theory holding for non-regular rings, also prevents too naive a form of the equivariant Borel conjecture from holding.

---

[24] Recall that if A is a nilpotent matrix, then $\mathbf{I} + tA$ is a typical element in the Nil term: $\mathbf{I} + t^i A$ contains an infinite number of linearly independent elements (distinguished by which covers they transfer nontrivially to). See Bass and Murthy (1967) for some examples.

[25] Compare to Stallings's result, Proposition 6.13.

Needless to say, we haven't proved any of this. The proof (see Quinn, 1988; Steinberger, 1988) uses the controlled h-cobordism. What occurs on a pure stratum is not merely a proper h-cobordism – that can be analyzed via Sieben-mann. It has control with respect to the lower stratum as one goes to ∞. As controlled K-theory is a homology theory, the

$$\bigoplus \text{Wh}(NH/H) \bigoplus K_0(NH/H) \bigoplus \text{Nil}_\pm(NH/H)$$

that one sees in the interior is modified by

$$H_*\left(S^1; \bigoplus \text{Wh}(NH/H)\right) \cong \bigoplus \text{Wh}(NH/H) \bigoplus K_0(NH/H),$$

leaving the Nil terms left over.

Exercise 6.21   Give an example of a closed aspherical manifold where there are equivariant homotopy equivalences not equivariantly homotopic to home-omorphisms because of a nontrivial Nil group.

For small groups like $\mathbb{Z}_p$ this doesn't make a difference, but even for $\mathbb{Z}_n$ or $\mathbb{Z}_p \times \mathbb{Z}_p$, the group Wh^{top} can be large (because of K_{-1} bubbling up from a fixed set of dimension 2 or because of Nil). In any case, these issues can be handled by assuming it away in the equivariant Borel conjecture:

Conjecture 6.22 (Modified equivariant Borel conjecture)   *Suppose $G \times M \to M$ is a tame action, and that it is an equivariantly aspherical manifold. If $f : M' \to M$ is an equivariant topologically simple homotopy equivalence, then f is equivariantly homotopic to a homomorphism.*

A negative aspect of this modification is that it loses the vanishing of White-head groups that is part and parcel of the usual Borel conjecture.

On the other hand, it is a possibly true (*prima facie*) rigidity statement.[26]

The better resolution of this difficulty is the Farrell–Jones conjecture, that makes a prediction of the structure of $\text{Wh}^{G,\text{top}}(M)$ in terms of the space of equivariant G-submanifolds of M of dimension 1. For example, if the action is semi-free, and suppose the fixed set contains no higher-rank abelian subgroups, the relevant Whitehead group should just depend on the $\text{Nil}(G)$, parameterized by the conjugacy classes of maximal cyclic subgroups of $\pi_1(F)$.[27]

We shall return to this later.

6.4 Cappell's UNil Groups

We are not out of the woods yet in understanding the equivariant Borel con-jecture because of a remarkable phenomenon discovered by Cappell (1973, 1974a,b).

---

[26] Since we will see that it is indeed false, perhaps it would be better to say frequently true.

[27] This description tacitly assumes no infinitely divisible elements.

This is a beautiful story worth telling in its own right, not just as an adjunct to the Borel story – so we shall delay the application to the equivariant Borel conjecture for a couple of sections and discuss a part of Cappell's work in its original context.

We are now back in the world of manifolds, with no group actions. For simplicity we will assume that all manifolds here are orientable and will only deal with one special splitting problem: Cappell's work is much more general.

We begin with a theorem of Browder (and its easy generalization by Wall, 1968).

Theorem 6.23 *Suppose M is a closed manifold (of dimension greater than 5) and V is a codimension-1 submanifold, dividing M into two parts, $M_\pm$, so that $\pi_1(V) \to \pi_1(M_+)$ is an isomorphism. Then any homotopy equivalence $f: M' \to M$ can be homotoped to one where f is transverse to V, and $f|_{f^{-1}(V)}$ is a homotopy equivalence.*

Corollary 6.24 *In the* PL *and* Top *categories, the question of whether a manifold is a connected sum only depends on the homotopy type of the manifold if it's simply connected (or one of the summands is).*

Let V be the separating subsphere in the connect sum decomposition, and make use of the Poincaré conjecture to assert that $f^{-1}(V)$ is also a sphere.

An analogue of this corollary (ignoring the possibility of taking a summand that is a counterexample to the Poincaré conjecture) without the simple connectivity was first proved in dimension 3. Stallings showed that a 3-manifold is a connected sum of non-simply connected pieces iff its fundamental group is a nontrivial free product (see, e.g., Hempel, 1976).

In dimension 4 this corollary is now known not to be true in PL (because of Donaldson's work), and in dimension 5 it is possible to fix the argument and establish the result. In general, we have the following theorem of Cappell (1974b, 1976a):

Theorem 6.25 *For manifolds whose fundamental groups have no 2-torsion, being a connected sum is homotopy-invariant. However, there are infinitely many manifolds homotopy-equivalent to $\mathbb{RP}^{4k+1}\#\mathbb{RP}^{4k+1}$ that are not connected sums.*

It is the second part of this theorem that will imply, for example, that the equivariant Borel conjecture still fails for certain involutions on the torus.

We now know, thanks to unpublished work of Connolly and Davis, that connected sum is homotopy invariant for *all* orientable manifolds of dimension 0 and 3mod 4.[28] It turns out that *this* positive result is a consequence of results proved about the equivariant Borel conjecture (or, perhaps better, the Farrell–Jones conjecture).

But let's do things in order.

The Browder–Wall splitting theorem is a consequence of the π–π theorem. Consider $M' \times [0,1]$ and glue on to $M' \times 1$ a normal Z, ∂Z cobordism of $f^{-1}(M_-, V)$ to a homotopy equivalence. This normal cobordism exists by the π–π theorem. We now have another surgery problem,

$$M' \times [0,1] \bigcup_{f^{-1}(M,V)} (Z, \partial Z) \to M \times [0,1] \bigcup_{M \times 1} M_- \times [1,2] \cong M \times [0,2],$$

relative to M_-. We can view this as a π–π problem, because by Van Kampen's theorem, the map $\pi_1 M_+ \to \pi_1 M$ is an isomorphism. When we solve it, we obtain an h-cobordism to the solution of the splitting problem. If we glue on an h-cobordism on the part mapping to $M_+ \times 2$ with negative torsion (of the above h-cobordism), we turn it into an s-cobordism – i.e. we have produced the desired s-cobordism.

To study the connected sum problem, it is easy enough enough to construct a normal cobordism of the homotopy equivalence to a split one.[29] The real problem is to somehow understand elements in the cokernel of $L(G) \times L(H) \to L(G * H)$ which will measure the difficulty in taking a normal cobordism to a connected sum, and modify it (via a Wall realization) to one where the surgery obstruction vanishes, and can be turned into a homotopy.

Cappell showed, on the one hand, that one can give a complete analysis in terms of an analogue of Nil, based on the bimodule $(\mathbb{Z}; \mathbb{Z}[G - e], \mathbb{Z}[H - e])$. These are bimodules with involution as in the Milnor duality formula. In this case, we essentially are dividing $\mathbb{Z}[G - e]$ (and $\mathbb{Z}[H - e]$) into pieces that are $\mathbb{Z}$ or $\mathbb{Z} \oplus \mathbb{Z}$ terms that are preserved or interchanged by the involution. When there is no 2-torsion, we are in the situation where everything is of the form $(\mathbb{Z} \oplus \mathbb{Z}, \mathbb{Z} \oplus \mathbb{Z})$ where there is nothing. However, terms of the form $(\mathbb{Z}, \mathbb{Z})$ give a very large group.

In this case, Cappell wrote down quite explicit elements in $L_2(\mathbb{Z}[D_\infty])$, where D_∞ is the infinite dihedral group. He showed that these are nontrivial by mapping to the finite dihedral groups D_{2n} (for n odd). More precisely he

---

[28] Their work, as well as work of Banagl and Ranicki, show that there are non-connected sums homotopy-equivalent to $\mathbb{RP}^{4k-1} \times S^3 \# \mathbb{RP}^{4k-1} \times S^3$ as well – so the phenomenon does arise in both of these sets of dimensions. The case of 3mod 4 was proved by Cappell in his original paper.

[29] Here we only need to use functoriality to build a map $L(G * H) \to L(G) \times L(H)$ that is almost a retraction of the maps induced by inclusion $L(G) \times L(H) \to L(G * H)$. (Why is it not a retraction? How far off is it?)

considered the Arf invariant in $L_2(\mathbb{Z}) = \mathbb{Z}_2$ and by first going to an odd-fold cover and then taking the Arf invariant (i.e., mapping to $L_2(\mathbb{Z})$). If splitting were possible, i.e. if the L-group were as small as predicted, these two elements would be the same. More precisely, $L_2(\mathbb{Z}) \to L_2(Z[D_\infty])$ would be an isomorphism, in which case passing to a finite-fold cover would just multiply the element by the index of the cover. Since the Arf element is of order 2, we would expect equality under odd-fold covers. But we do not obtain this by explicit calculation.

Some more details: recall that $L_{2k}(\pi)$ is built out of $(-1)^k$ symmetric quadratic forms over $\mathbb{Z}\pi$. Let $\pi = \mathbb{Z}_2 * \mathbb{Z}_2$, generated by involutions g and h. Here $gh = t$ is the translation in the usual view of D_∞ as the affine isomorphisms of $\mathbb{Z}$ (and g is then $x \to -x$ and h is $x \to 1 - x$).

Cappell's elements γ_k are defined on a two-dimensional quadratic form – generated by (e, f), with $\lambda(e, e) = \lambda(f, f) = 0$ and $\lambda(e, f) = 1$ (i.e. looking hyperbolic, i.e. trivial, from the λ point of view) and with $\mu(e) = g$ and $\mu(f) = t^k g t^{-k}$. Note that γ_k is essentially γ_1 pushed forward from a subgroup of index k. It is obvious that the augmentation, sending g and h to the trivial element, takes γ_k to the standard element of $L_2(\mathbb{Z})$ which has nontrivial Arf invariant. On passing to a large cover, one checks that the Arf invariant becomes trivial and we have a non-split example.

One can check that for different ks one gets different elements by examining the various transfers and augmentations and thus obtains Cappell's result that $L^2(\mathbb{Z}[D_\infty])$ contains an infinite $\bigoplus \mathbb{Z}_2$.

Now we know the full structure of this L-group[30] (and explicitly, not just as an abstract statement), namely:

$$L_0(\mathbb{Z}[D_\infty]) = \mathbb{Z}^3,$$
$$L_1(\mathbb{Z}[D_\infty]) = 0,$$
$$L_2(\mathbb{Z}[D_\infty]) = \bigoplus \mathbb{Z}_2,$$
$$L_3(\mathbb{Z}[D_\infty]) = \bigoplus \mathbb{Z}_2 \bigoplus \mathbb{Z}_4.$$

All unlabeled sums are infinite. They show that L-groups of nice small lattices can be infinitely generated. It is not shocking that they give rise to an infinitely generated group of counterexamples to the equivariant Borel conjecture – as we shall see in §6.5.

The results about homotopy invariance of connected sums for oriented manifolds of dimension 0 and 3 mod 4 for general fundamental groups is a consequence of first, the work of Cappell on the algebraic nature of the obstruction mentioned above, and second, some specific calculations that Connolly and

---

[30] Thanks to work of Banagl and Ranicki, and of Connolly and Davis.

Davis did – using the methods of Farrell and Jones[31] – so that the calculations done for the dihedral group end up sufficing for all groups. The first point is that the $\mathbb{Z}$-bimodule with involution basically only sees the number of elements of order 2 and the remaining number of elements. Unlike $L(\mathbb{Z}\pi)$ – that depends on the ring structure of π (and the involution) – UNil really looks at much less. It turns out that with cleverness one can reduce to the cases of $\mathbb{Z}_2 * \mathbb{Z}_2$ and, say, $\mathbb{Z}_2 * \mathbb{Z}_3$. The second case is algebraically tough[32] – however, the Farrell–Jones conjecture will reduce it to the former case,[33] and this group is a retract of the fundamental group of a two-dimensional hyperbolic orbifold, so it's a case that can be handled by the ideas of Farrell and Jones (1993a) and Bartels *et al.* (2014a,b).

6.5 The Simplest Nontrivial[34] Examples

Let us consider the simplest special case of the equivariant Borel conjecture, when M is a G-manifold with singular set of dimension 0.[35] In that case, the quotient is a non-manifold with just isolated singularities (corresponding to the nontrivial isotropy). This case doesn't require any controlled (or stratified) topology to analyze. As in §6.2, the key issues are all susceptible to analysis by means of proper topology and then one-point compactifying. We shall see that the $\mathrm{Wh}^{\mathrm{top}}$ theory is apt to have an especially simple form, because of the discreteness of the singular set, but that despite this there can be failures on very concrete manifolds, because of the nonvanishing of UNil.

Let W be the nonsingular part of M/G. We should then consider the proper topology of W. Its fundamental group Γ fits into an exact sequence $1 \to \pi \to \Gamma \to G \to 1$ where $\pi = \pi_1 M$.

The singular points correspond to subgroups of finite order in Γ. The isotropy group acts on the normal sphere to the fixed point, and then embeds a "space

---

[31] The more recent paper of Bartels and Lück (2012a) – which also is a further development of the Farrell–Jones ideas – includes enough examples to suffice for this purpose. It has the advantage of being published, while the work of Connolly and Davis has the advantage of only requiring ideas of negative curvature.

[32] The fundamental group contains a nonabelian free group of rank 2. However, the only virtually cyclic subgroups inside of it are dihedral – which, as we will later see, gives vanishing in the relevant dimensions.

[33] The basic example of this is the reduction of $\mathbb{Z} * \mathbb{Z}_2$ to $\mathbb{Z}_2 * \mathbb{Z}_2$. There are two conjugacy classes of maximal infinite dihedral groups each of which contributes its UNil elements to $L(\mathbb{Z} * \mathbb{Z}_2)$.

[34] By which we mean non-free – so that new complications arise that are not part of the ordinary Borel conjecture.

[35] This case is considered in detail in Connolly *et al.* (2015) which gives full justification of the somewhat heuristic descriptions given here. Recall that the singular set is the set of points where the isotropy group is nontrivial.

form" (i.e., manifold[36] quotient of the sphere) near an end of W corresponding to the deleted neighborhood of this point in M/G.

Indeed, a little reflection[37] shows that Γ acts on the universal cover of M, and that its singular set is discrete – all isotropy finite, injecting into G, and that by taking Γ orbits, the singular points are in a one-to-one correspondence with conjugacy classes of maximal finite subgroups of Γ. The key nontrivial observation here is that Smith theory (see, e.g., Bredon, 1972) guarantees that the fixed set of each element of prime (power) order (acting on the universal cover) is a single point, by our discreteness condition.

Moreover, these maximal finite subgroups are disjoint (except for the identity element), and (since M is compact) there can only be finitely many (conjugacy classes) of them. We shall call them $G_1, G_2, \ldots, G_k$ (or maybe use some other indexing set).

The Whitehead theory, according to Siebenmann (1970b), then fits into an exact sequence:

$$\bigoplus \mathrm{Wh}(G_i) \to \mathrm{Wh}(\Gamma) \to \mathrm{Wh}^p(W) \to \bigoplus K_0(G_i) \to K_0(\Gamma)m$$

which suggests – and one can correctly do so – extending the sequence both to the left and right and make Wh^p into a relative group. Indeed, for this special class of groups, i.e. where all elements of finite order lie in a unique conjugacy class of subgroup of finite order, the Farrell–Jones K-theory conjecture boils down to the statement[38] that:

(*) For such groups $\bigoplus \mathrm{Wh}(G_i) \to \mathrm{Wh}(\Gamma)$ and $\bigoplus K_0(G_i) \to K_0(\Gamma)$ are isomorphisms.

The reader might suspect (correctly) that the injectivity of these maps is (part of) a Novikov conjecture statement. In any case, conjecturally, these proper Whitehead groups vanish.

Needless to say, we don't expect

$$\bigoplus L(H_i) \to L(\Gamma)$$

to be an isomorphism; after all, that is not what happens in the torsion-free setting when there are no G_i!

---

[36] Because we are assuming the discreteness of the singular set.
[37] Using discreteness of singular sets!
[38] As was observed by Connolly, Davis, and Khan.

To simplify[39] the discussion,[40] let's ignore the issues of algebraic K-theory (i.e. how to decorate the L-groups), and use the $-\infty$ decoration here:

$$\bigoplus L_n^{-\infty}(G_i) \to L_n^{-\infty}(\Gamma) \to L_n^{-\infty}(W) \to \bigoplus L_{n-1}^{-\infty} \to L_{n-1}^{-\infty}(\Gamma).$$

After all, this is the sequence one would get for manifolds with boundary, and after crossing with a circle, we can put a boundary on these manifolds, and, furthermore, it will be essentially unique (certainly after crossing with another one!).

Henceforth we will just write L for this decorated version, which means that final results will have to take the change of decoration into account.

Let us now combine this with the surgery exact sequence:

$$\to S^p(W) \to H_n^{lf}(W; L(e)) \to L_n^p(W) \to .$$

Conjecturing that $S^p(W)$ vanishes (or, better, is contractible, if we spacify) would boil down to the statement that $H_n^{lf}(W; L(e)) \to L_n^p(W)$ is an isomorphism.

However, we would like to improve this since W is not an invariant of Γ, although its proper L-group is the cofiber of $\bigoplus L_{n-1}(G_i) \to L_{n-1}(\Gamma)$, which patently is. The idea is[41] to recognize that M with its G-action is the equivariantly canonical object that naturally arises, and on it there is a natural cosheaf of spectra which is $L(G_m)$, the L-group of the isotropy group at that point.

An analogous point is this: suppose we are interested in $S^p(M - A)$ where M is a compact manifold and A is a subspace,[42] then the normal invariants would *seem* to be the invariants of the hard-to-understand object $H_n^{lf}(M - A; L(e))$. However, thanks to excision, this group is isomorphic to the relative group $H_n(M, A; L(e))$. This latter group has much better functoriality. If the codimension of A is at least 3, then (modulo K-theoretic issues) $L_n^p(M - A) \cong L_n(\pi_1 M, \pi_1 A)$, also a group with much better functoriality.

---

[39] It is a little tricky trying to relate proper L-groups to more ordinary ones of groups (or rings). For a noncompact manifold with a simply connected end, the proper (h-)theory will be the reduced L^h-group of the interior. If W were $N \times \mathbb{R}$, then it would be $L^p(N)$ (with a shift, and here "p" means the algebra is based on projective modules, rather than on free modules). A key important case is the π–π case, where the proper L-group vanishes.

[40] Actually, the ideas of tangentiality that we discussed in §4.6 could allow us to put a boundary on, and work with, ordinary surgery of manifolds with boundary (recognizing the non-uniqueness of the boundary that we have put on). But we will not burden our discussion with this.

[41] This idea seems to have first been enunciated in algebraic K-theory by Quinn (1985b) in thinking about the K-groups of crystallographic groups. I was led to it by thinking about what an equivariant Novikov conjecture should say, and being inexorably led to the equivariant K-theory as the home of the equivariant signature operator – which also is essentially the same modification.

[42] Note that one does not need a submanifold.

As a consequence, although the space $M - A$ (importantly its homotopy type) depends on the exact embedding of A in M, $S^p(M - A)$ actually only depends[43] on the homotopy class of the inclusion map $A \to M$.

Back to our situation, we can write the homology group as $H_n(M/G; L(G_m))$. By doing this, rather than having a point with $L(e)$ as the relevant coefficient at the singularity, we put an $L(G_m)$ *there*, which replaces the $L(G_m)$ modification to $L(\Gamma)$ that takes place in the proper L-group. In short, the sequence becomes[44]

$$\to L_{n+1}(\Gamma) \to S^G(M) \to H_n(M/G; L(G_m)) \to L_n(\Gamma).$$

There are a number of ways of making this precise. On the face of it, the middle term doesn't quite make sense – G_m is actually a conjugacy class of subgroups, rather than a subgroup. From a stratified point of view, it should be viewed as the local fundamental group of the pure stratum near the point; in Davis and Lück (1998) a general theory is developed perfectly adapted to group action purposes, and essentially one uses the *einsatz*:

$$L(G_m) = L^G(G/G_m).$$

Note that G/G_m is a sensible thing to look at: it is the orbit corresponding to the given point in the orbit space.

The good news is that, with these modifications, one actually has a valid calculation (with the $-\infty$ decoration) for G-actions on M, aspherical actions such that there are no codimension ≤ 2 situations (note that the singular set below is the set of all points whose isotropy group is nontrivial):

$$\to L_{n+1}(\Gamma) \to S^G(M, \text{rel sing }) \to H_n(M/G; L(G_m)) \to L_n(\Gamma),$$

and

$$H_n(M/G; L(G_m)) \cong H_n^\Gamma(\underline{E\Gamma}; L(?)),$$

where $\underline{E\Gamma}$ denotes the universal space for proper Γ actions. More precisely, $\underline{E\Gamma}$ is a space that has a proper action of Γ and furthermore, given any X with proper Γ action, there is an equivariant map $X \to \underline{E\Gamma}$; moreover, this map is unique up to homotopy. (Note the similarity to $E\Gamma$ which has a similar characterization, except that only free Γ spaces are used.) We can characterize $\underline{E\Gamma}$ can be characterized as a proper Γ-space so that for all finite subgroups G of Γ the fixed set $\underline{E\Gamma}^G$ is contractible.

Once we reach this point, the formula for Wh^{top} in the general case has an entirely similar description. There is a very interesting point here – not visible

---

[43] This is indeed true, although, obviously, the above heuristic does not give a proof of this.

[44] We ignore here the tacit use of the fact that $S(*)$ is trivial.

when M is equivariantly aspherical, but of great use in trying to apply the equivariant h-cobordism theorem to more general G-manifolds.

The first point is that Wh^{top} (even relative to singularities) depends on more than the fundamental group of the top stratum. This is pretty clear: for a G-action on $M \times X$ for a free simply connected G-manifold X, the top stratum has the same fundamental group as that for M – indeed, it is all top stratum, since $M \times X$ has a free action, so its Whitehead group is $\text{Wh}(\Gamma)$. However, this point can be absorbed in the statement that this Whitehead group depends on the "equivariant fundamental group," which will then include the fact that, for all nontrivial subgroups of G, the fixed set is nonempty.

The second point is that it does not depend on more than this (i.e., the equivariant fundamental groupoid). This is truly remarkable and is a consequence of a theorem of Carter (and one of Hopf):

Let's think about an example – say M a manifold with a semi-free G-action with fixed set F. We have the sequence

$$H_0\big(\text{F}; \underline{\text{Wh}}(G)\big) \to \text{Wh}(\Gamma) \to \text{Wh}^{\text{top},G}(M\,\text{rel}\,\text{F}) \to H_0(\text{F}; \underline{K_0}(G)) \to K_0(\Gamma).$$

This is because the h-cobordism that we are considering on the top pure stratum is controlled over $\text{F} \times [0, 1]$ and controlled K-groups form a homology theory (see §§4.8 and 5.5.1).[45] To unpack this a little bit, the term $H_0\big(\text{F}; \underline{\text{Wh}}(G)\big)$ can be computed using an Atiyah–Hirzebruch spectral sequence whose E^2 term involves things like $H_0(\text{F}; \text{Wh}(G)), H_1(\text{F}; K_0(G)), H_2(\text{F}; K_{-1}(G))$ and so on.

The statement we made about the Whitehead group only depending on the equivariant "fundamental group" is therefore surprising because H_2 and higher all depend on F, not just its fundamental group. The reason that this statement is true is because of two facts:

(1) $K_{-i}(\mathbb{Z}G) = 0$ for $i > 1$ (Carter's vanishing theorem, Carter, 1980).
(2) $H_2(X) \to H_2(\pi_1(X))$ is surjective for all X (Hopf's theorem).

Naturality tells us that the part of $(\text{Wh}\Gamma)$ coming from $H_2(\text{F}; K_{-1}(\mathbb{Z}G))$ factors through $H_2\big(\pi_1(X); K_{-1}(\mathbb{Z}G)\big)$, but Hopf tells us that it actually always hits all of it.

Carter's theorem is the result of computation – there is no known purely conceptual explanation for this vanishing. Indeed, given current knowledge, one could conjecture[46] that its statement is true for all groups, not just finite ones.

---

[45] The infinite processes we can use to kill elements of the Whitehead group need to be controlled over F, which puts a condition on them (i.e. they are not arbitrary elements of $\pi_1(\text{F}) \times G$ – as we discussed in seeing that Nils enter).

[46] And this is a part of the Farrell–Jones conjecture that we will discuss later.

Hopf's theorem is quite simple: the Eilenberg–Mac Lane space $K(\pi_1(X), 1)$ can be obtained by attaching 3-cells and higher to X. (This argument was not actually available to Hopf, and, indeed, he needed to give a definition of $H_2(\pi_1(X))$ without having the concept of an Eilenberg–Mac Lane space!)

Now, let us turn to the construction of some counterexamples to the equivariant Borel conjecture as we have rephrased it: some equivariant simple homotopy equivalences to G acting on M which are not equivariantly homotopic to a homeomorphism. Indeed, the original examples of Cappell can be turned into such examples, albeit with nondiscrete singular set. Using the calculations of Connolly and Davis (2004), we can get similar examples with a discrete singular set, but for non-orientation-preserving actions.

Consider an affine involution on a torus with fixed points: such is necessarily of the form $\mathbb{T}_1 \times \mathbb{T}_2 \times \mathbb{T}_3$ where the first torus has a trivial action, the second is a product of the "complex conjugation action" (thinking of the circle as unit complex numbers) some number of times, and the final torus is the interchange of pairs of factors. We just use this to set notation.

Let Γ denote the group $\pi_1 \rtimes \mathbb{Z}$. This is the group that acts on the universal cover $\mathbb{R}^d$. Note that Γ always contains a subgroup isomorphic to $\mathbb{Z}_2 * \mathbb{Z}_2$. (Frequently it contains a subgroup like this that splits off, in which case, more transparently, $L(\mathbb{Z}_2 * \mathbb{Z}_2)$ is a split summand of $L(\Gamma)$.) This will be the source of nonrigidity.

Suppose that we are in the situation of an action of type $\mathbb{T}_2$, i.e. with isolated fixed points. In that case, indeed $\mathbb{Z}_2 * \mathbb{Z}_2$ is a split summand of Γ. (Here the $\mathbb{Z}_2$s must be given the orientation character of the action of the involution on M.)

We can act on the proper structures $S^p((M^d - F/\mathbb{Z}_2))$ by any element of $L_{d+1}(\Gamma)$. We shall use the nontrivial elements of $L_{d+1}(\mathbb{Z}_2 * \mathbb{Z}_2)$ coming from UNil. Cappell's elements live in $L_2(\mathbb{Z}_2 * \mathbb{Z}_2)$ with an orientation-preserving action, so it would be necessary to cross with a circle to act by these, but the elements constructed in Connolly and Davis (2004) can be used even in the isolated fixed-point situation.

We claim that the result of such an action produces a new proper structure, and indeed a new equivariantly homotopy-equivalent action. The proof of this is a straightforward application of functoriality:[47] the point being that these elements of $L_{d+1}(\mathbb{Z}_2 * \mathbb{Z}_2)$ survive the map $L_{d+1}(\mathbb{Z}_2 * \mathbb{Z}_2) \to L_{d+1}(\mathbb{Z}_2 * \mathbb{Z}_2, \mathbb{Z}_2 \sqcup \mathbb{Z}_2)$.

Actually, we can produce similar actions on hyperbolic manifolds. In order to do so, we note a key trick for showing that UNil groups split the L-groups, a trick that is similar to other transfer devices used in Chapter 4. Suppose, for concreteness, we have an involution on a hyperbolic manifold M. Suppose that γ is an invariant geodesic for the involution. Then there is a cover of M

---

[47] The original proof of this was by a counting argument, and just showed that this proper structure set was an infinitely generated group, but did not control individual elements.

associated to this subgroup of the fundamental group, and the involution lifts to that cover.

Note that there is a normal exponential isomorphism $\mathrm{Exp} \colon N\gamma \to \mathbb{H}/Z$, where $\mathbb{H}$ is the hyperbolic space, and Z is the group acting by translation associated to the geodesic γ. The normal bundle $N\gamma$ can be split as a product of trivial Euclidean bundles according to the eigenspace decomposition of the involution. The inverse of this map is Lipschitz (because of nonpositive curvature) and produces a map $S^{G,\mathrm{Bdd}}(\mathbb{H}/Z) \to S^{G,\mathrm{Bdd}}(\mathbb{R}^d/Z)$. We can split off the trivial summand, and then get a map map $\to S^{G,\mathrm{Bdd}}(\mathbb{R}^k/Z)$. where the action on the $\mathbb{R}^k$-direction is antipodal. This last structure set can easily be computed: at the infinity from the $\mathbb{R}^k$-direction, we have arbitrary control (by rescaling, since we are now in Euclidean space), and the action is free. The two fixed points can be deleted, at the cost of allowing proper control in those directions, but that mods out by $L_{d+1}(\mathbb{Z}_2 \coprod \mathbb{Z}_2)$. In any case, the UNil elements do survive.

The methods of constructing hyperbolic manifolds using quadratic forms explained in Chapter 2 give an ample supply of involutions to which to apply this construction.

With more effort, we are even led to speculate (and this is a theorem modulo the Farrell–Jones conjecture that we will get to later) that, in this case, the equivariant structure set is a sum of contributions associated to invariant unions of closed geodesics. (Free unions, though, contribute nothing, as indeed do ones where only odd-order isotropy arises. Indeed, a bit of thought reduces to geodesics that are invariant under some nontrivial involution.)

6.6 Generalities about Stratified Spaces

In §6.5 we dealt mainly with the situation of an isolated singular set, so that the quotient spaces of these group actions could be thought of as noncompact manifolds with some ends compactified by gluing in points. What happens when the singular set is higher-dimensional? One approach is via considering the quotients as *stratified spaces*.

For the purposes of the rest of this chapter, we will see that there is a reasonable classification theory for stratified spaces with respect to *stratified homeomorphism*, i.e. within a (simple) stratified homotopy type. Unfortunately, this rarely[48] will coincide with what one is interested in, in the situation of

---

[48] Or, fortunately, this occasionally will coincide with what we are interested in.

equivariant homotopy types. Consequently, our later sections will deal with the implications of the tension between "stratified" and "equivariant."

A *stratified space* X is a space with a filtration $X = X_n \supseteq X_{n-1} \supseteq \cdots \supseteq X_0$ by closed subsets (called *strata, stratum* in the singular). We shall always assume that, for each i, the *pure stratum* $X^i = X_i - X_{i-1}$ is an i-dimensional (ANR homology) manifold. Beyond this, there are various theories about how the strata are demanded to fit together. We shall use the notion of homotopically stratified spaces, introduced by Quinn (1988) (see also Hughes, 1996). The precise general definition need not bother us here – instead, we shall give some examples that are and some that are not.

Example 6.26 The one-point compactification, M^+, of a noncompact manifold M, is sometimes a stratified space for us, and sometimes not. The obvious stratification consists of a bottom zero-dimensional stratum consisting of the added point, and the remaining points form the top pure stratum.

The usual strong stratifications (such as Whitney stratifications) would require the existence of a compactification of M to a manifold with boundary: M^+ could then be viewed as the result of shrinking this boundary to a point, or gluing on the cone of the boundary to this compactification.

Unfortunately, this manifold with boundary is *not* a topological invariant of this situation (even when it exists). There is an indeterminacy associated to $\mathrm{Wh}(\pi_1 \partial)$.

Our assumption is that M is a *tame* in the sense of Siebenmann (1965). This is equivalent to the condition that $M \times S^{-1}$ has a compactification as a manifold with boundary. Essentially this condition means that complements of sufficiently large compact sets can be "pulled" closer from infinity. We refer to Siebenmann, and the predecessor work of Browder and Livesay (1973) for more information, and, in particular, how to recognize this.

However, many noncompact manifolds are not allowed (to be the nonsingular part of a compact stratified space): for example, a typical infinite cover of a compact manifold (such as infinite abelian covers of a surface of genus greater than 1), or any manifold with infinitely generated fundamental group or homology.

Example 6.27 A manifold with a *nice* submanifold (W, M) can be viewed as a two-strata space, with bottom stratum M, and the ambient manifold being the top stratum. The condition of tameness follows from the condition that M is *locally flat* in W, i.e. that each point in M has a neighborhood in W which is isomorphic to $(\mathbb{R}^w, \mathbb{R}^m)$; this does not guarantee that M has a topological bundle neighborhood.

There are some other submanifolds that are not locally flat, but still give us homotopically stratified spaces – they all have a nice homogeneity property. The basic source is the Cannon–Edwards theorem that the second suspension of any homology sphere Σ is a topological sphere[49] (see Daverman, 2007). As a result, if $W = M \rightarrow c\Sigma$ (or, more generally, the mapping cylinder of any Σ (block) bundle over M), one obtains a topological manifold in which M is embedded in a quite nontrivial, yet homogeneous, way. This is a rather exotic embedding from the conventional point of view, yet it gives a reasonable homotopically stratified space.

Example 6.28 If G is a finite group acting on M, a sufficient condition for M/G (stratified by orbit types) to be homotopically stratified is that, for $H \subset K$, the embedding of $M^K \subset M^H$ should be a locally flat embedding of manifolds. For this situation, the homogeneity property is quite remarkable: it includes some of the one-point compactification examples mentioned above! As in Example 6.26, there does not have to be a closed invariant "regular neighborhood" of the fixed set. The following result gives an example of how such actions occur naturally in converses to Smith theory.

Theorem 6.29 (See Weinberger, 1985a) *A submanifold Σ of the sphere is the fixed set of a (locally linear[50]) Q_8-action iff Σ has codimension a multiple of 4 and is a $\mathbb{Z}_2$-homology sphere. The top stratum of the quotient can be compactified as a manifold with boundary iff*

$$\prod \#\mathrm{tor}(H_i(\Sigma; \mathbb{Z})) \equiv \pm 1 \bmod 8.$$

In this case, there is always a PL locally linear action.

The group Q_8 can be replaced by any other group that can act freely on the sphere, but then the conclusions have to be modified. (The simplest modification is that for cosmetic reasons we wrote down a product of numbers that really should be an alternating product.) This is the simplest case where there is a nontrivial restriction on the homology of the fixed set that follows from algebraic K-theory. The actions can always be made PL locally linear in the complement of a point – indeed, that is a natural feature of their construction – the numerical obstruction is a Wall finiteness obstruction that doesn't arise in the noncompact setting. The local linearity at ∞ is a remarkable consequence of general features of homotopically stratified spaces (see Quinn, 1970).

---

[49] The deep part is that it's a manifold at all. Identifying the manifold with a sphere then follows from the Poincaré conjecture.

[50] If you wish.

Example 6.30 (Supernormal spaces) These are spaces modeled by a strengthening of the condition of normality that occurs in algebraic geometry. We mention them because their theory is more elementary than the general situation, but is quite beautiful and is a good place to start.

A stratified space X is *supernormal* if each *pure* stratum is dense in the corresponding closed stratum,[51] and near each point x of X_k and any $\varepsilon > 0$, there is a $\delta > 0$ such that any 1-manifold[52] in X^r for $r > k$, within δ of x, is null-homotopic within a ball of radius ε.

Note that any manifold with boundary can be thought of as a supernormal stratified space with two strata.

For an embedding of closed manifolds (W, M), supernormality is (by a nontrivial theorem; see Daverman and Venema, 2009) exactly the condition that one is in codimension greater than 2 and the submanifold is locally flat. If M is a subpolyhedron, supernormality (from the point of view of the top stratum) would follow if the Hausdorff codimension is greater than 2.

The Whitehead group for supernormal spaces[53] is just $\bigoplus \mathrm{Wh}(\pi_1(X_i))$ (i.e. just like the PL situation).

The surgery theory describes the *structure sets* (which are actually groups) $S(X) = \{(X', f)\colon f\colon X' \to X$ is a stratified homotopy equivalence up to stratified s-cobordism$\}$.[54] If Y is a union of strata of X, then we can also form $S(X \operatorname{rel} Y)$ which is defined in the same way, but we insist that $f|_{Y'}$ is already a homeomorphism.

A *stratified map* $f\colon X' \to X$ is a map that preserves *pure* strata, i.e. so that $f(X'^r) \subset X^r$. A *stratified homotopy equivalence* is a stratified map $f\colon X' \to X$ for which there is a stratified "inverse" $g\colon X \to X'$ such that the composites fg and gf are both stratified homotopic to the identity.

Theorem 6.31 (Cappell and Weinberger, 1991a) *If X is supernormal of dimension $n > 4$, with Σ its singularity set (i.e., the complement of the top pure stratum), then $S(X \operatorname{rel} \Sigma) \cong S^{\mathrm{alg}}(X)$, where $S^{\mathrm{alg}}(X)$ denotes the fiber of the assembly map – i.e. what surgery would predict had X been a closed manifold:*

$$\to L_{n+1}(\pi_1 X) \to S^{\mathrm{alg}}(X) \to H_n(X; L) \to L_n(X).$$

---

[51] This is just a convenience to ensure that our picture of the singularity set to correspond to the largest proper closed stratum.

[52] We cannot just use loops because we want to require normality in this definition. Normality is essentially the same condition with S^0s replacing the 1-manifolds in this definition.

[53] Ignoring low-dimensional difficulties.

[54] The equivalence relation can be taken to be homeomorphism if we only allow manifolds as strata. However, for our calculation to be correct as stated, it is necessary to allow ANR $\mathbb{Z}$-homology manifolds as strata, and then an s-cobordism theorem is not available. This is related to the discussion of functoriality in §4.7.

So the formal structure set of algebraic surgery theory has an interpretation even for (certain) non-manifolds (without using the artifice of thickening the space to be a manifold). In particular, the Borel conjecture for $\pi_1(X)$ then implies (and clearly is implied by!) the following:

Conjecture 6.32 *If X is an aspherical supernormal space, then any stratified space stratified-homotopy-equivalent to X rel Σ is homeomorphic to it.*

This is equivalent (as surely one would guess) to the statement that, assuming asphericity, $S(X) \rightarrow S(\Sigma)$ is an isomorphism. In this view, aspherical manifolds are topologically rigid because they have no singularities!

Remark 6.33 For the conclusions about the rel Σ theory, we only need the simple connectivity condition occurring in the definition for the parts of the top pure stratum X^n near x (i.e. not on the intermediate strata). This follows from "continuously controlled at infinity surgery" (Pedersen, 2000).[55]

The absolute theory is necessarily more complicated. For example, if M is a manifold with boundary ∂M, we can view it as a 2-stratum supernormal stratified space X. In the above notation, $S(X)$ then corresponds to the group $S(M, \partial M)$ (while $S(X \operatorname{rel}, \Sigma)$ is $S(M)$). The strata interact[56].

In Example 6.27, there is a "forgetful" map $S(W, M) \rightarrow S(W)$. *This is far from trivial.* (On the other hand, the forgetful – or, better, restriction – map $S(W, M) \rightarrow S(W)$ is tautologous.) The reason for this is that the closed stratum of a manifold that is a homotopically stratified space that is stratified homotopy-equivalent to (W, M) is actually a manifold, as we now explain.

It is quite easy to see that W' is an ANR homology manifold and that it has the disjoint disks property (DDP) (see §4.7 for this and the rest of this paragraph).[57] That it is a manifold if the top pure stratum is a manifold requires the theorem of Quinn that gives it a resolution, and then Edwards's theorem that DDP is then sufficient for manifoldness.

Now we can combine the maps $S(W, M) \rightarrow S(W) \times S(W)$ and the theorem above directly implies that this is an isomorphism if the codimension of M in W is at least 3.

Wonderful: we've calculated something. But what does it mean?

---

[55] This is not very hard, and the reader might want to try their hand at verifying this.

[56] Unlike the situation in K-theory, where the stratified object decomposed into pieces. That L-theory works differently could have been predicted by the fact that the involution given by turning h-cobordisms upside down does not preserve this decomposition. But, this is obvious, anyway, as above.

[57] So, if the reader had been content to accept that we could define $S(W)$ using homology manifolds, and that this only differs by some $\mathbb{Z}$s from the one defined using topological manifolds, then the forgetful map was easy to define!

It certainly includes the statement that if $M \subset W$ is a locally flat submanifold of codimension at least 3, then any manifold M' homotopy-equivalent to M embeds in any manifold W' homotopy-equivalent to W.[58] Moreover, this embedding "has the same homotopy theory" as that of $M \subset W$. For example, $M' \subset W'$ is homotopy-equivalent to $M \subset W$, as a pair, and $W' - M'$ is homotopy-equivalent to $W - M$.

The first observation is a completely natural (but perhaps surprising) statement to someone studying embedding theory, but the second one, while strengthening our conclusion (the embedding we produce of $M \subset W$ has even more properties than we might have asked for), is not particularly a natural one to someone who studies embeddings for a living.

For example, consider the two (homotopic) embeddings of $S^{k-1} \vee S^{k-1}$ in S^{2k-1}, according to whether the S^{k-1} are linked as in the Hopf link, or just embedded in two disjoint disks: the first has complement homotopy-equivalent to $S^k \times S^k$; and the other, $S^k \vee S^k \vee S^{2k}$, is completely different.[59]

We also note that there is also a third condition that comes out of isovariant homotopy equivalence regarding the normal bundles[60] of the submanifolds. More precisely, associated to a codimension-c locally flat[61] embedding there is a spherical fibration $S^{c-1} \to E \to M$. These spherical fibrations must match for M and M'. The proof of this goes by comparing the neighborhood systems near M and M' that are mapped to each other by f and g.

These three conditions serve to define the notion of a *Poincaré embedding*. A Poincaré embedding of M in W consists of a triple $((X, E), \pi, f)$, where (X, E) is a pair, $\pi \colon E \to M$ is a spherical fibration with fiber S^{c-1} and, denoting the mapping cylinder of a map by Cyl, $f \colon X \cup \mathrm{Cyl}(\pi) \to W$ is a homotopy equivalence. The stratified map $(W, M) \to (W', M')$ gives us an isomorphism of the underlying Poincaré embeddings, and the theorem that $S(W, M) \to S(W) \times S(M)$ is an isomorphism says that there is a unique isotopy class of embedding of M in W associated to any Poincaré embedding.

In general, we have to be careful in thinking through what we get out of a

---

[58] Actually, we should use simple homotopy equivalence. However, there are straightforward arguments that allow us to deduce the homotopy equivalence result from the above.

[59] The complements do have the same stable homotopy types, but this does not suffice for the application of (stratified) surgery.

[60] We are abusing terminology here, since locally flat submanifolds don't necessarily have bundle neighborhoods.

[61] With a little effort, this spherical fibration can be associated to non-locally flat embeddings. Combining this observation with the result about the stratified structure set quickly gives a proof (in codimension greater than 2) that topological embeddings can always be approximated by locally flat ones: see Daverman and Venema (2009), for a thorough discussion of such results. (In codimension 2, this is not true according to examples of Matumoto; the codimension-1 result is true, but would involve a little more work to deduce from these methods, since the embedding problem does not reduce to homotopy theory.)

calculation of $S(X)$ – it gives some useful information, but frequently not all the information we want: for instance, it won't classify for us the embeddings in a given homotopy class.

A full discussion of how to calculate $S(X)$ for a stratified X is outside the scope of our current exposition, but, *ignoring algebraic K-theory issues*, we can give a quick summary.

(1) There are spectra $\mathbf{L}(X)$ associated to a stratified space X; one has that $\pi_i \mathbf{L}(X) = \mathbf{L}(X \times D^i \text{relative to the boundary})$.[62] If Y is a union of closed strata of X, there is a restriction map $\mathbf{L}(X) \to \mathbf{L}(Y)$, whose fiber is $\mathbf{L}(X \operatorname{rel} Y)$. If $X \subset Z$ is an open inclusion then there is an induced map of $\mathbf{L}(X \operatorname{rel} \infty) \to \mathbf{L}(Z)$.

(2) $\mathbf{L}(X_n \operatorname{rel} X_{n-1}) \cong \mathbf{L}_n\big(\pi_1(X^n)\big)$.

(3) There is an exact sequence $\cdots \to S(X \operatorname{rel} Y) \to H(X; \mathbf{L}(\operatorname{loc}(X \operatorname{rel} Y))$, where H is the spectral cosheaf homology of the cosheaf that associates to a small open set U in X, the $\mathbf{L}$-space of $(U, U \cap Y \operatorname{rel} \infty)$.

Items (1) and (2) together say that $\mathbf{L}(X \operatorname{rel} Y)$ is built up out of the $\mathbf{L}$-spectra of the fundamental groups of the pure strata of X that are not in Y. They do not say exactly how they fit together – this is the issue of interaction mentioned above – and I will ignore it here, although *for our situation of discrete group actions*,[63] it turns out that there is no interaction after inverting 2 (see Chapter 13 of Weinberger (1994)).

Note the special case where (X, Y) is a supernormal pair, with Y the singularity set of X. In that case, all of the $\mathbf{L}(\operatorname{loc})$ are just $\mathbf{L}(R^x \operatorname{rel} \infty)$ (induced locally by the inclusion of a neighborhood of any manifold point in the neighborhood). In that case, the cosheaf homology is essentially the ordinary[64] $H_x(X; \mathbf{L})$ that arises in surgery theory. The global $\mathbf{L}$ term $\mathbf{L}(X \operatorname{rel} Y)$ is just the L-group of the top pure stratum, which has fundamental group $\pi_1(X)$ by Van Kampen's theorem. This explains (aside from K-theory[65]) Example 6.30.

If X is a manifold with boundary, then the homology term has the usual spectrum at the interior points, but is contractible (by the π–π theorem for the trivial group π) at the boundary points. If we work relative to the boundary,

---

[62] The L-groups of stratified spaces are sometimes written $L^{BQ}(X)$ in recognition of the paper Browder and Quinn (1973) that initiated their study.

[63] Acting preserving orientations.

[64] There is a twist in this group when the top pure stratum of X is nonorientable. The reason that these cosheaf homology groups become more conventional (generalized) homology groups is because of the great rigidity that $\mathbf{L}$ cosheaves have, referred to as "flattening" in Weinberger (1994).

[65] In this case $\operatorname{Wh}^{\operatorname{top}}(X \operatorname{rel} \Sigma) = \operatorname{Wh}(\pi_1 X)$, so the K-theory agrees exactly with the manifold situation.

then the spectral term will be $\mathbf{L}(\mathbb{R}^x)$ everywhere, and the global term will be the ordinary relative L-group. Both of these calculations are completely in accord with the classical calculations.

Finally, let us consider Example 6.26. We suppose that $X = M^+$. As we are ignoring algebraic K-theory, we can safely view M as interior$(W, \partial W)$. In that case, $S(X) \cong S(W, \partial W)$. As the structures of a point are contractible (i.e. they form a trivial group), for ease of exposition, we will compute $S(X \operatorname{rel} *)$ where $*$ is the compactification point. In that case the global spectral term is $L(\pi_1 M)$ (rather than $L(\pi_1 M, \pi_1 \partial W)$. However, the homology term is different – noting that at the cone point the cosheaf is $L(\pi_1 \partial W)$, it fits into the exact sequence

$$H_x(X, L(\operatorname{loc}(X \operatorname{rel} *))) \to H_x(X - *, L(\operatorname{loc}(X \operatorname{rel} *)))$$
$$\cong H_x(W, \partial W, L(e)) \to L_{x-1}(\pi_1 \partial W).$$

So the homology term absorbs the difference between the absolute and relative global L-groups. *In a stratified space, there is little difference between local and global problems*, i.e. what is local from the point of view of a k-stratum space is global from the point of view of $(k - 1)$-strata spaces.

The reader can gain some insight into this surgery sequence by thinking about the PL situation where the strata have regular neighborhoods, and the boundaries block fiber over the previous strata, and then use the theory of blocked surgery to give directly a proof of the "Verdier dual" exact sequence.

Let us now return to the situation of Γ acting properly discontinuously on X a contractible manifold, with all fixed sets contractible locally flat submanifolds, of codimension greater than 2. In that case:

(1) $\operatorname{Wh}^{\text{top}}(X/\Gamma, \operatorname{rel} \operatorname{sing})$ is the fiber of the assembly map $H(X/\Gamma; K(\Gamma_x)) \to K(\Gamma)$.

This group is thus related to the Nil terms and can vanish but can certainly be nontrivial. This would give one set of obstructions to the equivariant Borel conjecture, had we not already made the assumption that our maps are equivariantly-simple-homotopy equivalences.

Realizing elements of this group, one obtains counterexamples if the boundaries of the appropriate h-cobordism are not homeomorphic; if they are, then one can glue them together and get a counterexample for the group $\mathbb{Z} \times \Gamma$ (i.e. for $S^1 \times X/\Gamma$).[66]

---

[66] One can actually see that, for $\mathbb{Z}/p^2$ acting on $S^1 \times \mathbb{T}^{p(p-1)}$ (here the action is the one associated to the action of $\mathbb{Z}/p^2$ on the torus associated to the ring of integers in the cyclotomic field of p^2 roots of unity), there are a number of Nil terms in the topological Whitehead group, and that, when one realizes an h-cobordism with suitable torsion, the "other end" is not topologically simple homotopy-equivalent to the original manifold. This actually gives an infinite number of examples (for each p).

Now, assume that this vanishes:[67]

(2) $S(X/\Gamma \mathrm{rel\,sing})$ is isomorphic to the fiber of the assembly map

$$H(X/\Gamma; L(\Gamma_x)) \to L(\Gamma).$$

We have seen that this can be nontrivial because of UNil. On the other hand, if, instead of $\mathbb{Z}$ we deal with rings, R, in which 2 is inverted, there are no known counterexamples to an isomorphism statement[68] – this is not so useful directly for classification problems, but it is useful for understanding invariants of manifolds (and group actions).

Notation 6.34 We will usually denote the singularity set of a stratified space by Σ, unless otherwise stated.

In §6.7 we will discuss a form of equivariant Novikov conjecture and some evidence for it. This conjecture does not take into account the Nil and UNil phenomena that get in the way of rigidity as we have seen. It is interesting that these always seem to split off structure sets.

We will follow this with a discussion of the Farrell–Jones conjecture, which gives a specific statement about how all of the Nil and UNil contributions to $\mathrm{Wh}^{\mathrm{top}}$ and S can be explicitly computed in terms of the virtually cyclic subgroups of Γ. This will suffice to give an understanding (at least in theory) of what *isovariant* structures should look like (since vanishing is not always true). Finally, we will return to the equivariant Borel conjecture, and discuss the relation between the difference between equivariance and isovariance and embedding theory.

6.7 The Equivariant Novikov Conjecture

The Novikov conjecture describes the restrictions on the characteristic classes of the tangent bundles of homotopy equivalent manifolds. While it can be phrased as the injectivity of an assembly map, it has other interpretations and implications and analogues, as we saw in Chapter 5. Already in that chapter we discussed some equivariant aspects of the Novikov philosophy, e.g. vanishing of higher A genera for smooth actions of S^1 on the one hand and the higher-signature local formulas for homologically trivial actions on the other.

---

[67] Note that in the usual Borel conjecture we assume homotopy equivalence, not simple, and we deduce a vanishing of the Whitehead group from the conjecture.

[68] And, indeed, the Farrell–Jones conjecture implies that it is an isomorphism, with $L^{-\infty}$ as the version of L-theory used.

In this section, we will take seriously the issue of how to properly generalize the Novikov conjecture equivariantly. There are several possibilities that interact with each other: wisely did the authors[69] of "An equivariant Novikov conjecture" title their paper.

As in the non-equivariant case, we should be concerned with the issue of tangentiality of (equivariant) homotopy equivalences, and also with the restrictions that can be made on characteristic classes of tangent "bundles" of G-manifolds, as well as assembly maps in L-theory and C^*-algebra theory.

Rather similarly to the classical case, in the situation of equivariant homotopy equivalent compact G-manifolds, the equivariant Novikov conjectures give very similar information away from the prime 2.

We start by noting some of the obstacles to proceeding as we had before:

(1) We needed to put scare quotes around the word "bundles" as we laid out our path two paragraphs ago: in §6.2 we had seen that in the topological setting the tangent theory is not so naturally bundle-theoretic.

(2) Indeed, the local structure of a manifold is simply its dimension; one would want to generalize this to be something like a tangential representation (of the isotropy group) associated to each point. However, we have to confront the phenomenon discovered by Cappell and Shaneson (1981) that for many groups there are (linearly) different representations V and W that are equivariantly homeomorphic. So we don't have the analogue of dimension even if we had bundle theory!

(3) We also have to worry about what characteristic classes we can hope to use: in the topological case, Novikov's theorem on topological invariance of rational Pontrjagin classes led us to use the L-class. What shall we use here?

(4) As a final point to put us in the mood: in §6.6, when we were thinking about the equivariant Borel conjecture, in order to use ordinary manifold surgery techniques, we were led to consider the problem inductively, i.e. assuming that we already had a homeomorphism on the singular set (e.g. on fixed sets of all proper subgroups; abbreviated rel Σ). This led to an assembly map formulation involving

$$H_n(\underline{E\Gamma}/\Gamma; L(\Gamma_x)) \to L_n(\Gamma).$$

(5) However, we should be interested in formulations that are not rel Σ and also in restrictions on characteristic classes, etc., that are *a priori* not rel Σ, i.e. do not require precise analysis of what is occurring on fixed point

---

[69] Rosenberg and Weinberger (1990); correspondingly the title of the present section has an unwarranted definite article.

sets: the many examples we've already seen show that such information is
frequently difficult to get.

In achieving all of the above, we will also understand things like the higher-
signature localization formula of Chapter 5 as part of this picture. Applications
to closed (homology) manifolds will be given in Chapter 7.

Deviating from the way of the wise,[70] we will begin by dealing with the
last question first, starting with the simplest situation, the simply connected
case. After all, the key class that is relevant for the ordinary Novikov conjecture
is the L-class[71] of the manifold M that lies in homology (or $H_m(M; L^*(\mathbb{Z}))$)
which itself is a variation of the ordinary signature of the manifold – i.e. it's an
encoding of all the signatures of all of the submanifolds of the manifold (taking
into account their normal bundles).[72]

The equivariant version is, of course, the G-signature that we first met in
§4.10 and again in Chapter 5 when we studied homologically trivial actions. We
review some basic aspects of this invariant now[73] as a step towards considering
the topological characteristic class theory.

If G acts orientation-preservingly on X^{2k}, an even-dimensional oriented
Poincaré complex,[74] the middle cohomology $H^k(X; \mathbb{Q})$ admits a G-invariant
$(-)^k$-symmetric inner product pairing.[75] Let's go even further, and consider
the situation after extending scalars to $\mathbb{R}$. As such it has some signature-type
invariants that occur in the representation theory of G. If k is even, then one
chooses a G-invariant positive-definite inner product on H^k and diagonalizes
the cup product pairing in terms of this auxiliary pairing. The G-action preserves
both, and therefore preserves the positive- and negative-definitive parts, giving
an invariant in $RO(G)$. If k is odd, then one does the same, except that the
operator A describing cup product in terms of the auxiliary product is now

---

[70] See Avot 5:9 (*Ethics of the Fathers*, one of the books of the Talmud) regarding the wise
approach to answering a series of questions.

[71] Or, better, the $L^*(\mathbb{Z})$ orientation of a manifold. This is an intrinsic class in $H_m(M; L^*(\mathbb{Z}))$ that
defines the Poincaré duality between $H_m(M; L^*(\mathbb{Z}))$ and $[M : \mathbb{Z} \times G/\text{Top}]$ and refines the
L-class. Sullivan emphasized that the PL-block bundle away from 2 is a KO[1/2]-oriented
spherical fibration, which, away from 2, is this class.

[72] Or all the definable signature-type invariants of all the open subsets of M, in the case of the
controlled symmetric signature of M over M.

[73] See Wall (1968) and Atiyah and Singer (1968a,b, 1971) for early references, from different
points of view.

[74] Perhaps with boundary; in that case the relevant quadratic form will be singular, and one must
mod out by the null vectors, $\ker H^* \to (H^*)^*$, before following the prescription above.

[75] We use rational coefficients, which greatly simplifies our remarks throughout the section,
losing information only at the prime 2, thanks to Ranicki's (1979a) localization theory.
However, the theory at the prime 2 is indeed much deeper, and much more mysterious, as we
shall occasionally indicate.

skew-adjoint. A positive square root of AA^* gives a canonical complex structure and a representation ρ of G. The G-signature in this case is $\rho - \rho^*$.

One can view this more illuminatingly perhaps by taking the Wedderburn decomposition of the real group ring RG and considering the effect of the anti-involution $g \to g^{-1}$. Up to Morita equivalence one has pieces corresponding to $\mathbb{R}$, $\mathbb{C}$, and $\mathbf{H}$. Whereas $\mathbb{C}$ contributes for both k odd and even, $\mathbb{R}$ arises only for k even (every symplectic real vector space, the k-odd situation has a self-annihilating subspace of half the dimension), and the $\mathbf{H}$ case arises only for k odd (for similar reasons).

It is a nice fact that the maps[76] $L(ZG) \to L(QG) \to L(RG)$ are also isomorphisms away from the prime 2 (see, for example, Wall, 1968) and computed via these combinations of integer-valued invariants. This means that, for finite groups, *away from* 2, surgery obstructions can be computed as the difference of very simple-minded intrinsic invariants of domain and range (i.e. their G-signatures).

The fact that G-signatures are computed from the action of G on cohomology implies the following strong homotopy-invariance property:

Proposition 6.35 *If $f : M \to N$ is an equivariant map that is a homotopy equivalence, then $G - \mathrm{sign}\,(M) = G - \mathrm{sign}\,(N)$.*

A map as in the proposition is called, following Petrie (1978), a *pseudo-equivalence*. It is equivalent to asserting that $f \times \mathrm{id} : M \times EG \to N \times EG$ is an equivariant homotopy equivalence. It obviously makes more sense to use the equivalence relation generated by this notion. But in that case, it is exactly equivalent to the homotopy equivalences in the following *pseudo-category*.[77]

Motivation 6.36 Let $G = \mathbb{Z}_2$. A map between G-spaces $f : X \to Y$ might fail to be equivariant, i.e. $f(\mathrm{i}x \not\simeq f(x)$. If these two maps are not homotopic, then we have no chance of getting (say, up to homotopy) an equivariant map, but if they are homotopic, we are still not done. For example, let F be homotopy between f and $f(\mathrm{i})$. Then $F \circ \mathrm{i}$ is also such a homotopy, and we need F to be homotopic ($\mathrm{rel}X \times \{0, 1\}$) to $F \circ \mathrm{i}$. And then that homotopy G must be homotopic to $G \circ \mathrm{i}$, and so on. All of this still won't make you succeed, and you've traded a simple condition of equivariance for an infinite number of homotopies and higher homotopies, and the down-to-earth reader will surely want some justification of this ... Hopefully the pages that follow will provide some. In any case, this

---

[76] We can go further and add one more isomorphism to the real K-theory of $\mathbb{C}_\mathbb{R} * G$.

[77] The pseudo-category is a category: we use the perjorative "pseudo" to describe the morphisms that are *prima facie* odd. (Of course, mathematics often progresses through non-naive definitions and problems. As Gromov once said, "Naive problems are usually stupid.")

data is just an equivariant map from $S^\infty \times X \to Y$. The point being that blowing X up in this way makes it slightly easier to build maps (at least in theory).

Definition 6.37 The *pseudo-category* of G consists of G-spaces as objects (just like the usual equivariant category) but it has more morphisms. A morphism from X to Y consists of a G-equivariant map $EG \times X \to Y$. (This can be thought of as an EG-parameterized family of maps from X to Y that satisfy certain intertwining conditions with respect to the G-action and the parameter.)

A morphism in this category is thus an element of the *homotopy fixed set* $\text{Map}[X : Y]^{hG}$ (for the reader who remembers this notion from §4.9); this should be compared to the usual equivariant category where morphisms are elements of the *usual fixed set* $\text{Map}[X : Y]^G$.

One reason that this category is important is because pseudo-equivalences arise frequently. For example, any G-action on a contractible space is pseudo-equivalent to the action of G on a point (clearly!), but the fixed sets of such G-actions can be quite different for nontrivial subgroups of G, and thus these actions would not be equivariantly homotopy-equivalent.

If G acts on a space X then there is an equivariant fundamental group associated to the action:

Proposition 6.38 (Definition) *If G acts on a space X then the equivariant fundamental group associated to the action is given by the group of all lifts of the elements of G to the universal cover of X. This group, Π, fits into an exact sequence $1 \to \pi_1 X \to \Pi \to G \to 1$. It is an invariant of the pseudo-equivalence class of the group action on X (if G is discrete, it is the fundamental group of the Borel construction $X \times_G EG = (X \times EG)/G$).*

Proposition 6.39 *If G is a finite group and acts on an M-manifold, then we can define an invariant $\sigma_G^*(M) \in L^*(\mathbb{Q}\Pi)$. It is a pseudo-equivalence invariant.*

Indeed, the Borel construction $M \times_G EG$ is a $\mathbb{Q}\Pi$-Poincaré complex,[78,79]

Now, the ideas of controlled topology that we have discussed earlier assert that this invariant disassembles over M/G.

---

[78] First, note a piece of good news. Since $1/2 \in \mathbb{Q}$, there is no difference between symmetric and quadratic L-theories. I should point out, though, a slight subtlety. The finiteness condition on this chain complex is homological, so one only gets a projective chain complex. (If there is a G-invariant triangulation, this is direct, because all orbits are permutations, and permutation complexes are projective over $\mathbb{Q}G$. This suggests the true statement that for free actions the chain complex is defined in L^h and that one doesn't ever need all projective modules.)

[79] Probably one should point out that one need only invert in the coefficients the orders of the nontrivial isotropy groups. One might also point out that, using intersection homology sheaves, one can define these invariants, for example, for complex varieties with action.

Proposition 6.40   *There is an assembly map*

$$H_m(M/G; L(\mathbb{Q}G_x)) \to L_m(\mathbb{Q}\Pi)$$

and $\sigma^*_G(M)$ canonically lifts to the domain of this assembly map. We denote this lift by $\Delta(M)$.

This is completely analogous to the non-equivariant situation (aside from the coefficients being $\mathbb{Q}$ rather than $\mathbb{Z}$). To continue the analogy to the non-equivariant case, we should study the functorialities of this map and factor it through

$$H_m(M/G; L(\mathbb{Q}G_x)) \to H_m(\underline{E\Pi}/\Pi; L(\mathbb{Q}\Pi_x)) \to L_m(\mathbb{Q}\Pi).$$

This is indeed possible and is part and parcel of the interpretation of the left-hand side as controlled algebraic Poincaré complexes and the arrows as change of control spaces (or the forgetful map).[80]

Now we have a wonderful coincidence. The domain and range of this assembly

map are (away from 2) the same as for the assembly map that arises in the calculation of $S(M/G\,\mathrm{rel}\,\Sigma)$ considered in §6.6.

Corollary 6.41   *Away from 2, $\Delta(M)$ is a pseudo-equivalence invariant iff it is an isovariant homotopy-invariant (for maps that are homeomorphisms on the singular set!).*

This is highly significant, because, unfortunately, we do not have a well-understood theory of pseudo-equivalence (especially in the topological category). In addition, this corollary reduces a pseudo-homotopy invariance statement to a *tangentiality type* result in the equivariant Borel conjecture (see Ferry *et al.*, 1988).

Another important point that is almost implicit within the corollary is that $S(X, \mathrm{rel}\,\Sigma)$ is a summand of $S(X)$ (away from 2) for finite group actions. We record a somewhat more general statement that is proven by induction.

Theorem 6.42   *If G is a finite group tamely and acting orientation-preservingly[81] on a manifold M, then, inverting 2, the isovariant structure groups decompose:*

$$S(M/G) \otimes \mathbb{Z}[1/2] \cong \bigoplus S(M^H/(NH/H), \mathrm{rel}\,\Sigma), \otimes \mathbb{Z}[1/2].$$

---

[80] Thus the rel Σ isovariant structure sets have an equivariant functoriality (for finite groups acting orientation-preservingly) like manifold structure sets. The theory of functoriality for the isovariant structures themselves is much more complicated – Cappell, Yan, and I have been thinking about this from a number of points of view for years, with only fragmentary results.

[81] This includes the hypothesis that the fixed sets of all subgroups are orientable.

This decomposition is frequently true integrally, especially for odd-order groups, but it is not true in general. The right-hand sum is over components of strata of the quotient – so we would not count twice a component fixed by a subgroup that is also fixed by a larger subgroup.

The question of integral versions of this splitting is an important one. The difference between "yes" and "no" is often an element of $\mathbb{Z}_2$!

When one has an integral splitting, one knows that a "replacement theorem" holds: any manifold homotopy-equivalent to the fixed set is the fixed set of some action on an equivariant homotopy-equivalent manifold (and similarly for other strata). Sometimes it is even possible to arrange that the new action is on the same manifold, although there are situations (e.g. for some orientation-preserving involutions) for which replacement holds, but this strong form fails.

In any case, the results we have seen in Chapter 5 about higher-signature formulas for S^1-actions show that in codimension $2 \bmod 4$ replacement does not hold for *rational reasons*, and that strong replacement doesn't hold if the fixed set has codimension $0 \bmod 4$. So, our discussion has really required the finiteness of G.

The analysis of the group structure on these structure sets (for the finite case) is also facilitated by the following:

Observation 6.43 $S(M/G \operatorname{rel} \Sigma)$ is a (graded) module over $L^*(\mathbb{Q}G)$.

In particular we can view it as a module over $\mathrm{RO}(G)$ – and therefore apply the ideas of the localization theorem in equivariant K-theory (Atiyah and Segal, 1968).[82] For example:

Corollary 6.44 *If the action of G on M is free and pseudo-trivial, then, assuming the Novikov conjecture, the higher signature of M vanishes.*[83]

Very similar reasoning would give the localization of higher signatures to *twisted higher signatures* of fixed sets (where we twist the L-class of the fixed set by an appropriate characteristic class of the equivariant normal bundle; however, one would not obtain immediately the relevant converse statements from Chapter 5).

The reader should be able to deduce some non-pseudo-trivial results from the equivariant conjecture.

There are several directions in which we can go next, and we will go in many of them!

We should discuss the prime 2 and also generalize the Novikov conjecture

---

[82] Compare our discussion of ρ-invariants in §4.10.

[83] Indeed, if there is some element whose fixed-point set is empty, one gets the same conclusion (at least after inverting several primes).

from groups to metric spaces – after all, that was the route we had taken to Novikov's theorem on topological invariance of rational Pontrjagin classes, and we shall see that here it rewards us similarly – and we should discuss the index-theoretic version of these (knowing that the prime 2 will be a place of divergence as always), anticipating, at least, applications to other operators.

The characteristic class that we have introduced here, the equivariant controlled symmetric signature in $L^*(\mathbb{Q}\Pi)$, is certainly not the right thing to do. In the non-equivariant setting we would surely have wanted a $\mathbb{Z}$. However, I do not know any one method for defining the most refined "intrinsic" characteristic classes for group actions without taking their fixed sets, isotropy structure, and so on into account. This feels somewhat related to the realization problem: for manifolds all of $L^*(\mathbb{Z})$ arises as a signature, but none of the torsion elements of $L^*(\mathbb{Q})$. However, not every representation of G occurs as the G-signature of a G-manifold. Depending on the category or setting (e.g. smooth, PL locally linear, PL, topological), one gets different subtle phenomena on the interaction between G-signatures of the manifold and the fixed (or, better, the singular) sets. If the singular set is empty, then the G-signature is a multiple of the regular representation, but even if it's not, there is sometimes residual information available at the prime 2 connecting the G-signature to the fixed set – not the germ neighborhood of the fixed set. It is this information that is implicit in the surgery-theoretic formulation of the Novikov conjecture – since the singular sets for equivariant homotopy equivalences are stratified homotopy-equivalent, this information must be encoded – rather like the equality mod 8 of signatures of manifolds when there is a degree-1 normal map (and concomitant implications for characteristic class theory, such as equalities of Stiefel–Whitney classes). However, the intrinsic characteristic class theory has no room for this refinement and it seems that there are different options, and there is no *a priori* reason to imagine that they will capture the essence of Novikov phenomena at 2.

In particular, I see no reason to expect there to be a theory of intrinsic characteristic classes at 2, for which the pseudo-equivalence invariance is equivalent to the equivariant homotopy invariance. But, mathematics is more beautiful than it needs to be and maybe this is one of those opportunities.

Let us now continue our explorations by analogy to the non-equivariant case. The setting, as we described it, already has controlled aspects. It is completely straightforward to formulate bounded equivariant homotopy equivalence conjectures over metric spaces with proper group actions. By considering $M \times [0, \infty)$ as boundedly controlled over the cone cM, the equivariant Novikov conjecture (which is a theorem of cones of G-ANRs, by the non-equivariant proof) then gives:

Theorem 6.45 $\Delta(M)$ *is a topological invariant of the G-manifold M.*

This is extremely strong. We shall soon see that $\Delta(M)$ is essentially a topological version of the equivariant signature operator. This theorem therefore can be applied to the situation of M being a representation and it implies a celebrated result:

Theorem 6.46 (Based on Cappell and Shaneson, 1982; Hsiang and Pardon, 1982; Madsen and Rothenberg, 1988a,b, 1989)   *For G of odd order, linear representations of G are conjugate as topological group actions iff they are conjugate as representations. For all G, the Grothendieck group of representations under topological equivalence has the following partial calculation:*

$$\mathrm{RTop}(G) \otimes \mathbb{Z}[1/2] \cong \mathrm{RK}(G) \otimes \mathbb{Z}[1/2],$$

where RK *denotes the K-representations of G, and K denotes the real subfield of the cyclotomic field of all odd roots of unity.*

(To see why this fact about the equivariant signature operator is enough, one can read the introductions of Cappell and Shaneson, 1982, and of Madsen and Rothenberg, 1988a,b, 1989.)

For G with elements of order 2, one finds that the symbol of the equivariant signature operator is not a unit in the representation ring $R(G)$. This indirectly is related to the existence of nonlinear similarities.[84] It is also responsible for the different behavior that we have seen regarding the stabilization map

$$\mathrm{BPL}_{2k}(\mathbb{Z}_p) \rightarrow \mathrm{BPL}_{2k+2}(\mathbb{Z}_p)$$

(mentioned as Example 6.17 among the other trivialities in §6.2) which is highly connected for p odd (and k sufficiently large) but never a rational equivalence on π_2 (even for k large) when $p = 2$.

The beautiful description of the topological representation group[85] given above should not mislead you into thinking that the *size* of the bundle theory relevant to the topological category is smaller (regulated by RK). Not at all. It is the size of KO^G, but the image of (usually real) representation theory into this description is not the most naively expected one – and happily the kernel of this map is succinctly describable. It is a nice problem to analyze equivariant topological equivalence of bundles (sort of like the Adams conjecture does for fiber homotopy equivalence) – even stably.

The class $\Delta(M)$ is essentially a topological analogue of the equivariant signature operator – except that the latter is only defined when one has a Lipschitz

---

[84] Although nonlinear similarities only exist when G has elements of order 4.

[85] Yes, it's not a ring. Topological equivalence does not play nicely with tensor products. And, the group does indeed contain 2-torsion as Cappell and Shaneson showed.

invariant metric.[86] We shall ignore the details, since the equivariant homotopy equivalence (or pseudo-equivalence) of the higher equivariant symbol class is of interest even in the smooth case.

Theorem 6.47 (Rosenberg and Weinberger, 1990) *Let* Π *be the fundamental group of a G-action on a manifold with fundamental group* π. *The injectivity of the assembly map* $KO^{\Pi}(\underline{E\Pi}) \to K(C^*_R\Pi)$ *implies the pseudo-equivalence invariance of G-equivariant higher signature in* $KO^{\Pi}(\underline{E\Pi})$. *It also implies the vanishing of higher indices of equivariant Dirac operators on manifolds admitting equivariant positive scalar curvature metrics on spin manifolds with isometric G-action.*

The left-hand side (here using real C^*-algebras for some slight refinement, as emphasized in Rosenberg's early (e.g. 1991) papers is the domain for the Baum–Connes assembly map, and the injectivity part has all of the implications we would like. I will not bother repeating the details of this type of argument here, but rather will point out two nice advantages of the analytic version over the topological one:

(1) In the analytic situation, there is no trouble dealing directly with G compact, since the Baum–Connes conjecture is a statement about locally compact groups. For the situation where $\underline{E\Pi}$ is finite, one can deduce the relevant injectivity statement in the topological case by using the result of McClure (1986) for finite complexes X that $KO^G(X) \to \prod KO^H(X)$ is injective as H runs over the finite subgroups of G. It is clearly necessary to develop a theory of $\Delta(M)$ for G compact, rather than just finite. However, there are considerable technical difficulties to doing this related to the fact that the orbit G-spaces are homogeneous spaces and have interesting topology.

Indeed, this interesting topology also leads to the important point that the equivariant Novikov conjecture does not yield the information one is interested in about the variation of characteristic classes within an equivariant homotopy type for G-manifolds when G is positive dimensional. For instance, we have noted that there are interactions between higher signatures of manifolds and fixed sets that automatically hold and (are accounted for in L-theory but) not accounted for in equivariant K-theory.

(2) For G non-abelian (and connected!) there is a remarkable topological consequence of the above result. Lawson and Yau (1974) have shown that any smooth G-manifold has a G invariant positive scalar curvature metric.[87]

---

[86] I do not know an example of a C^0-group action that does not preserve a Lipschitz Riemannian metric, but I doubt that they always exist.

[87] Actually they did this for $G = SU(2)$; I have not checked that their argument works for all G, but presumably it does.

Consequently the manifold M has vanishing higher Dirac class in $KO(C_\mathbb{R}^* \pi)$ and assuming the Novikov conjecture in $KO^\pi(\underline{E\pi})$.[88]

This is nontrivial even in the situation of exotic spheres, because they are known to have a variety of different symmetry properties (as investigated in papers of Reinhardt Schultz).

We close with section by noting that the proof methods discussed so far apply to the situation where $\underline{E\Pi}$ is a non-positively curved locally symmetric manifold. In the analytic case, this is automatic: the machinery handles equivariance with almost no pain.[89] In the topological case, one needs, for example, the equivariant analogue of Ferry's theorem:

Theorem 6.48 (Steinberger and West, 1987[90]) *If $G \times M \to M$ is a homotopy locally-linear action of a compact group on a compact G-manifold,[91] then there is an $\varepsilon > 0$, such that if $f : M \to N$ is a G-map to a connected homotopy locally-linear G-manifold of no larger dimension, then f is equivariantly homotopic to a G-homeomorphism.*

(This theorem cannot be yet phrased in the full tame category including homology manifolds, because we do not know enough about homology manifolds to homotop CE maps to homeomorphisms when they should be!)

Appendix: Note on the Formulation of the Equivariant Novikov and Borel Conjectures

I've been blithely arguing by certain analogies with the unequivariant case. Here I would like to point out some dangers with doing this.

One moral is that the Novikov and Borel conjectures are considerably less well founded in the presence of infinite dimensionality (although they are frequently true even in this setting) and that infinite dimensionality is sometimes hidden in group action problems.

Another point that emerges is that, given the pseudo-invariance properties of our signatures, one should perhaps be led to dispense with the notion of the equivariant $K(\pi, 1)$. Better would be to consider the analogous objects in the pseudo-category: these are actually the $\underline{E\Pi}$ that we had been using[92] above without comment.

---

[88] Assuming the stated generalized Lawson–Yau theorem, one would get the stronger vanishing of the equivariant class in $KO^\Pi(\underline{E\Pi})$.

[89] Provided one sets up K-theory to be equivariant to begin with, and establishes the relevant forms of Bott periodicity and so on, making use of an equivariant Bott element.

[90] In high dimensions, but subsequent developments allow its restatement in the form given.

[91] And one can suitably relax this condition, as we have in Ferry's theorem in §4.6.

[92] Following Kasparov, and Baum and Connes, whose immediate goal had been to give models

A $K(\pi, 1)$ is the terminal object in the homotopy category of spaces with maps that are 1-equivalences: i.e. for maps $f: X \to Y$ with the property that given any $g: K^2 \to Y$ there is a map $g': K^2 \to X$ with fg' homotopic to g on the 1-skeleton K'. (Actually, the terminal objects are disjoint unions of $K(\pi, 1)$s but this hardly effects any of our conceptual thinking about the topology – everything happens on the components independently).

In the equivariant case, we should actually deal with the equivariant analogue of this notion. This is a G-space (with G compact – but we can easily change our mind and work with locally compact groups by defining an analogous notion in the universal cover) with the same universal property with respect to equivariant 1-equivalences, defined with respect to equivariant 2-complexes.[93] This boils down to the condition that for all $H \subset G$, the fixed set is an equivariant NH/H aspherical complex. Or, putting it all together, one wants the fixed set of any subgroup to be a disjoint union of aspherical complexes.

We have seen that the integral Novikov conjecture fails for groups with torsion – so it becomes reasonable to assume finite dimensionality. This would boil down to the equivariant Novikov conjecture failing for $G = \mathbb{Z}_p$ with the equivariant aspherical complex chosen to be S^∞. With finite dimensionality, we are led to consider the complex to be a point (which is $\underline{E\mathbb{Z}_p}$).

When $G = S^1$ and one uses the S^∞ model, even rationally equivariant homotopy equivalence fails. (Unlike the usual Novikov conjecture, which is not known to have any rational counterexamples using $E\pi$ in place of $\underline{E\pi}$; indeed, the rational injectivity statements are equivalent.)

Note that, in our formulation of the equivariant Novikov conjecture in this section, we used $\underline{E\Pi}$ when we had a G-action on a space M with fundamental group with fundamental group π. However, that space need not be the equivariant aspherical space associated to M. It accepts a G-map from M, but it might collapse different components, lose some fundamental group information, etc. In some sense $\underline{E\Pi}$ is the smallest model, and the one most likely to be finite-dimensional, and therefore the one most likely to have "correct" equivariant Novikov and Borel conjectures.

For simplicity let's consider what we can do when $G = \mathbb{Z}_p$ acts on a simply connected manifold. The only finite-dimensional equivariant aspherical spaces can be determined using Smith theory: they are ones that are contractible, have a $\mathbb{Z}_p$-action, with fixed set F that is aspherical and mod p acyclic. The rel Σ assembly map is typically not an isomorphism even in this case, if F

---

for $K(C^*\Pi)$ and thus could not have been misled by the possibilities suggested by equivariant algebraic topology and surgery.

[93] Note that the dimension of a G-cell is not the same as its non-equivariant dimension when G is positive-dimensional.

has torsion in its homology away from p – although it will be rationally.[94] *The infinite dimensionality forced, for example, by requiring the modeling of a disconnected fixed set causes even a rational failure of the injectivity of this assembly map. This means that some fundamental groupoid[95] situations give rise to wider variation of equivariant characteristic classes than one would have naively expected from the usual analogies between equivariant and ordinary surgery.*

6.8 The Farrell–Jones Conjecture

In §6.6 we were led to consider assembly maps

$$H(X/\Gamma; K(\mathbb{Z}\Gamma_x)) \to K(\mathbb{Z}\Gamma), \tag{6.1}$$

$$H(X/\Gamma; L(\mathbb{Z}\Gamma_x)) \to L(\mathbb{Z}\Gamma), \tag{6.2}$$

for X, e.g. a locally symmetric space on which Γ acts properly discontinuously.

In §6.7, we have seen that these maps are frequently injective by considerations of the relΣ tangentiality part of the equivariant Borel conjecture.[96] If we replace the ring $\mathbb{Z}$ by $\mathbb{Q}$ in the coefficients, then none of the Nil and UNil phenomena we discussed provide a problem, and one can[97] reasonably conjecture isomorphism.

Is there a moral to this?

Let's review our situation. We started by considering the less-refined assembly map:

$$H(B\Gamma; K(\mathbb{Q})) \to K(\mathbb{Q}\Gamma), \tag{6.3}$$

$$H(B\Gamma; L(\mathbb{Q})) \to L(\mathbb{Q}\Gamma). \tag{6.4}$$

However, for a finite group, one observes that the right-hand side behaves (completely) differently from the left-hand side; it has a much more number-theoretic nature.

There is a map, though, $B\Gamma \to X/\Gamma$, where the inverse image of a point $[x]$ in

---

[94] Using a Davis construction, this can be promoted to an equivariantly aspherical manifold where there is a failure of the tangentiality part of the equivariant Borel conjecture. Moreover, this cannot be attributed to equivariance versus isovariance, like our examples in §6.10, or Nil/UNil problems like our previous ones in §6.5.

This example, though, has no bearing on the form discussed in this section, or on the stratified Borel conjecture, discussed in Chapter 13 of Weinberger (1994); because a key incompressibity (namely π_1 injectivity) condition is violated.

[95] This captures both π_0 and π_1 issues.

[96] Make no mistake: this is an integral result, despite the fact that the version of the pseudo-equivalence version was only sufficiently precise away from the prime 2.

[97] We will not back out of this conjecture, as we have for some others in this book.

X/Γ is $B\Gamma_x$. We have essentially, in the target, "gathered up" the $H(B\Gamma_x; L(\mathbb{Q}))$ parts of $H(B\Gamma; L(\mathbb{Q}))$ and replaced them by $L(\mathbb{Q}\Gamma_x)$.

In other words, we can think formally along the following lines. The original Borel conjecture was that K- and L-theory are the simplest possible things (for group rings $R\Gamma$ consistent with $K(R[e])$ and $L(R[e])$. But, when we realized that this was wrong for finite groups, we just punted and said, OK, the correct conjecture should be the one that is correct for $R[G]$ for G finite – i.e. replace any part of the assembly map that maps through a finite group by one where the finite group acts trivially, at the cost of creating a cosheaf whose co-stalk at such a point reflects the correct answer.

Given that we got so much mileage out of doing this for finite groups, it's clear what to do[98] now that we have examples coming out of Nil and UNil. We know about counterexamples to the assembly maps (6.1) and (6.2) being isomorphisms among the class of groups that are virtually cyclic (i.e. have a cyclic subgroup of finite index). So we should "collect" all of these parts of the left-hand side together and make no predictions about their K- and L-theories – just simply predict the simplest possible answer consistent with assembly maps and calculations (left as a problem for the algebraically minded[99]) for these special groups.

The formal way to do this is to introduce a new classifying space $E_{vc}\Gamma$ for simplicial actions of Γ whose isotropy is virtually cyclic. There are equivariant maps of classifying spaces

$$E\Gamma \to \underline{E\Gamma} \to E_{vc}\Gamma.$$

Davis and Lück (1998) have given a nice formulation of the whole theory[100] by adding a final map to this:

$$E\Gamma \to \underline{E\Gamma} \to E_{vc}\Gamma \to E_{groups}\Gamma = \text{a point,}$$

where the last space is the classifying space of actions where any isotropy is allowed – it is a point, since Γ acting on a point is this classifying space. (After all, that space is now allowed, and surely everything has a unique map to it.)

---

[98] Except that I was shocked when Farrell and Jones took this step. I was certain that this could not be right because of what it implied for free abelian groups. And, indeed they waited until they had *proved* the conjecture that they asserted for many lattices that contain high-rank abelian groups before publicly making this conjecture. The class of virtually cyclic groups arises very naturally in the dynamical method that they had introduced into topological rigidity theory. In short, the genesis of this conjecture was a much less blithe process than I am pretending it to be. Nevertheless, if hindsight cannot be 20/20, what can be?

[99] This feels reasonable (but difficult) because the rings involved are (not necessarily commutative) finite-degree extensions of $\mathbb{Z}[\mathbb{Z}]$. It's the dimension-1 analogue of the dimension-0 issues considered as a major enterprise of the last century: computing $K(\mathbb{Z}G)$ and $L(\mathbb{Z}G)$ for G finite.

[100] Although I am not using their notation.

However, the domain of the assembly map for any "family" $\mathcal{F}$ is

$$H\big(\mathrm{E}_{\mathcal{F}}/\Gamma; K(R\Gamma_x)\big) \quad \text{and} \quad H\big(\mathrm{E}_{\mathcal{F}}/\Gamma; L(R\Gamma_x)\big)$$

for K- and L-theory, respectively.

Thus the map induced by the last forgetful map (it is forgetful because, when we go from a small family to a larger one, we are forgetting the special property that isotropy lies in the smaller family) is

$$H\big(\mathrm{E}_{\mathrm{vc}}\Gamma/\Gamma; K(R\Gamma_x)\big) \to K(R\Gamma), \tag{6.5}$$

$$H\big(\mathrm{E}_{\mathrm{vc}}\Gamma/\Gamma; L(R\Gamma_x)\big) \to L(R\Gamma). \tag{6.6}$$

These isomorphisms comprise the Farrell–Jones isomorphism conjecture.[101]

As they point out, when their conjecture is disproved by some group (or class of groups) they will be able to immediately generalize their conjecture by including the counterexample into a new one that has a larger family. Of course, it could be that the "final" conjecture would be the one where $\mathcal{F}$ ends up being the family of groups – but such a pessimistic conclusion is surely premature.[102]

An amusing situation arises if one applied this philosophy to the operator algebra context for understanding $K(C^*_{\mathrm{max}}\Gamma)$. We know that for Γ an infinite Property (T) group $K(C^*_{\mathrm{max}}\Gamma)$ is larger than the domain of the Baum–Connes assembly map (the trivial module C is projective in this setting; indeed, all finite-dimensional representations are isolated and give a very large cokernel to the assembly map). We are thus led to study the map

$$H(\mathrm{E}_{\mathrm{T}}\Gamma/\Gamma; K(C^*_{\mathrm{max}}\Gamma_x)) \to K(C^*_{\mathrm{max}}\Gamma).$$

The subscript T should be interpreted as the family of subgroups of Γ that are subgroups of a Property (T) subgroup of Γ. Whether this leads to any insights regarding the right-hand side, I do not know. It gives many new elements of that group.

Let us unravel what the Farrell–Jones conjecture means in some cases. We have to understand what $E_{\mathcal{F}}\Gamma/\Gamma$ looks like as a stratified space. First of all, there are nontrivial strata for subgroups π in $\mathcal{F}$. The fixed set of such a group is contractible, and the group $N\pi/\pi$ acts on this. At such a stratum we have $K(R\pi)$ or $L(R\pi)$. Strata corresponding to π and π' can intersect only if the groups generated by π and π' lie in the family $\mathcal{F}$.

We have already discussed the situation for torsion-free word hyperbolic groups in §5.5.3. The interesting situation there is K-theory and how the closed

---

[101] Where the L is decorated with -8.

[102] And, in this case, I would prefer the false conjecture that has been so fruitful over the previous decades to the one that is true yet meaningless.

geodesics, which are the maximal elements of the virtually cyclic family, contribute Nils (i.e. the fiber of the assembly map $H_*(B\mathbb{Z}, K(R)) \to K(R[\mathbb{Z}]))$. In L-theory, it's the usual assembly map (as it is for all torsion-free groups).

If there is torsion then the L-theory situation becomes interesting. For definiteness, assume that Γ comes from a $\mathbb{Z}_2$-action on a closed negatively curved manifold. Then the fiber of the virtually cyclic assembly map comes specifically from closed geodesics that are invariant under $\mathbb{Z}_2$, but, if the action is free, then it will not contribute either. However, each geodesic which is invariant under the involution will go through two fixed points, and give a UNil contribution via the inclusion of $L(\mathbb{Z}_2 * \mathbb{Z}_2)$ in $L(\Gamma)$.

For $\mathbb{Z}^n$, the trivial and finite families both give us $\mathbb{R}^n$ with free $\mathbb{Z}^n$-action. However, when we use the virtually cyclic family, the maximal subgroups correspond to primitive lattice points (up to sign) in $\mathbb{Z}^n$. However, there is a T^{n-1} family of geodesics in each of these free homotopy classes equal to the corresponding stratum in $E_{\text{vc}}\mathbb{Z}^n/\mathbb{Z}^n$. Each of these families produces a $H_*(T^{n-1}; \text{Nil}(R))$ contribution.

By the way, note that this description includes an analysis of $\text{Nil}(R[\mathbb{Z}])$ in terms of $\text{Nil}(R)$, but it is not simply the assertion that $\text{Nil}(R[\mathbb{Z}])$ is computed from $\text{Nil}(R)$ via the assembly map isomorphism. The description given by the Farrell–Jones conjecture is strong enough to enable an analysis of the action of $SL_n(\mathbb{Z})$ on the K-groups.

The story for, for example, $\mathbb{Z}^n \rtimes Z_2$, where the involution acts as multiplication by -1, is similar. In that case there are 2^n fixed points (on the torus), and many such closed geodesics and these tell the whole story (see Connolly *et al.*, 2014). If the involution had a positive eigenspace, then there would be tori of these interesting geodesics. They would then contribute the homology of these tori with coefficients in the Nils corresponding to the geodesics.

Finally, we note that the Farrell–Jones conjecture gives us a description of the isovariant structure set rel Σ: it is the relative homology group describing the difference between $\underline{E}\Gamma$ and $E_{\text{vc}}\Gamma$.

6.9 Connection to Embedding Theory

We now return to the equivariant Borel conjecture (or, more generally, to the problem of equivariant surgery classification). Our approach is via a profound connection between group actions and embedding theory that was already hinted at in §6.5.

Surgery theory does quite a good job (with the Farrell–Jones conjecture picking up much of the slack) of analysis of structures within an isovariant homotopy type, but does not do a very good job of classification within an equivariant homotopy type. The latter is what the equivariant Borel conjecture (or, surely we should say, question) asks about.

However, analogously, surgery shows that there is a unique isotopy class of embedding realizing a given Poincaré embedding, but it does not directly help with the problem of analyzing the embeddings in a given homotopy class.

Embedding theory has developed a set of geometric tools that enable good classification results in specific settings. We do not know of any way of mimicing these geometric tools (such as general position, multiple disjunction, and so on) directly.

However, *after the fact*, we can make use of the theoretical reduction of both embedding theory and isovariant classification within an equivariant homotopy type to homotopy theory to relate these problems to one another and get concrete and theoretical results. This section is devoted to developing the connections between the subjects, and the next will use this to give some specific analyses. The explicit results which we give there for some crystallographic manifolds are the subject of heretofore unpublished joint work with Sylvain Cappell. I also wish to acknowledge useful conversations with John Klein about the use of categorical techniques and explanations of the calculus of embeddings due to Goodwillie, Weiss, and him.

Recall the definition of a *Poincaré embedding* (see Wall, 1968). A Poincaré embedding of M in W consists of a triple $((X, E), \pi, f)$, where (X, E) is a Poincaré pair, $\pi: E \to M$ is a spherical fibration with fiber S^{c-1}, and $f: X \cup \mathrm{Cyl}(\pi) \to W$. In the PL and topological(ly locally flat) category, every Poincaré embedding can be realized by an embedding (that is unique up to concordance – by the relative version of this realization statement).

By analogy we can define a similar notion of an isovariant Poincaré complex. For notational simplicity we shall only discuss the semi-free case.

Definition 6.49 An isovariant G-Poincaré complex consists of a triple

$$((X, E), \pi)$$

with $\pi: E \to F$ an equivariant spherical fibration with fiber S^{c-1} having a free G-action, (X, E) with free G-action (i.e. covering a Poincaré pair) so that $Y = X \cup \mathrm{Cyl}(\pi)$ is a Poincaré complex.

We are interested in the possible isovariant Poincaré complexes (up to isovariant homotopy equivalence) within a given equivariant homotopy type. In other words:

Definition 6.50 If $G \times M \to M$ is a group action, then $\mathrm{Iso}(M)$ is the set of isovariant Poincaré complexes *with an equivariant homotopy equivalence to M* (up to isovariant homotopy type). This can be made into a Δ-set in the usual way, and we shall denote this by the same symbol. (The set of isovariant homotopy Poincaré complexes within the given equivariant homotopy type can be thought of as π_0 of the space $\mathrm{Iso}(M)$.)

Similarly we will denote the Poincaré embeddings of F in M which are homotopy equivalent as a pair to a given embedding as $\mathrm{PE}(M,\mathrm{F})$. Note that there is a map $\mathrm{Iso}(M) \to \mathrm{PE}(M,\mathrm{F})$ when F is the fixed set of the G-action on M.c Note that G acts on $\mathrm{PE}(M,\mathrm{F})$ by composing the map f with elements of G. Our main results concern the relation of $\mathrm{Iso}(M)$ to $S^{\mathrm{equi}}(M)$ and the relation of $\mathrm{Iso}(M)$ to $\mathrm{PE}(M,\mathrm{F})^{\mathrm{hG}}$.

Theorem 6.51 (Decomposition theorem) *For $G = \mathbb{Z}_p$, with p odd, acting tamely and supposing that the fixed set is of codimension greater than 2, there is an isomorphism*

$$S^{\mathrm{equi}}(M) \cong S^{\mathrm{iso}}(M) \times \mathrm{Iso}(M).$$

For applications to disproving the equivariant Borel conjecture, the above theorem is not even necessary. The point is that there is a total surgery obstruction to realizing elements of $\mathrm{Iso}(M)$ (or even $\mathrm{Iso}(M\,\mathrm{rel}\,\Sigma)$) that lies in a group[103] that is (assuming Farrell–Jones) trivial. As a result, even in the absence of the above theorem, $\mathrm{Iso}(M)$ would provide counterexamples to the equivariant Borel conjecture.[104]

It is necessary, though, for understanding the set of all counterexamples, when the dimension of the fixed set is relatively high compared to dim M.

The proof of the theorem is not purely stratified in nature, but relies on connections between the isovariant and equivariant categories that were pioneered by Browder (and continued to be studied by Dovermann and Schultz, 1990, Yan, 1993, and others).

This decomposition theorem is surely true in much greater generality (at least I think so). I hope to return to this in a later paper; below we will give a small extension of it.

We will not discuss the algebraic K-theoretic aspect of the decomposition theorem: essentially this is handled by the way that isovariant finiteness (when all strata are codimension 3 in one another) is equivariant in nature. We focus on the algebraic topology and the surgery.

---

[103] The delooping of the isovariant structure space.

[104] Ironically, we would be using the isovariant Borel conjecture to disprove the equivariant Borel conjecture in following such a route!

The first interesting ingredient is a variant of the Whitney embedding theorem, due to Browder, that asserts (in the semi-free case):

Theorem 6.52   *In the semi-free case, if* $\dim M > 2\dim F + 1$, *then every equivariant homotopy equivalence is equivariantly homotopic to an isovariant homotopy equivalence.*

In addition, we need a method for getting into the stable range that doesn't lose surgery obstructions. Again following Browder, we cross with the G-manifold $(\mathbb{CP}^2)^G$, where the superscript G denotes the product of #G copies of the projective plane, with the G-action given by permutation. Browder observed that:

Theorem 6.53   *If G is odd order, then, if taking the product* $\times(\mathbb{CP}^2)^G$ *does not change the number of strata in the quotient spaces, it induces an isomorphism on L-groups,*

$$L^{\mathrm{strat}}(M/G) \to L^{\mathrm{strat}}((M \times (\mathbb{CP}^2)^G)/G).$$

As a consequence of this and stratified surgery $\times(\mathbb{CP}^2)^G$ induces an injection of structure sets. Since existence of a structure underlying a given isovariant Poincaré structure is a surgery problem (i.e. lies in the delooping of the structure space) if the realization exists after crossing $\times(\mathbb{CP}^2)^G$ it will exist before crossing. However, by Browder's first theorem above, if the structure exists equivariantly, it exists isovariantly after crossing with $(\mathbb{CP}^2)^G$, explaining the above theorem.

Remark 6.54   The decomposition theorem holds at least in the greater generality of G is of odd order and acting semi-freely, working relative to the fixed point. (This suffices for our applications below.) Of course, the problem is that crossing with $(\mathbb{CP}^2)^G$ has more strata. However, working rel F will enable us to get around this as follows.

Note that the product map sends[105]

$$S(M/G, \mathrm{rel}\, F) \quad \text{to} \quad S(M \times (\mathbb{CP}^2)^G/G, \mathrm{rel}\, \text{singularities})$$

and we can study the existence problem relative to the singular set. The rel sing structure set can be thought of as the fiber of a *conventional* assembly map[106] –

---

[105] It is actually true that $S(M/G)$ can be decomposed as $S(F) \times S(M/G, \mathrm{rel}F)$, because the symmetric signature of the space form normal to F vanishes. (This is enough because of the way the symmetric signature of the link enters in the definition of L^{BQ}, the key object in stratified surgery. This vanishing can be deduced from the fact that the symmetric signature can be lifted under an assembly map to an odd torsion group, but the L-group has only 2-torsion.) For some detail, see Cappell and Weinberger (1995).

[106] With non-constant coefficient, when interpreted in the quotient.

interpreted as forgetting control – from the controlled free equivariant Poincaré complexes over the space mapping to the uncontrolled Poincaré complexes (which end up in L (the orbifold fundamental group)).

With this interpretation, there is a projection map

$$S\big(M \times (\mathbb{CP}^2)^G/G, \text{rel singularities}\big) \to \S(M/G, \text{rel } F)$$

whose precomposition with the product map

$$S(M/G, \text{rel } F) \to S\big(M \times (\mathbb{CP}^2)^G/G, \text{rel singularities}\big)$$

is an isomorphism – this map is a transfer associated to a fiber bundle with fiber $(\mathbb{CP}^2)^G$ and that it induces an isomorphism on L-groups is part of the proof of Browder's theorem (see Yan, 1993).

Now we turn to the map $\text{Iso}(M) \to \text{PE}(M, F)^{hG}$ alluded to above. The map is obtained by sending a typical vertex $\big(((X, E), \pi), \Phi\big)$, where $\Phi \colon X \cup \text{Cyl}(\pi) \to M$ is an equivariant homotopy equivalence, to the vertex of the homotopy-fixed-set $\big(X \cup \text{Cyl}(\pi)\big) \to (M \times_G \text{EG} \downarrow \text{BG})$ in the homotopy-fixed-set of G acting on $\text{PE}(M, F)$. Higher simplices are mapped similarly.

Theorem 6.55 *If M has boundary and each component of the fixed set touches the boundary, then*

$$\text{Iso}(M, \text{rel } \partial) \to \text{PE}(M, F, \text{rel } \partial)^{hG}$$

is a homotopy equivalence.

Without the boundary condition, this theorem is hopelessly false: on π_0 one can get uncountably many components on the right-hand side, while the left is clearly always countable. The problem is that one produces in the homotopy-fixed-set group actions on infinite-dimensional spaces that don't have a reasonable geometric interpretation.

On the other hand, the condition is not an unreasonable one, since it can be arranged through strategic puncturing of M at various fixed points.

The reader can wonder whether this theorem is ever of use, in that homotopy-fixed-sets involve maps of infinite-dimensional spaces into other objects. We close the section with some examples of how one can use this machinery, even in the absence of concrete information about the classification of embeddings. In §6.10 we will give some additional illustrations that have some more computational input that I hope are convincing that this approach is not completely worthless.

The proof of the theorem is quite simple and quite analogous to the old result of George Cooke about realizing homotopy actions by actions: a homotopy action in a map of groups $G \to \pi_0 \mathrm{Aut}(X)$, and the question is which of these are realized by group actions?

Theorem 6.56 (Cooke, 1978) *A homotopy action is realized by an action iff the induced map on classifying spaces has a lift:*

$$\mathrm{BAut}(X)$$
$$\downarrow$$
$$\mathrm{BG} \to \pi_0 \mathrm{Aut}(X)$$

Proof If there is an action, there's a lift. If one has a lift, then the associated X fibration over BG has as induced G-cover a space homotopy equivalent to X on which the G-action by covering translates is the desired realization. □

A warning, though, is that the space on which G acts could well be infinite and even infinite-dimensional when the cohomological dimension of G is infinite.

Similar is the following:

Proposition 6.57 *If X and Y are free G-spaces, then the map of mapping spaces*

$$[X, Y]^G \to [X, Y]^{hG}$$

is a homotopy equivalence.

This is a triviality from covering space theory and the homotopy equivalences between X/G and Y/G and their respective Borel constructions.

If we stare at what the right-hand side $\mathrm{PE}(M, \mathrm{F})^{hG}$ means, one sees an $\mathrm{F} \times \mathrm{BG}$ with a spherical fibration over it, together with some pair that is also given as a fibration over BG. The spherical fibration over $\mathrm{F} \times \mathrm{BG}$ can be thought of as a family of "spherical fibrations over BG" parameterized by F. A spherical fibration over BG is like the output of Cooke's theorem – it corresponds to a free action of G on a space of the homotopy type of $\mathcal{S}^{c-1}$ but it's not necessarily finite, i.e. corresponding to a homotopy lens space (or space form). This is a question that needs answering over each component of F once – which is why we need the boundary conditions.

But, if it is finite, then we have obtained the relevant equivariant spherical fibration over F. The total space of this is now included into a complement, which, if it were finite, would be exactly what we need for an isovariant Poincaré complex. The finiteness follows from:

(1) codimension greater than 2;
(2) the already established finiteness of the boundary of the regular neighborhood;

(3) the comparison of the relative chain complexes for $(X, \partial \mathrm{Nbd(F)})$ and (M, F) (this is a chain equivalence by excision);
(4) the fact that isovariant finiteness obstructions are equivalent to equivariant finiteness obstructions. This statement is pretty obvious in the PL case (because the relevant K-groups are sums of the K-groups of various strata) and [107] it is a consequence of Carter's vanishing theorem for negative K-groups, and the calculations of both of these obstruction groups for the topological case.

Some consequences of the above theorems are worth pointing out immediately – although they involve some massaging to get them for general finite groups (since the decomposition theorem wasn't proved in appropriate generality).

I conjecture that for semi-free actions, the decomposition theorem holds for all G. Indeed, I suspect that the phenomenon is extremely broad (and perhaps only requires a very small gap hypothesis).

(1) The pseudo-trivial orientation-preserving[108] G-action situation produces pairs (M, F) where the inclusion is a homology isomorphism at #G. As a result the homotopy-fixed-set analysis is straightforward, and one obtains that $S^{\mathrm{equi}}(M) \cong S^{\mathrm{iso}}(M \mathrm{rel} \, F) \times S(F) \times \mathrm{Emb}(F \subset M)$ and a complete reduction of the equivariant classification problem to embedding theory!
(2) For orientation-reversing involutions on the sphere, Chase showed in unpublished work that a mod 2 homology subsphere Σ of the sphere is the fixed set of an orientation-reversing involution of codimension greater than 1 iff Σ is isotopic to its mirror image – exactly the π_0 part of the homotopy-fixed-set condition. That the remaining part follows automatically follows from ideas of Dwyer (1989).
(3) If M is a G-manifold, then $S^{\mathrm{equi}}(M \times \mathcal{D}^i \mathrm{rel} \, \partial)$ is an abelian group for $i > 1$. The embeddings $(F \times \mathcal{D}^i \subset M \times \mathcal{D}^i \mathrm{rel} \, \partial)$ also form a group for $i = 1$ and is abelian for $i > 1$. These are the π_i of the spaces $S^{\mathrm{equi}}(M)$ and $\mathrm{PE}(M, F)$. Sometimes I like to refer to the embeddings $(F \times \mathcal{D}^i \subset M \times \mathcal{D}^i \mathrm{rel} \, \partial) = C_i(F, M)$ as the ith *concordance embedding group* of F in M. One obtains[109] an isomorphism

$$S^{\mathrm{equi}}(M \times \mathcal{D}^i \mathrm{rel} \, \partial) \otimes \mathbb{Z}[1/\#G] \cong S^{\mathrm{iso}}(M \mathrm{rel} \, F) \otimes \mathbb{Z}[1/\#G]$$
$$\times C_i(F, M) \otimes \mathbb{Z}[1/\#G].$$

---

[107] As we have already remarked in §6.5.

[108] Note that the $G = \mathbb{Z}_2$ case can be pseudo-trivial and orientation-reversing. In that case, the restriction to the boundary is not pseudo-trivial, which interferes with inductive arguments.

[109] Here, since we are inverting 2, one can rehabilitate the argument for odd-order groups to apply in general.

6.10 Embedding Theory

We begin by ignoring all the stuff about Poincaré embeddings and their connection to group actions discussed in §6.9.

It requires herculean effort to deduce from surgery even the most basic embedding theorem, that of Whitney:

Theorem 6.58 *If* $f: M^m \to W^w$ *is a continuous map, and* $w > 2m$, *then* f *can be approximated by an embedding.*

To do this, we would need to construct a spherical fibration (which can be done by the methods of Spivak, 1967), and a homotopy complement (which is very difficult, but clearly related to Spanier–Whitehead duality: see, for example, Spanier, 1981). All in all, a lot of work.

But embedding theory goes much further than this. Whitney proved a much deeper embedding theorem for when $\dim W = 2 \dim M$ using the famous "Whitney trick" that underlies the h-cobordism theorem and the process of surgery, and therefore underpins almost all that we know about high-dimensional topology. However, that embedding theorem is more subtle; the above is sharp as the "8 curve" in the plane cannot be approximated by an embedding.

The embedding of two kissing circles in the plane, then thought of as lying in $\mathbb{R}^3$, can be approximated by infinitely many *non-isotopic* embedded $S^1 \cup S^1$s distinguished by their *linking number*. So there is not a uniqueness theorem that goes with the above existence result (unless the dimension of the ambient space is even larger than what is demanded above).

Recall that the linking number of two disjoint oriented (compact) cycles X^x and Y^y in $\mathbb{R}^n$ (or S^n) with $n = x + y + 1$, namely $\mathrm{lk}(X, Y) \in \mathbb{Z}$, can be defined as the intersection number $\mathrm{int}(Z, Y)$, where Z is any chain bounded by X. This definition is not quite symmetric; viewing X and Z as cycles on the boundary of $\mathcal{D}^{n+1}$ we can define the linking number more symmetrically as $\mathrm{int}(Z, Z')$ where Z bounds X and Z' bounds Y. This then shows that

$$\mathrm{lk}(X, Y) = (-1)^{(x+1)(y+1)} \mathrm{lk}(Y, X).$$

Linking invariants and their generalization are fundamental to the theory of embeddings.

We will need variants for non-simply connected situations, and for more general targets. Note that the current definition only really involves knowing the vanishing of certain homology classes and certain (other) homology groups. If the cycles involved are simply connected, such a theory already arises in the proof of the h-cobordism theorem and in surgery theory – intersection (and self-intersection) numbers take values in (a quotient of) $\mathbb{Z}\pi$ – and one can occasionally define an associated linking theory.

If the cycles are non-simply connected there is more indeterminacy in their definition and we have to mod out by the influences of the fundamental groups of X and Y.

Theorem 6.59 *Suppose M is a connected oriented submanifold in W and that $w = 2m + 1$. Then the embeddings of M homotopic to the given one are in a one-to-one correspondence with*

$$\mathbb{Z}[\pi_1 M \backslash \pi_1 W / \pi_1 M] / \{g \not\equiv 1, g - (-)^m g^{-1}\}.$$

Addendum 6.60 If M consists of several components, then there are additional invariants that live in $\mathbb{Z}[\pi_1 M_i \backslash \pi_1 W / \pi_1 M_j]$. These have appropriate symmetry associated to interchanging i and j.

We shall only prove the theorem for the topological locally flat case (or PL case) and shall avoid thereby some arguments necessary for the smooth case (which are given in Whitney's well-known paper). We shall use the following basic theorem (concordance implies isotopy) that is an elementary consequence of the h-cobordism theorem:

Theorem 6.61 *If $i: V \subset W$ is an embedding with codimension greater than 2, then any proper embedding of $V \times [0, 1]$ in $W \times [0, 1]$ which restricts to i on $V \times \{0\}$ is equivalent to $i \times [0, 1]$.*

This is completely false in codimension 2, and in codimension 1 it is true for "incompressible" (i.e., π_1-injective locally two-sided) embeddings.

By the way, note that this theorem implies the Zeeman unknotting theorem: any locally flat embedding of a sphere in another with codimension greater than 2 is equivalent to the inclusion of an equator. (In other words, there's only one embedding.)

Suppose now we take two homotopic embeddings of M into W. We can homotop the map of $M \times [0, 1]$ into an immersion by Whitney's theorem. We are interested in the self-intersection of this immersion and will try to use the Whitney trick to remove them – completely analogously to what occurs in Wall (1968) in the description of the even-dimensional surgery groups – just taking into account the fact that M is not simply connected.

The self-intersection points (for a generic immersion) are all labeled by ± 1 according to orientation conventions. Moreover, choosing a base point and a path to each sheet of the intersection, we can get an element of the fundamental group by going along one sheet to the intersection and back to the other sheet. Note that there is an indeterminacy of which is the first or second sheet, and also of the paths from the base point – this gives only a well-defined double coset. Now, as usual, when two intersection points have the same group element and opposite signs, they can be cancelled.

The coefficient of the identity can be modified by changing the immersion near a point, or by dealing with embeddings of punctured versions of M and using the uniqueness of the embedding of S^{m-1} in S^{2m} to complete the discussion.

These kinds of invariants are relevant exactly at the "edge of the gap hypothesis." To go further, all of these linking invariants need to take values in homotopy groups of spheres, rather than $\mathbb{Z}$ (which equals π_0^s). This will suffice for getting through the metastable range. That this should be the case is pretty clear: if one considers embeddings of

$$S^n \cup S^n \subset S^{2n+1-k},$$

for n large, there is a natural invariant in π_k^s that turns out to determine the embedding. Using Zeeman unknotting, the complement of the first sphere is homotopy-equivalent to S^{n-k} (the simpler observation that the linking S^{n-k} included in the complement is a homotopy equivalence, by the Whitehead theorem and Alexander duality, or even by a Mayer–Vietoris argument, suffices for this purpose), and therefore the second sphere defines an element of $\pi_n(S^{n-k}) \cong \pi_k^s$. The relevant symmetry property can be proved similarly to the symmetry of the linking numbers in the stable range.[110]

In a less stable range of embeddings, e.g. for $S^3 \cup S^3 \subset S^6$ so that the corresponding invariant would take values in $\pi_3(S^2) \cong \mathbb{Z}$ via the Hopf invariant – and the two definable "Hopf linking numbers" can be different[111] (although they must agree mod 2 by the stable result).

Moreover, for embeddings of S^n in, for example, $\mathbb{T}^{2n+1-k}$, one would get an invariant in the "group"[112] $\pi_k^s[\mathbb{Z}^{2n+1-k}]$ with the coefficient of 0 being 0, and there being a symmetry condition connecting the coefficients of g and $-g$.

To illustrate the key ideas, let's work out some especially nice cases; for convenience, I will concentrate on crystallographic groups with holonomy $\mathbb{Z}_p$ an odd prime, acting with connected fixed set. We assume that p is odd so as not to get caught up in the surgery difficulties; no problems due to Nil or UNil – all of the isovariant structure sets vanish in this case, and this helps both for the existence of actions as well for their classification. It also helps with actually

---

[110] This uses the Pontrjagin interpretation of stable homotopy groups of spheres are framed cobordism.

[111] As John Klein pointed out to me.

[112] It's actually a tensor product, but if we were in the stable range, this notation would evoke a group ring.

doing the homotopy theory. As mentioned above, the $\mathbb{Z}$ gets replaced by π_i^s as we move forward, and for $p = 2$ we are not given much slack as $\pi_1^s = \mathbb{Z}_2$.

Theorem 6.62 (Cappell and Weinberger, unpublished) *If p is an odd prime, then (assuming $k > 1$ if $p = 3$)*

$$S^{\text{equi}}\left((\mathbb{T}^p)^k \times \mathbb{T}^{(p-2)k-1}\right) \cong \mathbb{Z}[\mathbb{Z}^{(p-1)k+(p-2)k-1} - \{0\}]^{\mathbb{Z}_{2p}}.$$

Here $\mathbb{Z}_p$ acts on $\mathbb{T}^p$ by permutation, and otherwise trivially. The extra $\mathbb{Z}_2$-action reflects the symmetry that linking numbers satisfy, so it gives a $+/-$ factor depending on some parities relating the coefficients of g and g^{-1}.

If one increases the size of the $\mathbb{T}^{(p-2)k-1}$ factor, then one moves deeper into the metastable range, and the one gets additional factors. One extra circle then gives another factor of $\mathbb{Z}[\mathbb{Z}_2^{(p-1)k+(p-2)k-1} - \{0\}]_{2p}^{\mathbb{Z}}$ where this corresponds to the π_1^s linking, etc. Throughout the metastable range we have that $S^{\text{equi}} \cong \text{Emb}(F \subset \mathbb{T})^{\mathbb{Z}_p}$. The original method will be explained in §6.11, but *morally* it follows from the fact that the Tate cohomology of $\mathbb{Z}_p$ acting on the embeddings is trivial.[113] Alas, this vanishing of Tate is computational in nature.

Remark 6.63 I believe that there is an example where the equivariant Borel conjecture fails for an isovariant Poincaré complex reason when the Tate cohomology is nontrivial. Indeed, I would not be much surprised if the analogous crystallographic actions for $\mathbb{Z}_2$ already include such examples, but I did not succeed at doing these calculations.

To go beyond the metastable range,[114] first of all, the homotopy theory becomes unstable (it should go without saying).

One also needs versions that are "multiple linking invariants" that arise from triple points and higher. This is because of the phenomenon of the Borromean rings (Figure 6.4): one can have three linked spheres that are pairwise unlinked in this range. And as one goes further into deeper ranges, there are higher and higher-order Borromean phenomena. Examples of this are the μ-invariants of Milnor (related to Massey products in the way that intersection numbers are related to cup products; see Milnor, 1954).

---

[113] There are convergence issues in the "obvious" spectral sequence argument that would lead to this conclusion. Note, however, that there is a similar issue that arises in trying to compare the equivariant maps from X to Y to $[X, Y]^G$ (the homotopy classes of maps that are homotopic to themselves after composing with elements of G). If the action of G on X is free, then the spectral sequence has better convergence properties, because X/G is finite-dimensional, and one does not really have to go to infinite dimensions, despite the implicit infinite-dimensionality of BG that arises in the definition of homotopy fixed sets.

[114] Embedding theory is essentially the homotopy theory of the map when $w \gg 2m$, the "stable range," because of the Whitney embedding theorem; the metastable range is when $w \gg 3m/2$. The next range is when $w \gg 3m/3$, etc. Each successive range requires yet higher-order information. The "calculus of embeddings" described below is one version of how to do this.

Figure 6.4

Let me explain this in the simplest case, in a somewhat non-classical way, relevant to embeddings of aspherical manifolds in one another (and therefore to the equivariant Borel conjecture).

For simplicity let's think about the classification of k-component linked spheres $\bigcup_k S^n \subset S^{n+c}$ in the sphere in codimension $c > 2$. This classification (even when the spheres have different dimensions) was established by Haefliger (1966). Now let's consider this from the Poincaré embedding point of view.

Firstly it is easiest to replace the given problem by the embeddings of $\bigcup_k \mathcal{D}^n \subset \mathcal{D}^{n+c}$ rel∂ (so-called "disk links"). Note that this is $\pi_n \mathrm{PE}(\mathcal{D}^c, k) = C_n(k \subset \mathcal{D}^c)$, where k denotes any k-point subspace of the disk.

Incidentally, it is interesting to note here how different is the (Δ)-space of embeddings of points in the disk to which we are led from the more naive "genuine embeddings." *That* space is a configuration space, and very well studied. In particular, it is a finite-dimensional space in this case – but PE is actually a function space.

The reason is because concordance implies isotopy. All of the embeddings classified in $C_n(\mathrm{F} \subset M)_{(n \geq 1)}$ are, abstractly, i.e. not relative to the boundary, just product embeddings[115] $\mathrm{F} \times \mathcal{D}^n \subset M \times \mathcal{D}^n$. What makes the element nontrivial is that on the top face $\mathrm{F} \times \mathcal{D}^{n-1} \subset M \times \mathcal{D}^{n-1}$ we have a nontrivial automorphism (relative to the boundary).

Thus, we are interested in the automorphisms of $\mathcal{D}^c$ that are the identity on the boundary and send k points to themselves, and the spheres S^{c-1} normal to these points to themselves, and finally map the complement to the complement.

---

[115] By the h-cobordism theorem.

Let us call this space $\text{Iso}(\mathcal{D}^c, k)$:

$$C_n(k \subset \mathcal{D}^c \cong \pi_{n-1}\text{Iso}(\mathcal{D}^c, k).$$

Note that there is a restriction map

$$\text{Iso}(D^c, k) \to \prod \text{homotopy equivalences } (\mathcal{S}^{c-1} : \mathcal{S}^{c-1}).$$

This map is null-homotopic because of the condition that the automorphisms restrict to the identity on the outer boundary. (We can study any factor on its own by filling in the other $(k-1)$ holes, and the geometry then gives an explicit null-homotopy.)

The homotopy fiber of this restriction is easily studied by obstruction theory. The complement we are discussing has the homotopy type of the wedge $\bigwedge_k \mathcal{S}^{c-1}$ and we have restricted these maps on a disjoint union of $(k+1)$ copies of $\mathcal{S}^{c-1}$ in this complement. The associated[116] spectral sequence for this situation has $E_2^{p,q} = H^p(\mathcal{S}^c, k+1; \pi_q(\vee_k \mathcal{S}^{c-1}))$. (It abuts to π_{q-p} (iso-invariant relative to the neighborhoods); one has to remove the

$$\Omega \prod \text{homotopy equivalences } (ph^{c-1} : \mathcal{S}^{c-1})$$

that comes from the injection of the Ω-base in the fibration.)

There are just two lines in this sequence: $p = 1$ and $p = c$. In particular, there is just one nontrivial differential. The homotopy groups that occur as coefficients are of $\bigwedge_k \mathcal{S}^{c-1}$. These are given by the Hilton–Milnor theorem (see Hilton, 1955). They are homotopy groups of $\mathcal{S}^{r(c-2)+1}$ where there is one sphere for each generator on the free Lie algebra in degree r on the generators of a vector space of dimension k. The Lie algebra operation is Whitehead product, and the nontrivial differential can easily be written down using Whitehead products (using the obstruction theory interpretation). For example, if c is even, working rationally the rank of this group is $kL_r - L_{r+1}$, where $L_r = (1/r)\sum \mu(d)k^{r/d}$ (the sum over divisors of r) is the number of generators of the free Lie algebra of degree r on a vector space of dimension k.

The above reworking of Haefliger's classical work can be modified for the Borel conjecture setting,[117] and, remarkably enough, many of the same features hold. For example, for $C_n(k \subset B\pi)$ the spectral sequence still has two lines, and one is taking cohomology of π with coefficients in various free Lie algebras on, for example, $(\mathbb{Q}\pi)^k$. This can be interpreted as multivariable "polynomial" invariants of the embeddings, which (together with symmetry properties) will rationally calculate $\mathcal{S}^{\text{equi}}$ (even outside the metastable range).

The whole story is very complicated, and while the ingredients now seem

[116] See Federer (1956).
[117] As Cappell and I did in our original approach to these calculations.

clear, computationally it currently looks a mess, and surely key aspects of structure elude us. More precisely, there is a calculus of embeddings due to Goodwillie, Klein, and Weiss (see Weiss, 1999; Goodwillie *et al.*, 2001) that puts these types of ingredients together, but it is via a sequence of complicated diagrams, and the analysis of the terms and how they are assembled has only been done in a few cases.[118]

Their theory is a descendant of the work of Whitney and Haefliger, and deals with genuine embeddings,[119] but can be adapted to deal with Poincareé embeddings. The approach goes like this:

One's first approximation to $\text{Emb}(F \subset M)$ might be the result of "gluing" together the spaces $\text{Emb}(\mathbb{R}^f \subset M)$ over all the submanifolds of M isomorphic to $\mathbb{R}^f$. (Note that when one such submanifold is included in another, the restriction map is a homotopy equivalence.) More explicitly, consider the category F of open subsets of F diffeomorphic to balls, and all smaller in diameter than some ε, say the injectivity radius of F, with morphisms being inclusions. One then can take the limit over F of $\text{Emb}(\mathbb{R}^f \subset M)$ as a guess for $\text{Emb}(F \subset M)$.

This doesn't quite work: What that actually gives, after doing the bookkeeping, is essentially the Smale–Hirsch description of immersion theory (Hirsch and Smale, 1959). Of course, this means that there is a global effect that immersion theory doesn't solve: the maps are only locally one-to-one, not globally one-to-one.

This might suggest taking a limit over the category of submanifolds of F isomorphic to two (in addition to the one) copies of $\mathbb{R}^f$ to prevent pairwise intersections. In this category there are morphisms where the two components "collide," i.e. are included in a single component of a larger set.

If we work modulo immersions (i.e. in the fiber of $\text{Emb} \to \text{Imm}$) then we can elide differences between unions of two points versus unions of two submanifolds each isomorphic to $\mathbb{R}^f$ and get some sort of description involving the configuration of pairs of points in F mapping into M. This is essentially the Haefliger theory in the metastable range (Haefliger, 1964).

But, we know that at the end of the metastable range the Borromean phenomenon begins! There are triple linkings not detected by pairs, so we need to go to the category of triples and further. This is the theory described beautifully in Weiss (1996) and developed in the papers surveyed in Goodwillie *et al.* (2001). The upshot is that we know that the k-tuple theory is not determined

---

[118] Although it has excellent theoretical implications, e.g. the theorem of Goodwillie and Weiss (1999) that many spaces of embeddings have finitely generated homotopy groups, or the calculations of spaces embeddings of knotted strings in high-dimensional spheres by Volic (2006).

[119] In other words, for points, it is configuration spaces that arise, rather than the concordance embeddings that arose in our analysis.

by the $(k - 1)$-tuple theory beyond a range, which requires serial elements of the categories we use for formulating "the simplest possible answer" to the problem of calculating embeddings. The main theoretical result of the calculus of embeddings is that in codimension greater than 2 this guess is correct.

However, the reduction of (π_0 of) embedding theory to Poincaré embeddings, i.e. the fact that isotopy classes of embeddings up to M in W (in codimension 3) are equivalent to those of M' in W' when M' is homotopy-equivalent to M and W', seems a mystery to these embedding-theoretical methods.[120] As we saw, it is the Poincaré embedding approach that links nicely with the categorical idea of homotopy fixed points. I am optimistic that the coming years will see progress on a useful synthesis of these points of view and their combination with the Farrell–Jones conjecture.

6.11 Notes

In §§6.1 to 6.3, the non-uniqueness of the isometry in a homotopy class is always a torus. This follows from the theorems of Borel explained in §6.1. The fact that the space of homeomorphisms homotopic to the identity is not even connected in high dimensions (except for the case of contractible manifolds relative to the boundary) is due to Hatcher (1978).

Although we make the choice here of "going cubist," i.e. dealing with blocked, rather than parameterized, structures, one need not do so. The way to go would then be obstructed by two issues. The most serious is that we do not understand Homeo(M), the space of homeomorphisms of a compact manifold, in high dimensions *except in a stable range* that is linear in the dimension of the manifold. This story is largely the story of pseudo-isotopy theory and Waldhausen's "algebraic K-theory of spaces." We refer the reader to Weiss and Williams (2001) and Rognes and Waldhausen (2013). The second obstacle is that even in the stable range we do not get contractibility because of non-trivial pseudo-isotopy spaces.[121] However, this is considerably illuminated by the Farrell–Jones conjecture that "blames" the whole difference on the (stable) pseudo-isotopy space of the circle.

---

[120] Indeed, these methods work more or less the same way in all categories: the essential difference is in their "base case" which is immersion theory, a subject governed by an h-principle, but with different homotopy theory in the different categories. However, the reduction of embeddings to homotopy is only true in PL and Top , not Diff. Presumably, one has to add the immersion of M to the embedding of M' to relate the two. In any case, I do not know how to do this.

[121] Pseudo-isotopy spaces are precisely the difference between blocked and parameterized structures.

The equivariant problems studied in this chapter have a long history – indeed pre-dating Borel. Originally, the philosophy of group actions was to relate all actions to "linear ones." For example, Smith showed that (in the terminology of this chapter) a pseudo-equivalence induces an isomorphism on the F_p homology of the fixed sets of all p-subgroups. (Borel introduced the Borel construction in his famous seminar on transformation groups (Borel, 1960) to give a more conceptual proof of Smith's theorems and extend their scope.) The theme then became to try to compare group actions on "standard" spaces like the disk, sphere, Euclidean space, projective space, etc., to their "linear models" if there were any (see e.g. Bredon, 1972). In low dimensions, there was the goal of geometricization (achieved through the work of Perelman, 2002, Thurston, 2002, and Boileau *et al.*, 2005). In higher dimensions, more and more of the early conjectures of this sort were disproved – first via isolated examples and subsequently systematically.

Among the early results in this "contrary" direction were $\mathbb{Z}_n$-actions on Euclidean space with no fixed points for all n that are not prime powers by Conner and Floyd (1959), an example of Floyd and Richardson (1959) of a group action on the disk with empty fixed set – subsequently developed into the theory of Oliver numbers (Oliver, 1975), and the theory of L. Jones (1971) of converses to Smith theory. Also of great importance was the spherical spaceform problem of determining which finite groups act freely (and to a lesser extent, the classification of these actions) on spheres (which was settled by Madsen *et al.* (1976): all subgroups of order p^2 and $2p$ must be cyclic, the first a fact from Smith theory, and the second a geometric result of Milnor (1957)) – which is different than the situation for free linear actions (where all subgroups of order pq must be cyclic: no metacyclic groups can act linearly).

We were left with a theory of enormous complexity, where all conjectures were false; the positive principles were the conclusions of Smith theory for p-groups, and converses to the combination of Smith theory (due to Jones, 1971) with the Lefshetz fixed-point theorem for non-p-groups (Oliver, 1976b), and a few standout classification results. The differences between the differentiable, PL, and topological categories became abundantly clear from the late 1970s through the 1980s (some of which are explained in §6.2 on trivialities, and some of which depend on the isovariant surgery and equivariant Novikov conjecture results that come later).

There are still a number of standout problems from the early days. My favorites: (Petrie's conjecture) if a homotopy $\mathbb{CP}^n$ has a smooth circle action, must it have the same Pontrjagin classes as $\mathbb{CP}^n$? Does every finite group act freely on a product of spheres? (Or more ambitiously, which groups act on which products?) And what are the possible fixed sets of PL $\mathbb{Z}_n$-actions on disks?

I think that the place for new progress in the theory is the world of aspherical manifolds, where rigidity suggests interesting problems. This chapter and the next give some initial results on the equivariant Borel conjecture, on the Nielsen realization problem, and so on. The following is another problem of the same sort.[122]

Conjecture 6.64 *If M is an aspherical manifold whose fundamental group has no center, then only finitely many groups can act effectively on it. If it has center of rank k, then it has a product of at most k-cyclic groups as a subgroup of bounded index.*

Turning now to more specific things mentioned in the body of the text. The construction of counterexamples to the Smith conjecture given here is surely folklore. In dimension 4, the PL Poincaré conjecture is not known, and in any case, the method we used here requires knots whose complement has fundamental group $\mathbb{Z}$. In dimension 4, thanks to the work of Freedman, any such knot is topologically trivial. However, Giffen's construction is very explicit, and is based on "twist-spinning" so one has no need for the Poincaré conjecture.

The theory of Cohen and Sullivan was the PL predecessor to the theory of resolution of homology manifolds. It also foreshadowed the work of Matumoto (1978) and Galewski and Stern (1980) on non-PL triangulations of topological manifolds, leading to the final result of Manolescu (2016) that there are topological manifolds of arbitrarily high dimensions that are not homeomorphic to polyhedra.[123]

The theory by Cappell and me of Rothenberg classes (Cappell and Weinberger, 1991a) measures the lack of homogeneity that might be present in a semi-free PL action whose fixed set is a manifold. It was established in the context of trying to understand the possible neighborhoods of fixed sets of semi-free group actions. It is very similar in spirit to the characteristic class theory $BSRN_2$ of abstract regular neighborhoods in codimension 2 invented by Cappell and Shaneson (1976, 1978).

Examples 6.11 and 6.12 are inspired by Milnor's (1961) counterexamples to the *hauptvermutung* for polyhedra: there are homeomorphic non-PL homeomorphic polyhedra. Milnor relied instead on the "stable classification theory" of manifolds by Mazur. This example has a beautiful irony: Whitehead torsion

---

[122] As far as I know, this is a conjecture of my own. I don't know whether I really believe it.

[123] For example, the topological manifold which is $S \times E_8$ where E_8 is the unique simply connected closed 4-manifold with quadratic form E_8 can be easily shown to be homeomorphic to a polyhedron, but not to a PL-manifold.

is trivial for homeomorphisms (a theorem of Chapman that follows easily from controlled topology, and also from the work of Kirby and Siebenmann (1977) showing that topological manifolds have handlebody structures).

Rothenberg (1978) developed the PL analogue of torsion for the equivariant setting. The torsions lie in a (group isomorphic to the) sum of Whitehead groups of the equivariant fundamental groups of the various strata (see §6.7). The upshot of this (see Examples 6.12 and 6.16) is that the equivariant torsion is *not* a topological invariant, and that locally linear G-manifolds are *not* equivariantly finite (and equivariant handlebody structures do not exist).

The result about fixed sets of Q_8-actions alluded to (and elaborated on as Example 6.18) is the following. A submanifold of S^n is the fixed set of a PL locally linear Q_8-action iff it is a mod 2 homology sphere of codimension a multiple of 4 and the product of the order of its integral homology groups is $\pm 1 \bmod 8$. It is the fixed set of a topological locally linear action irrespective of the orders of these groups. The necessity of this condition is due to Assadi (1982) who gave a thorough development of finiteness theory for fixed sets and the connections to numerical invariants and how restrictions on isotropy subgroups influence this problem. His work simultaneously extends aspects of the work of Jones and Oliver mentioned above (see also Oliver and Petrie, 1982).

The remaining result is due to Weinberger (1989) – see also Weinberger (1985a) – based on earlier joint work of Cappell and Weinberger (1991a) in order to even build actions on neighborhoods. (The extension of the action from the neighborhood to the whole sphere uses "extension across homology collars"– a result of Assadi and Browder, 1985, and Weinberger, 1985a.) The actual theory is more general – e.g. one can easily replace Q_8 by other groups, but the specific criteria will be different. I picked an example that was easy to state.

Quinn (1988) is a landmark in the application of controlled methods to stratified spaces. The paper is foundational: besides excellent results on concrete problems (e.g. to orbifolds) it puts everything in the right general context. In particular, it contains two very important results: the homogeneity of strata in a homotopically stratified space (which then implies many local linearity results for various constructions of group actions, which typically look locally linear aside from some limit set and can then be deduced to be locally linear everywhere[124]) and the topologically invariant h-cobordism theorem. (In the case of orbifolds, Steinberger (1988) proved essentially the same h-cobordism theorem, expressed very differently, based substantially on extending the ear-

---

[124] It also is among the motivations for the conjecture in Bryant *et al.* (1993) about the homogeneity of DDP ANR homology manifolds.

lier ideas of Chapman from the unequivariant case.) This theory justifies the comments in §6.3.

A precise version of the statement about the difference between Top and PL being algebraic K-theory can be found in Anderson and Hsiang (1976, 1977, 1980), which preceded the theory of Quinn (1979, 1982b,c, 1986). This theory can also be used for the purposes of the footnote in Example 6.17.

The problem of the stability of equivariant classifying spaces for neighborhoods is essentially equivalent to the issue of equivariant transversality in the topological setting. This is the point of view of Madsen and Rothenberg (1988a,b, 1989), and the work that they did on nonlinear similarity was part of a deep analysis of the category of locally linear G-actions when G is of odd order. Their approach to geometric topology required transversality and only worked for odd G – because of Example 6.17! The stratified surgery approach works more generally, but it lacks some of the depth of the Madsen–Rothenberg approach. In Chapter 7 we will see some phenomena where equivariant $K(\pi, 1)$ manifolds are really different for $\mathbb{Z}_2$ than for odd-order groups, essentially for reasons that boil down to this transversality issue (although perhaps translated significantly into other more algebraic language).[125]

The fact that, for groups of Oliver number $n_G = 1$, every finite polyhedron[126] occurs as the fixed set of a PL G-action is a modification of a trick of Assadi. It is a completely geometric argument, and I will give it here.

Proposition 6.65 *If G acts piecewise linearly on a disk with empty fixed set, then for any finite complex F, G acts on some disk with fixed set F.*

Proof Let $\mathcal{D}$ be the G-disk with empty fixed set. Consider $(F \cup x) * \mathcal{D}$, where x is a disjoint base point and $*\mathcal{D}$ denotes taking the join with $\mathcal{D}$. This has a G-action with fixed set $F \cup x$. Unfortunately, this join is not a PL disk – it is contractible. One can therefore take an equivariant thickening of this G-space (one essentially replaces each simplex with an equivariant handle (see, for example, Assadi, 1982). This produces a G-disk, denoted Δ, whose fixed set is an abstract regular neighborhood (i.e. a thickening) of $F \cup x$. Let's denote this by $\mathrm{Nbd}(F) \cup \mathrm{Nbd}(x)$. Now take another join $\Delta * \mathcal{D}$. This produces another G-disk, whose fixed set is again $\mathrm{Nbd}(F) \cup \mathrm{Nbd}(x)$ with a key difference: the fixed set is now entirely located on the boundary sphere.

Restrict the action to $\partial(\Delta * \mathcal{D})$ and remove an equivariant regular neighborhood of $\mathrm{Nbd}(x)$. This gives an action on a disk with fixed set $\mathrm{Nbd}(F)$ entirely

---

[125] An equivariant signature class for these manifolds turns out not to be an orientation, or, more fundamentally, its restriction to small balls is not a unit in a suitable ring.

[126] And any finite-dimensional compact ANR occurs for "tame topological actions." For instance, the end-point compactification of any locally finite tree (which will frequently have a Cantor set at infinity) is a fixed set.

included in the interior of this disk! Now recall that regular neighborhoods are mapping cylinders – so collapse the mapping cylinder lines down to F. This is still a disk! (Since these cylinder lines are all interior to the disk.) The G-action has fixed set F.   □

Moving on to §6.4, the UNil theory of Cappell applies to all amalgamated free products of groups $A *_B C$, where B injects into A and C. It is equivalent to an appropriate codimension-1 splitting theorem; see Cappell (1976a,b).

Cappell showed that his UNil groups are 2-primary in three senses. First of all, UNil has exponent a power of 2. (It follows from Ranicki's (1979a) localization theorem and the vanishing theorem we will soon assert that it has exponent 8, but Farrell showed that it's actually of exponent 4 in general.[127]) Second, if one studies $L(R\pi)$, and $1/2 \in R$, then UNil vanishes. It is for this reason that when we discuss the L-theory of groups with torsion, we can reasonably conjecture that

$$H_*(\underline{E\Gamma}/\Gamma; L(R\Gamma_x)) \to L(R\Gamma)$$

is an isomorphism for all Γ, if $1/2 \in R$, while for $R = \mathbb{Z}$, we need to replace $\underline{E\Gamma}$ by $E_{vc}\Gamma$ in the Farrell–Jones conjecture. The third and final vanishing theorem of Cappell is that, if B is square-root-closed in both A and C, then UNil vanishes. This condition means that if, for example, $a^2 = b \in B$, then $a \in B$. So, for the case of connected sums, the square-root-closed condition applies iff the fundamental group has no 2-torsion.

UNil, as mentioned in the text, depends relatively little on the groups A and C, but rather significantly on the group B. The work of Connolly and Davis referred to in the text gives complete information for B trivial. It is clear that the case of B finite should be studied next, especially in light of the Farrell–Jones conjecture.

As we turn more seriously towards the equivariant Borel conjecture, it is important to refer to the early work of Connolly and Kosniewski (1990, 1991) on this problem. Their work (based on the ideas of Farrell and Hsiang that will be explained in Chapter 8) gave a number of cases of odd-order group actions on tori where the equivariant Borel conjecture is true and some counterexamples based on Nil (if one did not assume topological simplicity). They raised the issue of whether the gap hypothesis[128] could be another source counterexample. I had pointed out to them in a letter that UNil was another source counterexample. Thanks to the important work of Connolly and Davis (2004),

---

[127] And for the infinite dihedral group there are elements of order 4 as Banagl and Ranicki (2006) and Connolly and Davis (2004) show.

[128] That is, when the fixed set of some subgroup was more than around half the dimension of the manifold (or some other stratum it is included in).

Banagl and Ranicki (2006) and Connolly *et al.* (2014), involutions satisfying the gap hypothesis can have their equivariant structure sets analyzed (at least when the fixed set is discrete[129]). In a subsequent paper Connolly *et al.* (2015) deals with the analysis of equivariant structure sets when the singular set is discrete, assuming the Farrell–Jones conjecture holds. This covers many cases in light of the verification of that conjecture by Bartels and Lück (2012a) for all CAT(0) situations.

That there are failures of the equivariant Borel conjecture because of the gap hypothesis as well was first shown in Weinberger (1986). That a tighter connection to embedding theory should exist was explained in Weinberger (1999a). That paper defined concordance embedding groups, proved some of the theorems in §§6.9 and 6.10, and suggested that there might be a "Sullivan conjecture for equivariant structure sets." Shirokova's unpublished University of Chicago thesis showed that the counterexamples given for actions on the torus could be generalized to all finite group actions where the singular set was of the right dimension. In particular, she realized the role of double cosets in the relevant linking theory.

The first precise classification results (where the result was not just that the structure set vanishes) were those arising from joint work with Cappell in the situation of $\mathbb{Z}_p$ odd acting affinely on the torus. The method used an equivariant analogue of Farrell's fibering theorem (see Chapter 7) to reduce it to understanding monodromies, and then calculating with the Federer spectral sequence (and an isovariant analogue). After the event, it seemed that the results could be explained very well by the fact that the Tate cohomology of $\mathbb{Z}_p$ acting on the embeddings vanished. The realization that this would follow from the "Sullivan conjecture" mentioned above and some conversations with John Klein led to the treatment given here.

The conjecture that assembly maps:

$$H_*(\underline{E\Gamma}/\Gamma; K(R\Gamma_x)) \to K(R\Gamma),$$
$$H_*(\underline{E\Gamma}/\Gamma; L(R\Gamma_x)) \to L(R\Gamma),$$

could be isomorphisms was made by Quinn (1985b) (see also Quinn, 1987a), recognizing that they were false because of Nils and UNils. The *h*-cobordism theorem in Quinn (1970) and the stratified surgery theorem in Weinberger (1986) relate these to rigidity, as Quinn points out (at least regarding *K*-theory). Of course, the issues regarding Nil and UNil were only confronted by the Farrell–Jones conjecture.

---

[129] The case they deal with explicitly. However, most of their paper directly generalizes to the case asserted.

In §§6.5 and 6.6, besides equivariant rigidity, other motivations for stratified surgery were the nonlinear similarity problem and the development of intersection homology.

The main positive results about nonlinear similarity were the case of odd p-groups, proved by Schultz (1977) and Sullivan (unpublished) and then the general odd-order case by Hsiang and Pardon (1982) and Madsen and Rothenberg (1988a,b, 1989). The Hsiang and Pardon approach can be compared to a daring commando raid, while Madsen and Rothenberg's was like a major strategic effort aimed at much broader objectives. Meanwhile for even-order groups, Cappell and Shaneson showed that nonlinear similarities exist (and gave a type of stable classification, as we had mentioned). Hambleton and Pedersen (2005) gave a solution of the problem for all cyclic groups.

Intersection homology (Cheeger, 1980; Goresky and MacPherson, 1980, 1983) gave a perspective from which many stratified spaces (e.g. complex varieties) could be viewed as being like manifolds (e.g. satisfying Poincaré duality). It became natural from that point of view to wonder whether surgery theory could be extended to that setting. In a piece of work that briefly preceded stratified surgery, Cappell and I showed how to extend surgery to the "supernormal even-codimensional stratified spaces" (Cappell and Weinberger, 1991b). In that setting all the usual theorems about manifolds naturally extend (such as the Novikov conjecture for stratified homotopy equivalences).

For §§6.7 and 6.9, the equivariant Novikov conjecture was first studied in Rosenberg and Weinberger (1990). We realized how closely it fit into the framework used by Kasparov – at least in many cases. Our interest was for both its topological and its differential geometric implications. Gong (1998) and Hanke (2008) amplify each of these directions, respectively. One point that we did not appreciate at the time is the one made in the appendix to §6.7 – i.e. that the equivariant Novikov conjecture has systematic failure when one does not have the relevant injectivity of fundamental groupoids of fixed sets.

The connection between the equivariant Novikov conjecture and equivariant surgery was a fortuitous conclusion. That the L-groups break up (for finite group actions) away from 2 is a general phenomenon (Lück and Madsen, 1990a,b; Cappell *et al.*, 1991). At 2, Lück and Madsen (1990a,b) give a general result for locally linear G-manifolds. Cappell *et al.* (2013) describe some integral splitting results of a "replacement theorem" sort. The relΣ theory has good equivariant functoriality, which leads to a good formulation of such isovariant surgery in terms of assembly maps. The not relΣ theory has some functoriality as well – hopefully Cappell, Yan, and I will write a paper about this in due course – but currently it is a difficult and complicated set of examples.

This work on functoriality is based, as is the functoriality relevant to the

Atiyah–Singer theorem (Atiyah and Singer, 1968a,b, 1971), on periodicity theorems. The first periodicity theorem for isovariant structure sets was due to Yan (1993) for odd-order groups, and was based on the method of Browder explained for the decomposition theorem in §6.9. Weinberger and Yan (2005) proved a similar theorem for general compact groups. Unlike the Browder method, this requires stratified spaces rather than G-manifolds. We therefore have not yet been able to prove a decomposition theorem in general.

The pseudo-category is introduced here to foreshadow other uses of homotopy fixed sets in §§6.9 and 6.10. On the other hand, the pseudo-category is very powerful in the theory of group action. The celebrated Sullivan conjecture (now a theorem of H. Miller, 1987; see also Carlsson, 1991; Lannes and Schwartz, 1986) says that the space of pseudo-maps from a point to X is p-adically equivalent to fixed set of $\mathbb{Z}_p$ acting on X (see also Dwyer and Wilkerson, 1988).

The approach we have chosen to give in §§6.8–6.10 for the Farrell–Jones conjecture is the one they gave "after the fact." As I emphasized in Footnote 98, Farrell and Jones were motivated by the role that geodesics played in their proofs of Borel conjecture statements. It was only when they analyzed pseudo-isotopy spaces for non-positively curved closed manifolds (or at least some locally symmetric spaces of that sort) that they were willing to make this bold conjecture.

One point that I think is significant is that the Goodwillie–Klein–Weiss calculus of embedding idea that occurs in §6.10 can be described similarly. Recognizing that an h-principle fails for two points (or larger finite sets), one again finds the simplest functorial expression compatible with true calculations and discovers (following Weiss, 1996) – in their case – a theorem.

Thus, the Farrell–Jones conjecture, the Goodwillie–Klein–Weiss calculus of embeddings, the Sullivan conjecture, and its variant for structure sets are all of one spirit. Given the ubiquity of h-principles (see Gromov, 1986), this somewhat more sophisticated variant might be of help in other circumstances where h-principles fail.

The explicit calculations are influenced by ideas of Kearton, Hacon, Mio, al Rubaee, and Habeggar. I refer to Goodwillie *et al.* (2001) for a survey.

7

Existential Problems

7.1 Some Questions

The topics of this chapter can be motivated from several points of view. The first one is straightforward enough: what kind of existence theorems can be counterparts of the uniqueness statement that is the Borel conjecture?

Another point of view is that, given the Borel conjecture says that the fundamental group of an aspherical manifold determines it, it should follow that all properties of aspherical manifolds should be properties of their fundamental groups alone, and it is interesting to then inquire to what extent we can create such a dictionary.

A third point of view, like in Chapter 6, is to take the geometrical reasoning of the Borel conjecture seriously, and try to elaborate on the connection between the Riemannian geometry and topology.

Let me now be more specific, starting, as we often do, with some theorems of Borel (see Borel, 1983) regarding compact group actions on compact aspherical manifolds. First, he provided a lot of information about the identity component in the following theorem.

Theorem 7.1 *If M is a closed aspherical manifold, and G is a connected Lie group acting effectively on M, then $\pi_1(G) \to \pi_1(M)$, induced by the inclusion of an orbit, is injective, with image lying in the center of $\pi_1(M)$.*

Corollary 7.2 *Under the above conditions G is a torus, and if, in addition, $\pi_1 M$ is centerless, it must be trivial.*

For, in a noncommutative compact Lie group, the maximal torus does not inject on the fundamental group. And, the centrality of the image of the orbit is because there is a continuous map $G \times M \to M$.

This theorem concentrates one's attention on finite groups.

Theorem 7.3 *If $\pi_1 M$ is centerless, then for any G-action on M the map $G \to \text{Out}(\pi_1(M))$ is injective.*

Note that this also includes the previous corollary (since $\text{Out}(\pi_1(M))$ is discrete). Note that Mostow rigidity implies that, for $M = K\backslash G/\Gamma$, we have $\text{Isom}(M) = \text{Out}(\pi_1(M))$. So the isometry group in this case is as large as can be. Of course, since this is a finite group, there are many other metrics whose isometry groups are as large as this one.

Among questions that we will focus on in this chapter are the following:

(1) If the fundamental group of an aspherical manifold M has center, does the manifold admit a circle action? (Conner–Raymond conjecture.)

(2) If G is a finite subgroup of $\text{Out}(\pi_1(M))$ for an aspherical manifold whose fundamental group is centerless, is G realized by a group of homeomorphisms of M? (Nielsen realization problem.)

(3) In what senses is the symmetric metric on $K\backslash G/\Gamma$ the most symmetric one?

The discussion can start on general spaces. The first important negative examples are due to Raymond and Scott (1977); they show that for certain three-dimensional nilmanifolds, there are finite subgroups of $\text{Out}(\pi_1(M))$ that are not realized by any group actions at all on any space with the given fundamental group!

These are based on an algebraic obstruction: they show that (in their situation) there is no group extension

$$1 \to \Gamma \to \pi \to G \to 1$$

(where G has the given action on Γ) because of an obstruction that lies in $H^3(G, Z(\Gamma))$. This prevents the group action from being realized on any space with fundamental group Γ and has nothing to do with asphericity!

If the center of Γ is trivial, then an extension always exists, and the phenomenon lies deeper. For example, there is always then a G-action on some space homotopy equivalent to the $K(\pi, 1)$ on which G acts in the desired way (i.e. the regular G-fold cover of $K(\pi, 1)$). However, this will typically be an infinite-dimensional space – e.g. if π has elements of finite order.

This is the result of building the group action as a free one rather than allowing fixed points. (If π has torsion, this will be infinite-dimensional, although, by taking a skeleton, we can always avoid that.)

When there is a global fixed point to the action, then the map $G \to \text{Out}(\pi_1(M))$ lifts to $\text{Aut}(\pi_1(M))$. For actions that are "as aspherical as possible" (i.e. associated to the action of G on $\underline{E}\pi/\Gamma$, i.e. so that on the universal cover all finite subgroups have contractible fixed sets), the converse holds.[1]

The opposite extreme is where *no nontrivial element of G lifts to* $\mathrm{Aut}(\Gamma)$. In that case the group[2] π automatically satisfies Poincaré duality. The Nielsen conjecture being true would then boil down to the following statement.

Conjecture 7.4 *An aspherical Poincaré complex is homotopy-equivalent to a manifold if it has a finite sheeted cover that is.*

(We have tacitly used the fact that a finite complex satisfies Poincaré duality iff a finite sheeted cover does.)

This is a special case of a conjecture of Wall.[3]

Conjecture 7.5 *If π is a group satisfying Poincaré duality, then there is a closed aspherical manifold with fundamental group π.*

This is very natural from the point of view that every uniqueness statement, like the Borel conjecture, should have an existence statement that goes along with it. Thus, one should ask whether every $K(\pi,1)$ which could conceivably be a manifold is one.[4,5]

A similar question would be:

Conjecture 7.6 *An aspherical manifold is a product of two manifolds iff its fundamental group is a nontrivial product.*

This is a consequence of the Wall conjecture, as we leave as an exercise.

One might be so bold as to make similar conjectures about fiber bundles and so on. The stage is set for the problems we plan to address in this chapter. Section 7.2 begins with the Wall conjecture.

---

[1] One can construct actions on finite aspherical complexes where there is a map $G \to \mathrm{Aut}\left(\pi_1(M)\right)$, but there are no fixed points for the action. This requires G not to be a p-group. The manifold case is more difficult. In recent work with Cappell and Yan (Cappell *et al.*, 2020) we show that indeed, for each non-p-group, one can find actions without fixed points that lift to $\mathrm{Aut}(\Gamma)$.

[2] That is, $K(\pi, 1)$ does.

[3] Wall did not conjecture this: he asked it as a question (Wall, 1979). We are here commiting a standard historical crime of attributing the positive answer to a question as a conjecture of the proponent if it lasts more than a few minutes. This is especially venal in my case, since I do not believe this conjecture.

[4] I had originally planned on using Wall's conjecture as the first existential problem, but I decided instead to follow Borel's trail.

[5] Exercise: Show that the question of whether the universal cover of an aspherical manifold is $\mathbb{R}^n$ only depends on the fundamental group.

7.2 The "Wall Conjecture" and Variants

As Poincaré duality would follow from the existence of an aspherical homology manifold, it is much more reasonable[6] to conjecture that that is what exists in the presence of Poincaré duality.

Expanding the Borel conjecture in this way, i.e. in the uniqueness statement, is actually equivalent to the version if one allows ANR DDP (Absolute Neighborhood Retract Disjoint Disk Property) homology manifolds in the class of objects among which the manifold is unique. (If there were a non-resolvable homology manifold homotopy-equivalent to M, namely a $K(\pi, 1)$ manifold, then there would be a manifold homotopy-equivalent to $M \times \mathbb{T}^4$ with different p_1.) And, if one works in this setting, one would at least expect to get the uniqueness of the homology manifold up to s-cobordism.[7]

However, the Nielsen part would be expected, because the DDP homology manifold would have a cover that is (s-cobordant to) a manifold, and that would make it a manifold. The following seems reasonable to me.[8]

Conjecture 7.7 *The question of whether a torsion-free group is the fundamental group of a closed aspherical manifold only depends on the coarse quasi-isometry type of the group.*

This seems to me quite believable, at least modulo the Borel conjecture. I will explain some of the evidence below and give some heuristic. In particular, we will see, following Bartels *et al.* (2010), that it's true for hyperbolic groups.

For hyperbolic groups, something slightly stronger is suspected (the following is an analogue of the "Cannon conjecture", which we will soon get to).

Conjecture 7.8 *The question of whether a torsion-free hyperbolic group is the fundamental group of a closed aspherical manifold only depends on the boundary of the group.*

Let me remind the reader a little about the theory of hyperbolic groups[9] (Gromov, 1987). The property of a group being hyperbolic is a property of its Cayley graph. Perhaps the simplest description would be that all closed curves in the Cayley graph bound "disks" of area that grows linearly in the length of the curve.[10]

---

[6] At least in my view.

[7] Unfortunately, the surgery exact sequence in Bryant *et al.* (1993) is only completely proved there for homology manifolds that satisfy an orientability condition; I believe it is true in full generality. But, in any case, we would not have uniqueness up to homeomorphism because we don't have any s-cobordism theorem for homology manifolds.

[8] At least in dimensions other than 4, where I have no feeling for what is reasonable.

[9] We already had a brief orientation on this in §2.4.

[10] Or, equivalently, for any Riemannian metric on any compact manifold with that fundamental

However, the more traditional, and probably more intuitive, definition is in terms of "thin triangles." Every triangle in the graph (i.e. a union of three geodesics) is uniformly thin: there is a constant δ such that each side is within δ of the union of the other two sides.

This condition is typical of trees and (the universal cover) of negatively curved Riemannian manifolds. Euclidean spaces of dimension greater than 1 are not hyperbolic, and, for example, hyperbolic groups never contain a $\mathbb{Z}^2$. Part of their joy is that they exist in great profusion. One can add large random relations to non-elementary (i.e. not virtually $\mathbb{Z}$) hyperbolic groups to get new ones; there are gluing or combination theorems for certain amalgamated free products. And, from some point of view, almost all groups are hyperbolic.[11]

The boundary of a hyperbolic group consists of equivalence classes of geodesic rays. An important property of hyperbolic metric spaces is that any "quasi-geodesic," that is, a path that is uniformly embedded in the space, is a finite distance from a geodesic. As a result, this notion of boundary is a coarse quasi-isometry invariant. For trees, this is just the end-point compactification (and usually consists of a Cantor set). For the usual examples of co-compact lattices, the boundary is a topological sphere.

For torsion-free word hyperbolic groups, the boundary is one dimension smaller than the group. Its homology is usually, as in the case of the free group, infinitely generated in at least one dimension. If it is not, then one can show that the group is actually $\mathbb{Z}$ in one dimension, k, and the hyperbolic group is then a Poincaré duality group of dimension $k + 1$.

In dimension 2, the above conjecture is known to be true. According to Eckmann (1986)), all two-dimensional Poincaré duality groups are fundamental groups of surfaces. In dimension 3, in light of the geometricization theorem, this problem is closely related to Cannon's conjecture that torsion-free hyperbolic groups whose boundaries are S^2 are fundamental groups of closed hyperbolic 3-manifolds. Once one gets the 3-manifold in the above statement, the hyperbolic structure will be automatic.[12]

Dimension 4 is out of reach, but in higher dimensions this last conjecture is true.

group, any null-homotopic curve bounds a disk with area linearly bounded by the length of the curve.

[11] This last makes one extremely pleased with the result that the Baum–Connes conjecture is true for hyperbolic groups (Lafforgue, 2002; Mineyev and Yu, 2002), even with coefficients (Lafforgue, 2012) and so is the Borel conjecture – and indeed the whole package – (Bartels and Lück, 2012a).

[12] One cannot hope in dimension 4 to hyperbolize aspherical manifolds with boundary an S^3 because of complex hyperbolic manifolds and the Gromov–Thurston examples (discussed in §2.3).

Theorem 7.9 (Bartels *et al.*, 2010) *Two torsion-free hyperbolic groups with the same boundary of dimension greater than 5 are simultaneously fundamental groups of closed aspherical manifolds or are not.*[13]

Corollary 7.10 *So, if the boundary is a sphere, the hyperbolic group is the fundamental group of a closed aspherical manifold.*

This is as per the conjecture by Wall. (Consider the other group to the fundamental group of a closed hyperbolic manifold.)

To see this, cross with a circle, and then use the Borel conjecture to guarantee that there exists a homology manifold realizing these Poincaré complexes. (This uses the total surgery obstruction, and the result of Bryant *et al.* (1993) that, whenever the total surgery obstruction vanishes, there is a homology manifold realizing the object if one is in dimension greater than 5.) One can then take the cover corresponding to the group $\mathbb{Z}$ (as subgroup) and compactify by gluing to each of these homology manifolds the common boundary. It is not hard to prove that there is stratified homotopy equivalence between these homology-manifold-stratified spaces and thus an element of the relative to the singularity set structures. That group vanishes (consider $S^{\text{strat}}(S^{n+2}\text{rel}\,S^n)$ as a typical example), so these covers are h-cobordant, and thus have the same local index.[14]

Is there any good reason to believe Wall's conjecture regarding its aspect that goes beyond the Borel conjecture?

One can try to guess the analogue of pseudo-equivalence (see §6.7) and then consider the Borel and Wall conjectures in this setting. The following seems like a reasonable choice (to me).

Definition 7.11 A space X is "haspherical" if the map $X \to \mathbf{B}\pi_1(X)$ is a $\mathbb{Z}$-homology isomorphism.

Note that the Borel conjecture implies that, if X is haspherical and π is torsion-free, then X is rigid, i.e. has vanishing structure set if it's a manifold (and if it has a boundary, then working relative to the boundary).

So, we now can ask the Wall question: if X is a Poincaré complex and haspherical, is X homotopy-equivalent to a manifold?

---

[13] If the boundary is a sphere, then the $L^*(\mathbb{Z})$ orientability holds automatically (and indeed a normal invariant can be constructed from the action of the group on the boundary). In general, the Quinn invariant is detected by the controlled at infinity homotopy type of the universal cover of $\mathbf{B}\Gamma$. Recent work of Ferry *et al.* (2019) gives information in low dimensions: in particular, it also handles the case of S^4 as boundary.

[14] The original proof, which loses a dimension, is to consider the universal cover of the homology manifolds, compactify these, and glue them together to obtain a connected homology manifold with each of these universal covers being open subsets. Ferry *et al.* (2019) use a different variant by crossing with S^1.

Proposition 7.12 *Assuming the Borel conjecture for* π, *then* X, *as above, is homotopy-equivalent to an ANR homology manifold, but not necessarily a manifold.*

It is an interesting question to inquire if any π has an haspherical X that is not a manifold. For π a product of fundamental groups of surfaces of high genus, the answer is no (using the Atiyah–Kodaira fibration and their application to the Novikov conjecture, just as we did in §5.3). However, we will presently see that it's doable with $\pi_1 X$ free abelian, so that the space X is an integral homology torus.

That X is homotopy-equivalent to a homology manifold follows from the argument above. Now, for the counterexample, we start with a torus $\mathbb{T}^k$. We take a regular neighborhood of the 2-skeleton and do a Wall realization applied to an element of the form $x \otimes \mathbb{T}^k$ for a nontrivial element x of $L_0(e)$ on the boundary, ∂, of this regular neighborhood. (This produces a normal cobordism $V \colon \partial \to \partial'$ whose surgery obstruction – as a map to $\partial \times [0, 1]$ – is $x \otimes \mathbb{T}^k$ in $L_k(\mathbb{Z}^k)$.) Split the torus along ∂ and glue in a copy of V, by a homeomorphism to the boundary of the regular neighborhood, and by a homotopy equivalence to the complement.

This is almost X. The trouble is that we have not controlled its integral homology. If we could arrange for our normal cobordism V to be a $\mathbb{Z}$-homology equivalence, we would be done. But that is exactly what the Cappell and Shaneson (1974) homology surgery theory is for. We exactly need to know for this that our element vanishes in $\Gamma_k(\mathbb{Z}[\mathbb{Z}^k] \to \mathbb{Z})$. For k odd, this is trivial, because that group is trivial for general reasons about odd Γ groups (they inject in the L-group of their target, which is here $L_{\mathrm{odd}}(e) = 0$). After arranging for the normal cobordism to be an integral homology h-cobordism, we obtain the desired haspherical Poincaré complex. It obviously has a vanishing total surgery obstruction, and is homotopy-equivalent to a homology manifold, but it is not resolvable.

Once we have examples for odd-dimensional tori, we can cross with a circle and get examples in even dimensions as well.

Remark 7.13 One could have asked a different question that might seem more natural: *If a group* π *satisfies Poincaré duality over* $\mathbb{Z}$, *is it the fundamental group of a haspherical (homology) manifold?*

The answer to this is no. Apply the Baumslag *et al.* (1980) construction to a finite simply connected Poincaré complex that does not have any normal invariants (i.e. whose Spivak fibration does not have a lifting – see §3.8). This will be an aspherical $\mathbb{Z}$-Poincaré complex, which cannot be a homology manifold, because, if it were, you can see that the Spivak fibration would have to be reducible.

Another analogue of the Wall conjecture was suggested by Mike Davis (2000). Accepting the idea that one should only ask for $\mathbb{Z}$-homology manifolds from $\mathbb{Z}$-Poincaré duality, the question becomes[15] *If π is a group (with suitable finiteness properties) that satisfies $R\pi$ Poincaré duality, for a ring R that is a subring of $\mathbb{Q}$, then is there an R-homology manifold with fundamental group π whose universal cover is R-acyclic?*

This conjecture has at least one thing to recommend it over Wall's: Wall's conjecture is so hard[16] partly because it is currently very hard to come up with new $\mathbb{Z}$-Poincaré duality groups that are not manifolds by their very construction. For $\mathbb{Q}$ there is a very natural source – namely, any uniform lattice that has torsion. (The action of π on G/K has finite isotropy, but inverting the orders of these groups restores the Poincaré duality that one would have had in the free situation.)

The bad news is that this conjecture is very badly false.

Theorem 7.14 (Fowler, 2009) *If π is a non-torsion-free uniform lattice, and π contains an element of odd order ($\neq 1$), then there is no ANR $\mathbb{Q}$-homology manifold with fundamental group π and $\mathbb{Q}$-acyclic universal cover.*

It is an open problem whether the same holds in the presence of only 2-torsion. (You will soon see the issue when we sketch the argument in the next paragraph.) However, lest one conjecture that Davis's conjecture is missing a torsion-free hypothesis, Fowler has given examples where the $\mathbb{Q}$-homology manifold exists despite the existence of torsion. Interestingly enough, his construction is a Davis construction (see §2.3).

Here's a sketch of why Fowler's theorem is true. Suppose X^n is a $\mathbb{Q}$-Poincaré complex. Then it has a symmetric signature[17] in $L^n(\mathbb{Q}\pi)$. This is a rational homotopy invariant.[18] If X were a $\mathbb{Q}$-homology manifold, then one could lift the symmetric signature back to $H_n(B\pi, L^*(\mathbb{Q}))$ under the assembly map:

$$H_n(B\pi, L^*(\mathbb{Q})) \to L^n(\mathbb{Q}\pi).$$

---

[15] Davis actually asked a slightly different question, and only for torsion-free groups, that Fowler (2012) disproved.

[16] If it's false, that is!

[17] The reader might wish to review some of the discussion in §6.7 at this point.

[18] Actually, there is a slight technical issue. In the $\mathbb{Z}$ case, the symmetric signature is defined up to sign unless one chooses an orientation. We have let this go without saying. In the case of $\mathbb{Q}$, the fundamental class can be sent to any nonzero multiple under a rational homotopy equivalence. This can change, e.g. the 1×1 quadratic form (1) to (k) for some positive integer k. This can change the symmetric signature in $L^*(\mathbb{Q})$ by an element of order 2 or 4. One can live with this issue (say by ignoring the prime 2) or avoid it by keeping track of fundamental classes (like careful people keep track of orientations).

The point is this. When π is torsion-free, this map should be an isomorphism[19] (as part of the Borel package), but when π is not, the right-hand side has additional elements coming from

$$H_n(\underline{E\pi}/\pi, L^*(\mathbb{Q}\pi_x)).$$

The question, then, is where does $\sigma^*(X)$ lie with respect to these extra pieces? Actually it's pretty clear what element it is: we have the equivariant symmetric signature of G/K that lies in the same group and is clearly equal to a lift of $\sigma^*(X)$. At that point, we can use localization theorem technology borrowed from Atiyah and Segal (1968) or (what Fowler does) use a proof of the equivariant Novikov conjecture in this case by going to a g-equivariant symmetric signature of the universal cover G/K in its "bounded L-theory." One gets an obstruction in this way from the ρ-invariant (see §4.10) of the lens space that is normal to a generic point in a stratum of $\pi \backslash G/K$. It is here that one needs a condition. For the free involution on a sphere, the ρ-invariant happens to vanish and the proof breaks down, but for odd primes, the formula in Atiyah and Bott (1967, 1968) shows that the ρ-invariant is never zero and the proof is complete.

Remark 7.15   In many cases where there is 2-primary torsion, the above argument can be applied as well. (In many cases, a more elementary argument using algebraic K-theory suffices Fowler (2012) – a necessary condition is that all the singular strata corresponding to cyclic subgroups have Euler characteristic equal to 0.)

Even the torsion-free case of the Davis question strikes me as unlikely, despite the failure of the ideas above to disprove it.[20] My main reason for hesitation is that we do not have a good theory for surgery on R-homology manifolds for $R \neq \mathbb{Z}$. The assembly map being an isomorphism does not imply that there is a unique homology h-cobordism class of homology manifolds with the given R-homotopy type (as far as I know). Even for manifolds, local surgery theory has a more complicated normal invariant set than ordinary surgery theory (see Taylor and Williams, 1979b).

Despite the falsity of these many variations on the Wall conjecture,[21] we will continue to exploit and expand these ideas in the following section.

---

[19] And this is a case that is actually known, again by the remarkable paper of Bartels and Lück (2012a).

[20] Obviously, when the Borel package is in place, the method above gives no restriction.

[21] By the way, there is a form that is as well founded as the Borel conjecture: *Suppose X is a Poincaré space whose non-empty boundary is a manifold M (i.e. (X, M) is a Poincaré pair). Then if X is (h)aspherical, it is homotopy-equivalent relative to the boundary to a (unique) manifold, rel M.*

7.3 The Nielsen Problem and the Conner–Raymond Conjecture

The best evidence for the free Nielsen problem comes from Borel conjecture via the Wall conjecture. When there are fixed points, the situation is much more complicated.[22]

We will, following Block and Weinberger (2008) and Cappell *et al.* (2013), concentrate on $G = \mathbb{Z}_2$. In this case, Smith theory determines the 2-adic equivariant homotopy type of the action. Indeed, any action will be pseudo-equivalent to the action of $\mathbb{Z}_2$ on $\underline{E\pi}/\Gamma$ (in the notation of §7.1).

(This observation explains why, whenever G is a p-group, a lifting to Aut(Γ) guarantees a global fixed point, because the lift to $E\pi$ has a fixed point by Smith's theorem. However, for $\mathbb{Z}_n$, with n composite, there are fixed-point free actions on Euclidean space (see Bredon, 1972) and this argument fails.[23])

In Block and Weinberger (2008) it is observed that, for very low-dimensional fixed-point sets, the fact that the fixed set is an $\mathbb{F}_2$-homology manifold implies that it's a manifold (see Bredon, 1972) and that the map is an equivariant homotopy equivalence. Cappell *et al.* (2013) deal more directly with the fact that it's a pseudo-equivalence. By being more careful, we can achieve:

Theorem 7.16 *There are closed manifolds W with word hyperbolic group fundamental group Γ such that $\mathbb{Z}_2 \subset$ Aut(Γ) but for which there is no involution realizing this homotopy involution. Indeed, this involution is not realized on any closed ANR homology manifold homotopy-equivalent to W.*

We take our inspiration from the Gromov and Piatetski-Shapiro method[24] (see method three in §2.2.3). In other words, we will build two involutions on aspherical manifolds with (incompressible[25] aspherical) boundary W_1 and W_2 so that we have an equivariant homotopy equivalence

$$h \colon \partial W_1 \to \partial W_2.$$

---

[22] See Farrell and Lafont (2004) for examples of fixed sets of automorphisms of aspherical manifolds that don't have integral Poincaré duality, so the extension does not correspond to an aspherical orbifold.

[23] Thus, for X a finite-dimensional aspherical complex, the lifting condition suffices for prime powers. And, as we mentioned in Footnote 1, conversely, for $\mathbb{Z}_{pq}$, with p and q distinct primes, in Cappell *et al.* (2020), we construct an aspherical manifold with $\mathbb{Z}_{pq}$ in Aut(π), but which has a fixed-point free action by combining the argument in Bredon (1972) with a Davis construction.

[24] We could have been inspired by the way we produced haspherical homology manifolds that are not homotopy-equivalent to manifolds by realizing a surgery obstruction by gluing, or the construction of non-resolvable homology manifolds (Bryant *et al.*, 1993). These, in turn, were inspired by the work Jones (1973) on patch-space decompositions for Poincaré spaces. Of course, we have to recognize a common thread in all these examples.

[25] i.e. π_1-injective

We will arrange that h is homotopic to a homeomorphism h' (this would be automatic if the Borel conjecture were true), so we can form the manifold

$$W = W_1 \cup_{h'} W_2.$$

It is homotopy-equivalent to the result of gluing using h – but that will only give a Poincaré complex with an involution, not a manifold. Since h is not equivariantly homotopic to a homeomorphism, there isn't an obvious reason why W should have an involution in this pseudo-equivalence class (i.e. realizing the same automorphism of π_1) and it will be our problem to eliminate this unlikely possibility.

The involutions and equivariant-homotopy equivalence $h\colon \partial W_1 \to \partial W_2$ will be the UNil counterexamples to the Borel conjecture discussed in the previous chapter (see §6.5). Then the W_i will be built using cobordism theory and relative hyperbolization (see §§2.3 and 2.4).

The UNil obstruction to the Borel uniqueness for ∂W_1 gives rise to a UNil obstruction to the existence of an involution for W. By choosing the initial ∂W_1 to be hyperbolic, the remaining parts of the construction can be done carefully enough to give word hyperbolicity of W.

Remark 7.17 We can use pseudo-equivalences on the boundary to similar effect in making this construction. We have so far only examined carefully examples that come out of equivariant homotopy equivalences and using UNil. Presumably there are also examples that come from Nil (i.e. the simple homotopy condition) or via embedding theory. Possibly pseudo-equivalence allows for phenomena where one would get actions on CW-complexes that cannot be realized on manifolds because there is no $\mathbb{F}_p$ homology manifold $\mathbb{F}_p[\pi]$-homology-equivalent to the fixed set. Among other advantages, these should give rise to examples for p odd, for example.

Now for a few more details.

To obtain an aspherical manifold, one can start with an affine involution on the torus.[26] Such a manifold always bounds equivariantly. We can even make it bound explicitly an equivariant aspherical manifold, so that it is incompressible. This is W_1. We then do the equivariant Wall realization to the free part (as in §6.5). This will produce a smooth involution on the torus with an equivariant null-cobordism (gluing on the null-cobordism of the affine torus). This can be hyperbolized relative to the boundary (Davis *et al.*, 2001) to produce the null-cobordism with involution W_2.

It is a diagram chase involving the equivariant total surgery obstruction

---

[26] Note that this always obstructs the word hyperbolicity of $\pi_1 W$.

(and Cappell's splitting theorem) to see that this equivariant homotopy cannot be realized by an involution on an ANR homology manifold; see Block and Weinberger (2008). If the fixed set is of dimension ≤ 2, this suffices.[27]

However, one can get around this by considering the algebraic mapping cone of the pseudo-equivalence $M \to W$. It gives an element in $S^{\text{alg}}(W \times_{\mathbb{Z}_2} E\mathbb{Z}_2)$. (One should be careful – this mapping cylinder often is not chain equivalent to a finitely generated free chain complex. But, using Wall's (1965) homological criteria for finiteness, it is projective.) The existence analogue will be in the delooped version of this – which naturally has a map to $S_{w-1}(B\pi)$ which has a map to the UNil, as before. (See Cappell *et al.*, 2013 for some more discussion.)

To achieve word hyperbolicity, one starts with an involution on a hyperbolic manifold M inducing the correct orientation character and with fixed set of codimension greater than 2. This can be done as in §2.2 (together with the observations in §6.7 about the splitting off of UNil factors in this situation). (See especially §2.2.3 on grafting, where involutions with codimension-1 fixed sets are constructed; it is easy to modify this to increase the codimension.)

Then we consider $W_1 = M \times [0, 1]$ hyperbolized, to achieve acylindricalness. We then build W_2 as before (see Belegradek, 2006, for why this is incompressible and relatively hyperbolic, with the boundary as the maximal "parabolic"). Then, as in Belegradek (2006), the gluing theorem of Dahamani (2003), or even the more basic one of Bestvina and Feighn (1992), shows that W has word hyperbolic fundamental group.

Problem 7.18 Does there exist a counterexample to Nielsen when W is genuinely negatively curved? Even non-positively curved is not obvious to me.

Now let us turn to the Conner–Raymond conjecture. Recall that this is the question of whether every closed aspherical manifold X, with nontrivial center, $\mathbb{Z}(\pi_1 X) \neq 0$, has a topological circle action.

The X is quite simple: note that W as constructed above has a homeomorphism H inducing the relevant involution on π. Then X is the mapping torus of H,

$$\pi_1(X) \cong \pi_1(W) \rtimes \mathbb{Z},$$

where the automorphism of $\pi_1(W)$ is H_*. Let t be the generator of $\mathbb{Z}$.

Proposition 7.19 *The center $Z(\pi_1 X) \cong \mathbb{Z}$ generated by t^2.*

---

[27] Strictly speaking, one should use the equivariant analogue of taming theory (see Ancel and Cannon, 1979; Ferry, 1992) to replace the action by one where the fixed set is embedded locally flatly. Then the action would be necessarily equivariant homotopy-equivalent to the action of the Poincaré complex W.

As a result, because of the work of Borel (explained in the introduction) any circle action on X must have orbits in the homology class of some nontrivial even power of t. In particular, they must represent nontrivial one-dimensional rational homology classes.

We claim that, in fact, X has no circle actions. Our proof will be based on a lovely theorem of Conner and Raymond.[28]

Theorem 7.20 (Conner and Raymond, 1971) *If X is a connected space[29] with a circle action so that the orbits are nontrivial in $H_1(X;\mathbb{Q})$, then there is a space Y with a $\mathbb{Z}_n$-action such that X can be identified with $(Y \times \mathbb{R})/n\mathbb{Z}$, where $\mathbb{Z}$ acts diagonally on $Y \times \mathbb{R}$ factoring through the $\mathbb{Z}_n$-action on Y and by translation on $\mathbb{R}$. The action of the circle on X is via the left action of $\mathbb{R}/n\mathbb{Z}$.*

The n is related to how divisible the orbits are as elements of $H_1(X;\mathbb{Z})$. Note that one should be a bit careful: the orbits do not all represent the same element $H_1(X;\mathbb{Z})$ unless one views them as immersed circles (i.e. as being given via the orbit map $S^1 \to X$) rather than just as subsets.

To get a feeling for the theorem, let's just consider the case of free actions. (The orbit condition implies that all isotropy is finite, so this is not far off.) In that case X can be described as a principal S^1 bundle over X/S^1. However, these bundles frequently have the homologically trivial fibers. The condition for which this is not the case is that the Euler class in $H^2(X/S^1;\mathbb{Z})$ must be of finite order (i.e. vanish rationally). This leads to a finite cyclic cover of X/S^1 on which the bundle is trivial. That finite cover is Y, and X/S^1 is $Y/\mathbb{Z}_n$.

We note that this theorem is extremely general, and does not apply, for example, to the setting of manifolds. After all, X/S^1 can well be a non-manifold (recall the examples of Bing (1959) mentioned in §6.1), so we will be forced to allow Y to be an ANR homology manifold. (Note that $Y \times \mathbb{R}$ is a cover of X, so we do obtain that Y is an ANR homology manifold from the hypothesis that X is a (ANR homology) manifold.)

That is why we modified the Nielsen problem in our treatment above to exclude the action of $\mathbb{Z}_2$ on any ANR homology manifold in the homotopy type. The proof of the theorem comes about by eliminating any other possibilities

---

[28] Frequently the proofs and disproofs of conjectures are based on the work of the ones who formulate the problem. This might engender a feeling of irony in those of a competitive spirit, yet for those of us who think of mathematics as a magnificent cooperative endeavor, nothing is more natural. Surely, the milestones that are marked by being able to confirm or refute the beliefs of those who have thought profoundly about a subject should be the result of walking further down the road that their insights paved. In this case, Cappell, Yan, and I were surely "walking in the footsteps of giants."

[29] We will suppress the point-set topological hypotheses in this theorem; suffice it to say, that the theorem holds in great generality.

of what the $\mathbb{Z}_n$-action in the Conner–Raymond theorem can look like for the manifold X. These details are not particularly hard.

We note that the manifold X has the following interesting property:

Remark 7.21 As constructed, X has Riemannian metrics g_n so that the indices $[\mathrm{Isom}(\tilde{X}, \tilde{g}_n), \pi_1(X)] \to \infty$ but no metric for which this index is infinite.

This is simply because X has arbitrarily large self-covers (in the topological category; i.e. not by Riemannian "self"-covers) associated to odd-order cyclic quotients of the HNN map $\pi_1(X) \to \mathbb{Z}$ (because odd powers of H are pseudo-isotopic to H).

We will later see that this is indeed unusual: for M homeomorphic to a compact locally symmetric manifold (of non-positive curvature) there is a[30] $C(M)$ so that any Riemannian metric g on M with $[\mathrm{Isom}(\tilde{M}, \tilde{g}), \pi_1(M)] > C(M)$ is actually isometric to a locally symmetric metric (and therefore has G as its isometry group, so the index is uncountable). Farb and I had conjectured[31] that for quite general aspherical manifolds there is a "magic number theorem" (see §7.7 below), but this remark puts an upper bound on the extent to which one can reasonably conjecture that phenomenon (e.g. it might be good to assume that $\pi_1(M)$ is centerless).

Question 7.22 If a closed aspherical manifold X has fundamental group with nontrivial center, can it have a sequence of (Lipschitz) Riemannian metrics g_n, so that $[\mathrm{Isom}(\tilde{X}, \tilde{g}_n), \pi_1(X)] \to \infty$?

In §7.6 we will continue this discussion.

One can also ask whether the Conner–Raymond conjecture is virtually true? The examples that are constructed using failures of Nielsen realization and the Conner–Raymond theorem are virtually products with circles, so for them this is trivially true.

Is there any form of the Conner–Raymond conjecture that is closer to the Borel conjecture? Here is one that I know.

Conjecture 7.23 *An aspherical manifold M has nontrivial center in its fundamental group iff there is a connected topological group G that acts on M so that the orbits are not null-homotopic (i.e. the map $G \to M$ is not null-homotopic).*

That the center is nontrivial if there is such an action is obvious.

Regarding the converse, there is a universal case of this conjecture, namely that $G = \mathrm{Homeo}_0(M)$ (where the subscript 0 indicates the identity component).

---

[30] Which only depends on the volume of M in its standard locally symmetric metric.

[31] Farb and Weinberger (2008).

By the Borel conjecture, the blocked version of this space would be homotopy-equivalent to the space of self-homotopy equivalences of M, whose identity component is a $K(Z(\pi_1 M), 1)$ (presumably a homotopy torus[32]).

So the question is whether $\mathrm{Homeo}(M) \to \mathrm{Homeo}(\tilde{M})$, comparing essentially fiber bundles and block bundles, is, for example, a $\mathbb{Q}$ homotopy equivalence, or at least as far as π_1. Farrell and Hsiang (1978a) explain why this should be true[33] (at least in a stable range, using work of Waldhausen) if one knows that assembly maps in algebraic K-theory are isomorphisms.

Although the above heuristic does not make sense in low dimensions, e.g. dimension less than 7 or 8, nevertheless, I have no reason to doubt (and no good reason to believe) the conclusion.

7.4 Products: On the Difference that a Group Action Makes

Once one gets used to the Borel conjecture for manifolds, and the even larger Borel package extending its reach in various algebraic and geometric directions, one gets used to things like the following:

- Two simply connected manifolds are homeomorphic iff the results of crossing them with any compact aspherical manifold are.
- And that the simple connectivity we assumed is just to avoid algebraic K-theory difficulties (such as Whitehead torsion issues).

We shall see that frequently such statements are indeed consequences of the Borel conjecture, but that it is not quite true for all aspherical "objects." In particular, we will see that this is not true for aspherical homology manifolds that are not resolvable, if there are any. (Or, to vary the point somewhat: it is true for haspherical manifolds but not haspherical homology manifolds. We can cross with the "fake homology tori" constructed in §7.2 and create some interesting homeomorphisms.)

And, the point then becomes even more evident and significant in the equivariant setting. Odd-order locally linear group actions behave like manifolds, but beyond the locally linear setting, or when there's 2-torsion, in some ways these orbifolds act like non-resolvable homology manifolds (or even more extremely).

---

[32] I do not know whether the center can be a group like, for example, $\mathbb{Q}$.

[33] See the discussion referred to in §5.5.3 (and the notes in §6.11). Roughly, the reasoning goes like this: The A-theory assembly map governs topological concordance space theory (which forms the obstructions on a simplex-by-simplex basis to turning a block bundle into a fiber bundle). Rationally, that assembly map is equivalent to the algebraic K-theory assembly map, which is an isomorphism assuming the Borel conjecture in K-theory.

Let's put a little flesh on this skeleton.

Shaneson's thesis from the modern perspective,[34] a restatement of Farrell's:

$$S^s(M \times S^1) \cong S^s(M \times [0,1]) \times S^h(M),$$

$$L^s_{n+1}(M \times S^1) \cong L^s_{n+1}(M) \times L^h_n(M),$$

where the last statement, modulo decorations, is the isomorphism

$$H_n(S^1; L(\pi_1 M)) \cong L_n(\mathbb{Z} \times \pi_1 M),$$

which looks just the Borel conjecture for $\mathbb{Z}$ (with coefficients in the group ring $\mathbb{Z}[\pi_1 M]$). The composition

$$S^h(M) \to S^S(M \times S^1) \to S^h(M)$$

(the left arrow is induced by taking the product with S^1, and the right is applying Farrell's fibering theorem) is the identity, which establishes the injectivity (aside from K-theory) of $\times S^1$. We can apply this n times to get an injectivity statement for taking the product with $\mathbb{T}^n$.

These results are steps in (and analogues of) the relations[35] of $S^{\mathrm{Bdd}}(M \times \mathbb{R}^n \downarrow \mathbb{R}^n)$ to one another via taking the product with $\mathbb{R}$. Essentially all that is affected is the decoration. Aside from the K-theory issues (that effect only the prime 2) these maps are all isomorphisms.[36]

Corollary 7.24 *If Z is a non-positively curved manifold,[37] then taking a product with Z is injective on structure sets (aside from change of decoration[38])*

We use the diagram

$$
\begin{array}{ccc}
S(M) & \to & S(M \times Z) \\
& & \downarrow \\
S^{\mathrm{Bdd}}(M \times \tilde{Z} \downarrow \tilde{Z}) & \to & S^{\mathrm{Bdd}}(M \times \mathbb{R}^n \downarrow \mathbb{R}^n)
\end{array}
$$

where the bottom arrow uses the inverse of the exponential map to see that the top arrow is an injection (modulo decoration).

---

[34] What a wonderful example of terrible history! Shaneson's thesis essentially helped create the modern perspective wherein statements about structure sets and L-groups are viewed as essentially equivalent. Perhaps the best thinking covers itself up (in this way) and (is so successful it) becomes invisible.

[35] Chapman proved that a bounded structure over $\mathbb{R}^n$ can always be "wrapped over a torus" and then is transfer invariant, i.e. isomorphic to any of its finite covers.

[36] Note also that bounded structures are the same as controlled structures, and then by the yoga of controlled topology one should get $H^{lf}_*(\mathbb{R}^n; S(M))$ – except that this only works with $-\infty$ decoration.

[37] The proof uses nonpositive curvature very weakly: it just requires a Lipschitz homeomorphism h of the universal cover with $\mathbb{R}^n$ that has the property that $d(x, y)$ can be bounded in terms of $||h(x) - h(y)||$ (i.e. h must be "effectively proper" or equivalently a "uniform embedding").

[38] And for Z an n-manifold, one loses no more than one does for the n-torus.

The attentive reader might have noticed that this "logarithm" was the key to the proof of the Novikov conjecture for such Z, and therefore come to the conclusion that this injectivity is part and parcel of this package.

And, indeed for manifolds, it is.

Let's think about the bottom line in the diagram where Z is now just the universal cover of an aspherical homology manifold. One might not be able to find the relevant kind of logarithm map, but it still is reasonable to believe that $S^{\mathrm{Bdd}}(M \times Z \downarrow Z) \cong H_Z^{\mathrm{lf}}(Z; S(M)) \cong S(M)$ (with a shift of decoration).

However, what is unreasonable is to expect that this isomorphism is implemented by taking a product with Z. In the manifold case we saw this by unpeeling one R at a time.

In the homology manifold this can't go on all the way down to a point(!), for then it would be a manifold. And, indeed what happens is this. The effect of crossing with a (homology) manifold X in surgery is governed by the symmetric signature $\sigma^*(X)$. It is the image under assembly of a controlled symmetric signature that lives in a group isomorphic to $H_Z^{\mathrm{lf}}(X; L^*(\mathbb{Z}))$.

The usual way to see that multiplying with something is injective is to show that that thing is a unit, or maps to a unit under some map. For a manifold, at least, the image of controlled symmetric signature in $H_Z^{\mathrm{lf}}(X, X - x; L^*(e)) \cong L^\circ(\mathbb{Z}) \cong \mathbb{Z}$ is 1, and therefore if the map from controlled to uncontrolled has good enough properties; that is, assuming the Novikov conjecture, one can expect this product phenomenon.

However, when X is a homology manifold, then this image of the controlled symmetric signature is some number that is $1 \bmod 8$, and it determines whether or not X is resolvable. So, if X has local index equal to 9, crossing with X can kill 3-torsion in a structure set. And, if the local index is 17, then 17-torsion can die, but the 3-torsion is safe, and so on.

In the setting of $\mathbb{Z}$-homology manifolds then, crossing with something aspherical doesn't have to be integrally injective (modulo decorations), but it does have to be (assuming the Novikov conjecture[39]) rationally.

It's interesting to ponder a Poincaré complex P whose total surgery obstruction is of order 17 and that we cross with an aspherical homology manifold X with local index 17 whose canonical Ferry–Pedersen reduction is a stably trivial bundle. Then $p \times X$ will exist, and, if one believes the Borel package,

---

[39] Suspect the sanity of someone who wants to start by considering $\mathbb{Z}_2$ as the first nontrivial example in studying group actions: it is frequently much more difficult than any odd-order group, as we had noted in Fowler's theorem (§7.2), for example, or the nonlinear similarity problem (§6.7). And, indeed, the issue here is quite similar! Needless to say, I can imagine some situations where $\mathbb{Z}_2$ is "the first case" (because one wants only one singular stratum, and vanishing K-groups, etc.; or among people for whom nontrivial means "the first case not yet handled through the efforts of all mathematicians over the course of the previous millennia").

it will approximately fiber over X; however, the local structure will not be a product. (The same thing, of course, happens in the manifold setting when one has a non-simply connected Poincaré complex, which is only finitely dominated, and its total surgery obstruction vanishes in $BS^{-\infty}$ and crosses with a high-dimensional torus.)

Now, let's turn to the equivariant situation. We have so far seen that this venue is richer in phenomena because of Nil and UNil, and, if we choose to be equivariant rather than isovariant, also because of embedding theory. But now we will see that the situation is richer for yet another reason: the local structures that are present are richer (and more geometrically apprehensible) than what occurs in (homology) manifold theory and we can lose the rational injectivity of crossing.[40]

So, let's think about the smallest group, $\mathbb{Z}_2$, and the smallest nontrivial aspherical universal cover R with the involution $X \to -X$. We will denote this by $(\mathbb{R}, -)$. What happens here?

The issue occurs almost immediately.

Let's consider S^n with a free involution. Then taking the product with $(\mathbb{R}, -)$ can be thought of as (at least, it seems closely related to) the "suspension map"

$$
\begin{array}{ccc}
S^{\mathbb{Z}_2}(S^n) & \to \quad S^{\mathbb{Z}_2}(S^n \times (\mathbb{R},-)) \cong & S^{\mathrm{Bdd}}(\mathbb{R}\mathrm{P}^{n+1} - \text{a point}) \\
\downarrow & & \downarrow \\
S(\mathbb{R}\mathrm{P}^n) & \longrightarrow & S(\mathbb{R}\mathrm{P}^{n+1}).
\end{array}
$$

The boundedness in the top line is over $\mathbb{R}$ and $[0, \infty)$. Boundedness over these turns out to be equivalent to propriety. The vertical lines are isomorphisms (any proper homotopy $\mathbb{R}\mathrm{P}^{n+1}$ – a point can be compactified, and the boundary is necessarily a homotopy sphere, and therefore a sphere).

This "suspension map" was analyzed directly by Browder and Livesay (1973) in the early days of surgery. They were interested in the kernel and cokernel in the bottom line to get an approach to the structures of $\mathbb{R}\mathrm{P}^n$. In any case we know that $S(\mathbb{R}\mathrm{P}^n)$ is finite[41] iff $n \not\equiv 3 \bmod 4$, and is the sum of $\mathbb{Z}$ and finite group for $n = 3 \bmod 4$. In any case, it surely is not rationally injective.

Bolstered by this we can decide to directly compute

$$
S^{\mathbb{Z}_2}(S^n) \to S^{\mathbb{Z}_2}(S^n \times (S^1, -)).
$$

Note that $S^{\mathbb{Z}_2}(S^n \times (S^1, -)) \cong S(\mathbb{R}\mathrm{P}^{n+1} \# \mathbb{R}\mathrm{P}^{n+1})$, so there is a very significant

---

[40] Despite expecting rational injectivity of appropriate assembly maps!

[41] For simplicity and connections to the classical literature in this paragraph and the next, S denotes a topological manifold structure, rather than homology manifold structures. (Thus $S(S^n)$ is trivial, not $\mathbb{Z}$.)

lack of surjectivity because of UNil, but also the $\mathbb{Z}$ coming from $L_0(\mathbb{Z}_2)$ for $n = 3 \bmod 4$ also dies.[42]

This "anomalous" product is the key to at least two interesting geometric phenomena.

The first is nonlinear similarity (Cappell and Shaneson, 1981), discussed earlier in this book – the fact that for some even-order groups, like $\mathbb{Z}_{4k}$, when $k > 1$, there are distinct linear representations that *are* conjugate via homeomorphisms. They begin with representations that are not conjugate because of ρ-invariants, and then after crossing with $(\mathbb{R}, -)$ they become conjugate.

The second is a very nice result of Hambleton and Pedersen (1991) related to the classical spherical spaceform problem: namely, which groups act freely on some sphere. The answer, due to Madsen *et al.* (1976), is that a finite group so acts iff all subgroups of order p^2 and $2p$ are cyclic. The first condition is homotopy-theoretic, and follows from Poincaré duality of the putative quotient, but the second condition, due to Milnor, is essentially surgery-theoretic.[43]

The question then arises for groups that act freely, properly discontinuously, and cocompactly on $\mathbb{R}^n \times S^k$, must every finite subgroup satisfy the $2p$ condition? (The p^2 condition is indeed automatic since it comes out of cohomological considerations: see Cartan and Eilenberg, 1956.) Here the answer is affirmative, and basically the reason also involves crossing with $(\mathbb{R}, -)$ a few times.

Formally, the point for group actions is this. When one does a similar analysis for a controlled symmetric signature, now what arises is a more complicated local group rather than $L^*(\mathbb{Z})$; the isotropy enters, and one has – at least! – things like ρ-invariants entering. Away from 2 the information is essentially the same as the equivariant signature operator, and that has the property that for odd-order groups (acting smoothly – so we are dealing with a representation theory problem!) one obtains a unit, but for even-order groups, it frequently is a 0-divisor, with the kinds of implications just mentioned.

In summary: usually homology manifolds and manifolds behave very similarly, but they don't with respect to transversality problems. Bundle structures are much rarer.

Similar issues arise in the orbifold setting. For odd-order groups the equivariant signature (for locally linear actions) is locally a unit, i.e. is an orientation – but for even-order groups it's frequently a 0-divisor. For ANR homology manifolds that are not resolvable, the symmetric signature is neither a unit nor a 0-divisor, so one tends to see phenomena that are intermediate between the differences between odd-order and even-order group actions.

---

[42] Taking the cover associated to $\mathbb{Z}_2 \subset \mathbb{Z}_2 * \mathbb{Z}_2$ gives a computational proof of the bounded vanishing mentioned in the previous paragraph.

[43] In its modern formulation: see J. Davis (1983).

7.5 Fibering[44]

We recall a theorem of Browder and Levine (1966) that was the predecessor of Farrell's thesis:

Theorem 7.25 *A closed manifold M with $\pi_1(M) = \mathbb{Z}$ is a fiber bundle over S^1 iff its universal cover has finitely generated homology.*[45]

Farrell's theorem gives one generalization of this – what happens for fibration over S^1 if π_1 is not $\mathbb{Z}$.

But an alternative generalization asks what we can say more generally about the structure of manifolds whose universal covers have finitely generated homology?

Note that for many fundamental groups there are no such manifolds.[46]

Example 7.26 If M is a closed manifold and $\pi_1 M^n$ has infinitely many ends (e.g. is a nontrivial free product other than $\mathbb{Z}_2 * \mathbb{Z}_2$), then H_{n-1} of the universal cover of M is infinitely generated.

Indeed the universal cover has infinitely many compact separating codimension-1 submanifolds that are not homologous (as in Figure 7.1).

More generally, the following is a consequence of Quinn (1972, 1982b,c, 1986) – and see Block and Weinberger (1997).[47]

Proposition 7.27 *If $B\pi$ is a finitely dominated complex and M is any closed manifold with fundamental group π and whose universal cover has finitely generated homology, then (i) $B\pi$ is a Poincaré complex and (ii) the universal cover of M is a Poincaré complex.*

It is obviously not necessary for $B\pi$ to be finitely dominated for such an M to exist. Any finite group is a counterexample! Nevertheless, let us provisionally make this assumption.

---

[44] See also our discussion above in §5.1

[45] They proved this theorem in the smooth and PL categories in dimension greater than 5, but it is now known to be true in the topological category in all dimensions (for the "usual reasons").

[46] For some groups there are no such finite complexes or even having finitely generated homology through some fixed dimension. This is a variation of the usual FP_k hierarchy (see e.g. Brown, 1982) for groups, which seems worth further study.

[47] The argument is basically this. A finite complex X is a Poincaré complex iff the map from the boundary of the regular neighborhood of X in a high-dimensional Euclidean space is a homotopy sphere. This implies that, for a fibration of finite complexes, $F \to E \to B$, as these fibers for E are the join of the fibers for E and B, and it therefore follows that E is Poincaré iff both F and B are. In our situation, $E = M$, and the fibration is associated to the classifying map of the fundamental group, and we have assumed that all three are finite complexes. Thus, all three must be Poincaré complexes. Since M is a manifold, its universal cover automatically has a normal invariant, and consequently the fiber is homotopy-equivalent to a topological manifold (in all dimensions).

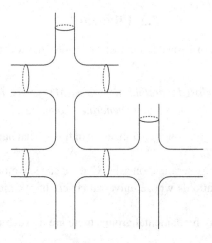

Figure 7.1 Part of the universal cover of a manifold with free fundamental group; a set of separating hypersurfaces as in the picture are linearly independent whenever no subset bounds a compact region.

Conjecture 7.28   *If M is a closed manifold with $\pi_1 M$ of finite type and whose universal cover has finitely generated homology, then there is an aspherical ANR homology manifold X and a UV^1 approximate fibration $M \to X$.*

Let's review some of the definitions and motivate the conjecture. We shall see that it is a natural analogue and consequence of the Borel conjecture. (Indeed, it also implies the Borel conjecture, so I guess that means it's equivalent to it.)

The first statement that there should be an aspherical ANR homology manifold with the same fundamental group as M follows from the existence version of the Borel conjecture (with Wall conjecturing that this should even be homotopy-equivalent to a manifold). The manifold is then homotopy-equivalent to a fibration over X with a Poincaré space as fiber.

If X is not resolvable, then obviously M cannot fiber over it. (The local index of M would have to be divisible by the local index of X.)

However, approximate fibration is somewhat less. A map is an approximate fibration if the usual condition for a fibration, namely that for any square

$$
\begin{array}{ccc}
A & \to & M \\
\cap & & \downarrow \\
A \times [0,1] & \to & X
\end{array}
$$

there is a diagonal lift, $A \times [0,1] \to M$, but now only demanding that the diagram commute up to ε. (In other words, for each ε we want a lift.) This condition is quite close, when X is a manifold, to being a block bundle (see Quinn, 1982a, 1987b).

Let $E(\tilde{M} \downarrow X)$ be the total space of the universal cover fibration over X, i.e. the product $(\tilde{M} \times \tilde{X})/\pi$ which has a natural fibration structure over X.

The obstruction to homotoping the map $M \to E(\tilde{M} \downarrow X)$ to a controlled homotopy equivalence over X (which, parenthetically, is equivalent to homotoping it to an approximate fibration) is precisely an element of a certain (looping[48] of a) structure set of X (because the homotopy fiber involved is simply connected) by making use of the theorem of Bryant *et al.* (1993). (See §4.9 for a blocked discussion and also §5.1.) Consequently, by the Borel conjecture again, we get the conclusion of the conjecture.

Conversely, if for example X is an aspherical manifold for which the Borel conjecture fails, then gluing together a counterexample in $S(X \times \mathcal{D}^4)$ to $X \times \mathcal{D}^4$ builds a manifold that is homotopy equivalent to $X \times S^4$ and therefore has finitely generated homology in its universal cover. However, the obstruction to making it approximately fiber over X is exactly the nontrivial element in $S(X \times \mathcal{D}^4)$ that starts the construction, completing the proof. (This requires some thought, perhaps, but it follows from the proof of Siebenmann periodicity given in Cappell and Weinberger, 1987.)

Question 7.29 Is there any natural construction of manifolds whose universal covers have finite type that might be a useful source of groups that satisfy Poincaré duality that are not immediately forced to be fundamental groups of manifolds? Note that any Poincaré $B\Gamma$ is the fundamental group of such a manifold – but this doesn't count, since we don't have a construction of these, as I've already lamented – the boundary of a regular neighborhood of $B\Gamma$ embedded in any Euclidean space would be such a manifold.

One source *might be* even a construction of manifolds M where $\mathrm{Out}(\pi_1 M)$ might or might not be trivial, and where there is a finite subgroup G for which no element of order p lifts to $\mathrm{Aut}(\pi_1 M)$. In that case the nontrivial extension $1 \to \pi_1 M \to \Gamma \to G \to 1$ produces a Poincaré duality group which is not obviously a manifold. On the other hand, if we knew the Novikov conjecture for π (even rationally), this would not be a candidate for a counterexample to the Wall conjecture based on the local index. And, in any case, I don't know any interesting examples of this sort.

Remark 7.30 Note that by striving for a general discussion, we were inexorably led to approximate fibrations. However, when X is a manifold, nothing we have said precludes M from actually fibering over X. Nevertheless, since

---

[48] By $\dim M - \dim X$.

there are sphere block bundles that do not correspond to sphere bundles, the total space of one of these over a closed aspherical manifold would provide such an example.

Remark 7.31 There is a slightly different line of reasoning that could lead to the above conjecture. (And naturally leads to a version for arbitrary regular covers in place of the universal cover: but this version is obstructed by (a sequence of) Nil groups in general. What follows is a somewhat different recombination of the ingredients in §5.1.)

Let us consider what a stratified version of the Borel conjecture could be – in the first interesting case, where our space has two strata. The question is which of these can be rigid?

This is actually kind of complicated. Here is an example that has nothing to do with the Borel conjecture:

Example 7.32 Suppose that $f: M \to N$ is an approximate fibration whose homotopy fiber is $\mathbb{CP}^{2k}$, then $(\mathrm{Cyl}(f), \mathrm{rel}\ \partial)$ is a rigid stratified space.

Interestingly Cappell, Yan, and I observed that this is not true in the slightly more general situation where the homotopy fiber is a general simply connected manifold F with signature equal to 1! There is an interesting contribution of the monodromy map $\pi_1 N \to \mathrm{Aut}(F)$.

So let's consider just the situation relative to a singularity[49] (see §§6.5 and 6.6). There is a fibration:

$$S(X \operatorname{rel} \Sigma) \to S(X - \Sigma) \to \mathrm{Fiber}\left(H\left(\Sigma; L(\text{local holink})\right) \to L(\text{global holink})\right).$$

(The loop of this fibration is

$$\mathrm{Fiber}\left(H_{+1}\left(\Sigma; L(\text{local holink})\right) \to L(\text{global holink})\right)$$
$$\to S(X \operatorname{rel} \Sigma) \to S(X - \Sigma),$$

which just means that the extensions of a structure of $X - \Sigma$ over Σ correspond to the ways of making the end of $X - \Sigma$ into an approximate fibration over Σ.)

It's easy enough to compute $S(X - \Sigma)$: it is essentially a manifold with boundary where one is not working relative to the boundary. In the mapping cylinder case, it vanishes. (It boils down to $S(M \times [0,1) \operatorname{rel} M \times \{0\}) = 0$.) This calls attention to the issue of whether

$$H\left(\Sigma; L(\text{local holink})\right) \to L(\text{global holink})$$

---

[49] Actually, there is a stabilization necessary here when the homotopy link (holink) is not simply connected, because of decoration issues. We shall ignore this.

is an isomorphism. If $\pi_2 \Sigma = 0$, then the L-sheaf is "flat," i.e. is a formal consequence of the short exact sequence[50]

$$1 \to \pi_1(\text{local holink}) \to \pi_1(\text{global holink}) \to \pi_1(\Sigma) \to 1$$

and then we would expect this fiber to be trivial at least when Σ is haspherical (by comparison to the aspherical case – as there is a map to the corresponding twisted assembly map over $B\pi_1(\Sigma)$).

All of this suggests that for $\text{Cyl}(f)$, if the map is an approximate fibration,[51] and the base is aspherical, one might conjecture rigidity. A little thought shows that the reasoning discussed in Chapter 4 would prove the split injectivity. Which makes the conjecture of rigidity (despite its clear falsity due to Nil and UNil issues!) plausible. This is a special case of the twisted Borel conjecture in §5.1.

The connection between the neighborhoods and approximate fibrations is very close in the topological category. As we discussed earlier, there is no direct regular neighborhood theory. What there is, is the "teardrop neighborhood theorem" of Hughes *et al.* (2000). It says that there is a deleted neighborhood of $X - \Sigma$ that has a canonical approximate fibration structure over $\Sigma \times (0,1)$. (Then Σ is glued in in the obvious way, and, when drawn appropriately, the open sets in the relevant topology look like teardrops.)

Crossing with $(0,1)$ has the effect of taking a loop space, and one is thus led geometrically to the conjecture of existence and uniqueness of approximate fibration structures when the target is aspherical.

Remark 7.33 Note that, once we have decided that Σ is aspherical, then the Borel conjecture would give *its* rigidity, so $S(X \text{ rel } \Sigma) \to S(X)$ would be an isomorphism. Thus, we have two circumstances where we have vanishing of $S(\text{Cyl}(f), \text{rel } \partial)$ – namely when the fiber is $\mathbb{CP}^{2k}$ or when the base is aspherical. When the fiber is $\mathbb{CP}^{2k}$ then any manifold M' homotopy-equivalent to M gives rise to a structure N' on N, so that M' approximately fibers over N'. In the situation of aspherical base, the base is itself rigid, so the result is that M' fibers over N itself.

Remark 7.34 Approximate fibrations arose naturally when we considered rigidity in the topological category. As we have emphasized many times in these notes, the topological setting is the natural one for rigidity.

Had one worked in the PL category one would have been led to consider block

---

[50] See Chapter 13 in Weinberger (1994).
[51] Actually, all one needs is that all "fibers" have the same π_1 (i.e. an approximate fibration with respect to 2-complexes). In that case, stratified surgery does not directly apply, but nevertheless controlled surgery would still give a suitable rigidity.

bundles instead. In that case, we would have trouble with decorations, which also reflects the way there are K-theory obstructions in Farrell's theorem.[52] While the whole Whitehead group enters in Farrell's theorem, in the problem of approximate fibering over the circle (and this is part of the Borel package) it is only the Nil part: writing, in the product situation, $\mathrm{Wh}(\mathbb{Z} \times \pi) = \mathrm{Wh}(\pi) \times \tilde{K}_0(\pi) \times \mathrm{Nil}_\pm(\pi)$. The $\tilde{K}_0(\pi)$ part arises from trying to put a boundary on the infinite cyclic cover. However, any such cover, with boundary or not, is the infinite cyclic cover of an approximate fibration over the circle (see Hughes and Ranicki, 1996, for a discussion of wrapping). The Whitehead part is entirely irrelevant to the problem. The h-cobordisms between different "almost fibers" can all be mapped to points in an approximate fibration.

Now let us turn to the possible structure of manifolds whose universal covers have finitely generated homology but have torsion in their fundamental group.

Conjecture 7.35 *For such a manifold to exist it is necessary and sufficient that* $\mathrm{HR}_i(\pi)$ *is zero for all but one dimension, and in that dimension it is isomorphic to* $\mathbb{Z}$.

Here HR is the "Rips homology" of the discrete metric space π (made into a metric using the word metric). It is defined as the limit of the locally finite homology of the nerve of the covering of π by balls of radius k, as $k \to \infty$ (see §4.8 or Block and Weinberger, 1997, or Roe, 2003). For groups of finite type, this is equivalent to Poincaré duality (see, for example, Brown, 1982).

The work we've already done on the Nielsen theorem indicates that there are groups satisfying the conditions of the conjecture where there is no $E\pi$-manifold. If one takes the counterexample to Nielsen and crosses it with S^3, one easily obtains a manifold whose universal cover is homotopy equivalent to the sphere but we cannot attribute this to, for example, an approximate fibering, as (we try to) in the finite type situation.

Of course, there are situations where $E\pi$ geometrically exists. For example, this is the case when π is a uniform lattice in a connected Lie group G. In that case, the Borel package gives one a nice characterization:

Theorem 7.36 *If M is a manifold whose fundamental group is that of a uniform lattice in G and whose universal cover has finitely generated homology, then there is a π-equivariant UV^1 approximate fibration $\tilde{M} \to G/K$ iff appropriate Nil and UNil obstructions vanish. Of course, if the fundamental group is, in addition, torsion free, then conclusion holds unconditionally.*

The need to handle such obstructions is easy to see. Suppose we take one of

---

[52] That the version of Farrell's theorem for approximate fibrations only involves Nil was observed by Ferry in the late 1970s.

Cappell's manifolds $M = \mathbb{RP}^{4k+1} \# \mathbb{RP}^{4k+1}$ terms that is not in itself a connected sum. Suppose that there were a D_∞ equivariant approximate fibration f of the universal cover to $\mathbb{R}$ (with the usual action where one involution is $x \rightarrow -x$ and the other is $x \rightarrow 1 - x$). Then $f^{-1}(1/3, 2/3)$ would descend into M, and would be a copy of $S^{4k} \times (0, 1)$ – as there is a unique manifold element in $S^p (S^{4k} \times (0, 1))$. Taking a slice, M is thus decomposed into a connected sum. Similarly, it is not hard to realize Nil obstructions on some manifolds with fundamental group $\mathbb{Z} \times F$, where F is a finite group, and then there would be no $\mathbb{Z} \times F$ equivariant map of the universal cover to $\mathbb{R}$ (with trivial F-action, and $\mathbb{Z}$ acting by translation).

The proof of the above theorem is now identical to the discussion of the torsion free case above.

7.6 Manifolds with Excessive Symmetry[53]

In this section we will describe some theorems in Riemannian geometry that have a philosophical relation to the topological issues explored in the previous few chapters. We will not do more than give the slightest hints of the arguments.

Recall again Borel's theorem:

Theorem 7.37 *If M is aspherical and $\pi_1 M$ is centerless, then any group G that acts on M injects into* $\mathrm{Out}(\pi_1 M)$.

If M is compact locally symmetric (with no virtual hyperbolic surface factors) then Mostow rigidity implies that $\mathrm{Isom}(M) = \mathrm{Out}(\pi_1 M)$. Then we can assert that, for action on M, there is a semiconjugacy (i.e., an equivariant map that's not necessarily a homeomorphism) homotopic to the identity

$$H : M \rightarrow M$$

to an action by isometries. In this sense, the locally symmetric metric is maximally symmetric among all metrics.

Of course, there are other metrics that are equally symmetric. $\mathrm{Isom}(M)$ is a finite group, and any metric equivariant with respect to this group has the same symmetry. Is there any way we can be more demanding and perhaps characterize the locally symmetric metric?

Another point to note is this. Instead of just considering *metrics on M*, we can also consider manifolds N with $\pi_1 N \cong \pi_1 M$. If the G-action on N is trivial

---

[53] Benson Farb has criticized the title of this section and suggested "Manifolds that only have the slightest bit more symmetry than they need to have" as an alternative.

in Out(π), then one can factor the natural map $N \to M$ through N/G.[54] In particular, if the homology class of N is nontrivial in $H_n(B\pi \cong M; \mathbb{Q})$, i.e. if N is essential, then G is finite.[55] If the map were degree 1, we would have Borel's injectivity for this larger class of manifolds.

This connects to our higher-signature localization discussion, since the fixed sets would have no choice but to be lower dimension for a nontrivial action, but such wouldn't be able to have this relevant higher signature. It's not shabby getting this far just from homology considerations without using any high technology.

In short, we see that there is a rather larger class of manifolds for which Isom(M) provides a bound on their symmetry.

We will warm up with the following result:

Theorem 7.38 (Farb and Weinberger, 2005)   *If $N \to M$ is a map of nonzero degree, M an irreducible locally symmetric manifold, and* Iso(N') $\cong$ Iso(M') *for every finite sheeted cover, then:*

(1) *if M is arithmetic, then N is isometric to M;*[56]

(2) *if M is not arithmetic, then there are such N not homotopy-equivalent to M (above dimension 3, for trivial reasons) and there are N diffeomorphic to M that are not locally symmetric.*

The key to the theorem comes from the existence and uniqueness theorems for harmonic maps in a given homotopy class when the target is compact and non-positively curved. The difference between the arithmetic and non-arithmetic cases is because of a theorem of Margulis: *Every non-arithmetic lattice is included in a maximal lattice that includes all of the lattices that are commensurable to it*. This makes part of (2) very simple. We can take N to be $M\#$ several $(S^2 \times S^{n-2})$.

For the arithmetic case, one makes use of many conjugates of π in the group $G(\mathbb{Q})$ which produce many extra isometries that finite covers have. Ultimately this transfers the whole action of G from the universal cover of M to that of N (by taking limits).

Indeed, a little work makes the same conclusion follow from the hypothesis that M and N have the same dimension and fundamental group. The map of universal covers is smooth, and, given the action of G, every point of M is

---

[54] This uses the finiteness of Out(π), the existence of a lattice Γ in G such that $\pi \lhd \Gamma \to$ Out(π), all of which follows from Mostow rigidity, and the fact that G/K is an $E\Gamma$.

[55] And we can be more quantitative: depending on how divisible $[N]$ is in $H_n(B\pi)$, we can bound #G.

[56] Actually homothetic to M; one can rescale the metric!

a regular value, so the map is a finite sheeted cover, and the conclusion then follows.

One can push these ideas further in a number of ways:

(1) One doesn't need that all of the isometries of all of the finite covers extend. One just needs that there are many that do.
(2) One can, in the spirit of the previous paragraph, remove essentiality[57] types of conditions (like asphericity) and end up with statements about fibering, i.e. that the harmonic map that one produces has no singularities, and that the domain manifold Riemannianly fibers over the target.

It turns out that the correct setting for these results is the following (that follows from the same ingredients together with some Lie theory):

Theorem 7.39 (Farb and Weinberger, 2008) *Suppose M is a compact aspherical Riemannian manifold whose fundamental group has no normal abelian subgroup and is not virtually a product of manifolds. Then if $[\mathrm{Isom}(\tilde{M}) : \pi]$ is infinite, then π is a uniform lattice in a semisimple group G, and M is isometric to $K\backslash G/\pi$.*

If M is not assumed aspherical, then one must assume that $\mathrm{Isom}(\tilde{M})$ is not a compact extension of π, and one obtains that a finite cover of M is a Riemannian fiber bundle over $K\backslash G/\pi$.

The condition about compact extension is to avoid situations like the following. Suppose π has, for example, a dense representation in a compact group H, then the quotient manifold under the diagonal action $(K\backslash G \times H)$ has a large isometry group for its universal cover (i.e. containing π with infinite index), even if one gives $K\backslash G$ a highly non-symmetric metric, that is merely π-invariant.

Normal abelian subgroups truly are the enemy. A three-dimensional solv-manifold is abstractly a torus bundle over a circle. One can consider families of flat structures on the torus, parameterized over the circle, with a small compatibility condition (and don't even all have the same volume) and obtain metrics that have excessive symmetry and do not come from locally symmetric metrics or fiber over anything.

It seems very reasonable to try to improve the condition "$[\mathrm{Isom}(\tilde{M}) : \pi]$ is infinite" to something more quantitative. This is most salient in the "no normal abelian subgroup" situation. The example on the Conner–Raymond conjecture gives an example of an obstacle involved in removing this condition and replacing the condition in the conclusion by something that takes the abelian subgroup explicitly into account.

---

[57] A manifold is called essential when it represents a nontrivial cycle in the group homology of its fundamental group.

Conjecture 7.40 ("Magic number" conjecture) *For each group π that has no normal abelian subgroups, there is a number $C(\pi)$ such that any Riemannian aspherical manifold with fundamental group π such that $[\mathrm{Isom}(\tilde{M}): \pi] > C(\pi)$ is isometric to a locally symmetric Riemannian manifold.*

This conjecture is true when π is a lattice, or a word hyperbolic group. It seems that a counterexample would have to be rather exotic.

Our discussion also makes the following seem possible (although, I confess, unlikely[58]).

Conjecture 7.41 *An aspherical manifold M has Lipschitz Riemannian metrics for which $[\mathrm{Isom}(\tilde{M}): \pi_1 M]$ is arbitrarily large iff $\pi_1 M$ has a normal abelian subgroup.*

7.7 Notes

The paper of Borel on symmetry of aspherical manifolds (that we began our discussion with in §7.1) was unpublished for many years, but appeared in his collected works. The theorem itself was published by Conner and Raymond much earlier. I recommend the book by Lee and Raymond (2010) for many results about Lie group actions on aspherical manifolds, which builds on and reviews the excellent work of Conner and Raymond.

The Nielsen problem can be viewed as a variant of the Borel conjecture: if one believes that a homotopy equivalence gives rise to a canonical homeomorphism in the homotopy class, then one would have been led to the Nielsen problem. In the classical setting of surfaces, Nielsen proved it for cyclic groups. As mentioned in the text, the general case was first proved by Kerckhoff (1983).

Borel's theorem, of course, only applies to closed aspherical manifolds: after all, there are many finite group actions on Euclidean space and $\mathrm{Out}(\pi) = e$! However, it is natural to ask about locally symmetric manifolds of finite volume, i.e. quotients by non-uniform lattices. I had been interested in this question for many years, and the result that Borel's theorem holds for these was proved by G. Avramidi (2013). However, as he points out, the proof leaves open many questions: for example, given an arbitrary action of a finite group A on $K\backslash G/\Gamma$, is the dimension of the fixed set the same as in the classical action? Even for A a p-group this is open.

The results of Borel underscore a problem for understanding G-manifolds. For ordinary closed manifolds, we make a lot of use of the comparison of M

---

[58] And therefore this problem's demise will be a measure of how much we have to learn.

to $B\pi_1(M)$ (at least in the torsion-free case and to related things when π has torsion). The fact that for nonabelian compact Lie groups, and more generally, there is no very nice model for $K(\pi_1^G(M), 1)$ where $\pi_1^G(M)$ is the map from the orbit category of G, $\mathrm{Orb}(G) \to \mathrm{Gpd}$, the category of groupoids (since fixed sets can be disconnected or empty) makes it much harder to understand equivariant structure sets even when we have a theoretical analysis via a surgery sequence.

In §7.2 the connection of the Wall conjecture to the Borel conjecture requires Ranicki's total surgery obstruction (Ranicki, 1979b). This in turn is related to the idea that "coherent Poincaré transversality" gives a reduction of the Spivak fibration (Levitt and Ranicki, 1987). Actually, the (integral form of the) Novikov conjecture therefore gives the reduction of the Spivak fibration, but there is a surgery obstruction in principle – but surjectivity of the assembly map would give this, which is clearly part of the Borel conjecture. When Bryant *et al.* (1993) came out, it became clear (it seems to me) that it was more reasonable to ask for homology manifolds instead of manifolds. Davis (2000) suggested that one ask the question about aspherical R-homology manifolds.

This was disproved by Fowler (2009) in his thesis, which we have followed in spirit in the text.

Thanks to intersection homology there are other settings where one can ask for Poincaré duality. It would be interesting to find groups that naturally act on IH-acyclic "Witt spaces" or something similar. Then the Wall conjecture would somehow resolve the space (although perhaps in a non-local way).

Moving to §7.3, hyperbolic groups were first studied by I. Rips, who showed that they act properly discontinuously and cocompactly on a finite-dimensional contractible complex (the Rips construction). This result was published and then much elaborated by Gromov (1987), who explained their stability properties, their boundaries, isoperimetric inequalities, rationality of their word zeta functions, etc. They have become a much studied class of groups and are central to geometric group theory. I recommend Coornaert *et al.* (1990), Ghys and de la Harpe (1990), and Alonso *et al.* (1991) as good references (although there are a number of others).

As hyperbolic groups are generalizations of the idea of the fundamental group of a closed hyperbolic (or negatively curved) manifold, relatively hyperbolic groups are a generalization of hyperbolic manifolds with cusps. (These are never hyperbolic in dimension greater than 2, because they contain nontrivial abelian subgroups.) The basic paper is Farb (1998) which establishes many of their properties. We use the fact that relative hyperbolization (see Davis *et al.*, 2001) can be made hyperbolic relative to the boundary (Belegradek, 2007). That the pieces glue together to form a hyperbolic group is based on "combination theorems." The original combination theorem was Thurston's uniformization of

Haken 3-manifolds. For hyperbolic and relatively hyperbolic groups, Bestvina and Feighn (1992) and Dahamani (2003) provide analogues. (Drutu and Sapir, 2005 provide the quasi-convexity necessary for applying these theorems.)

The Conner–Raymond conjecture grew out of their work on injective torus actions (all actions on aspherical manifolds are injective). Their understanding of group actions on aspherical manifolds led to many examples, including, for example, the first examples of closed manifolds that have no symmetry (Conner *et al.*, 1972).

As discussed earlier, pseudo-isotopy theory is a deep subject connecting higher algebraic K-theory to groups of homeomorphisms and diffeomorphisms of manifolds. Besides the paper by Farrell and Hsiang (1978a) already referred to in the text, I recommend Cohen (1987) and Weiss and Williams (2001) as useful surveys, although Cohen's (1987) description of the then "recent" results was a bit optimistic.

The material on products in §7.4 should be well known, but doesn't seem to be. That products are likely not isomorphisms when the controlled symmetric signature is not a unit (in a relevant) ring is obvious in retrospect and also gives rise to failure of equivariant transversality – as discussed at the beginning of Chapter 6, first appearing in the form of lack of stability of equivariant classifying spaces – and some forms of transversality for homology manifolds (if one asks for bundle neighborhoods).

That bounded over $\mathbb{R}$ boils down to the proper theory is because tame ends of manifolds can be "wrapped up" and have an automatic periodicity. The relevant geometry is part of Siebenmann (1970a), which gave an alternative approach to Farrell's thesis for the problem of fibering over the circle.[59]

The Browder–Livesay theory is an elegant one wherein this particular non-simply connected problem is boiled down to a simply connected problem. The quadratic form $\langle u, Tv \rangle$, where $\langle , \rangle$ denotes cup product and T denotes the involution on the 2-fold cover, plays a major role. The orientation reversing nature of T interchanges the usual $\pm$ symmetry. In more modern L-theory this is called a change of "antistructure" and there is now a much more general theory of Browder–Livesay groups, associated to quadratic extensions of rings. Given our discussion, it should not surprise the reader that Cappell and Shaneson relied on such a Browder–Livesay theory in their calculations leading to the existence of nonlinear similarity.

The discussion on fibering in §7.5 is surely folklore and I am not sure who noticed what, and when. Ferry had told me decades ago that approximately

---

[59] Farrell's original approach did not place the algebraic obstruction all at once in Wh: it lived in several pieces. The connection between the two approaches is given by the formula for the Whitehead group of a twisted extension $\mathbb{Z}[\pi \rtimes \mathbb{Z}]$.

fibering over the circle is much less obstructed than fibering, i.e. that it's controlled by the Nil part of the fibering obstruction, and that the homotopy fiber does not need to be a finite complex. When considering the twisted analogue of the Borel conjecture, one quickly realizes its connection to block bundles, except that it doesn't get the decorations right. So, for "K-flat" groups, one gets general block fibering theorems (like those relevant to our question about spaces with finitely generated homology in their universal covers). Farrell and Jones (1989) point this out, and I had pointed out such things based on thinking about approximately fibered neighborhoods and possible stratified rigidity in Weinberger (1994). See Farrell *et al.* (2018) for recent results about approximately fibering compact manifolds over aspherical ones.

That one can prove results about approximate fibrations over ANR homology manifolds was the struggle in Bryant *et al.* (2007).

The teardrop neighborhood theorem of Hughes *et al.* (2000) is a variation on the periodic structure that can be given to a tame end, referred to in the notes in S6.11.

In §7.6, the proof of the main theorem characterizing Riemannian manifolds with excessive symmetry is a combination of the Myers–Steenrod theorem, which tells us that the isometry group of any Riemannian manifold is naturally a Lie group, the theory of harmonic maps (in order to build canonical maps to model spaces), and the Conner conjecture (a theorem of Oliver, 1976a), which asserts that the quotient of a finite-dimensional contractible space under a compact group action is contractible. This enables one to get information about isotropy groups and use homological algebra.

The use of harmonic maps to rigidify homotopy theory and make maps automatically equivariant arose earlier in work of Schoen and Yau (1979b). They also play a role in Frankel's proof of a conjecture of Kazhdan, of which Farb and Weinberger (2008) give an alternative proof.

These techniques are somewhat extended to noncompact manifolds in Farb and Weinberger (2010), except that the issues in general are much more complicated. As a result, attention is concentrated on moduli space (of curves). We give a new proof of some theorems of Ahlfors by showing that no complete Finsler metric on moduli space with finite co-volume has even a single point in its universal cover at which it is symmetric, i.e. possesses an involution with an isolated fixed point. We also obtained that, for any such metric, the symmetry of moduli space is never excessive. Avramidi (2014) strongly improved on this by showing that there are no unexpected isometries at all in any finite volume metric on any finite cover of moduli space.

On the other hand, he also gave a very simple construction of complete infinite volume metrics on moduli space (and non-uniform locally symmetric spaces) that do have excessive symmetry, so the finite co-volume conditions in Farb and Weinberger (2010) were necessary.

The fibration in the non-aspherical situation was significantly extended by van Limbeek (2014). Melnick (2009) has extended some of these results to Lorentzian manifolds.

8

Epilogue: A Survey of Some Techniques

This last chapter is a short epilogue to our ruminations. Having discussed many variations on the Borel conjecture and hopefully gained some appreciation for the problem, we now briefly discuss some of the very significant attempts that have been made to prove both it and its noncommutative geometric cousin, the Baum–Connes conjecture. To do an adequate job would take two or three more volumes of the length of this one[1] and clearly (indeed tautologously) that cannot be done here. Instead, we will give a breezy overview of some milestones, giving detail precisely for the parts that are easiest or that connect directly to earlier discussions.

The first four sections of the chapter will be devoted to the Borel and Farrell–Jones conjectures and we will then turn to the Baum–Connes conjecture.

8.1 Codimension-1 Methods

The first results on the Borel conjecture grew out of the study of codimension-1 submanifolds of homotopy-equivalent manifolds. We first discussed this idea in the setting of the Novikov conjecture in §4.4 (splitting theorems).

Geometrically, these ideas arose first for 3-manifolds in the classical theory of Haken manifolds, i.e. of irreducible 3-manifolds that contain an incompressible surface. A major result of that theory is that a connected irreducible[2] 3-manifold has a hierarchy iff it has an incompressible surface.

An incompressible surface is a two-sided codimension-1 submanifold that is one-to-one on π_1. A hierarchy is an inductive structure so that you start with an incompressible surface, cut open your manifold along it, and then find a new surface there, and keep on going. The key point, though, is that the process

---

[1] May they soon be written.
[2] That is, one in which each embedded S^2 bounds a ball.

terminates, and you end up with a ball. You then think of the 3-manifold as obtained by the reverse process of constantly gluing 3-manifolds to themselves along pieces of their boundary.

This sequence of incompressible surfaces enables inductive proofs. Thus, Waldhausen (1978) showed that, for such 3-manifolds, the universal cover (of their interior) is $\mathbb{R}^3$ and any homotopy equivalence which is already a homeomoprhism on the boundary is homotopic, relative to the boundary, to a homeomorphism.

The proof of the splitting theorem in dimension 3 is a consequence of the basic theorems of Papakyriakopoulos, namely the Dehn lemma, loop theorem, and sphere theorem. The combination of Dehn lemma and loop theorem tells us that, if the fundamental group of the boundary of a 3-manifold M does not inject, then it is for the obvious reason that there is an embedded D^2 in M that intersects the boundary in an essential curve. The sphere theorem asserts that $\pi_2(M)$ is nontrivial (for an oriented 3-manifold) iff there is an essentially embedded S^2. This last gives a quick proof of the basic result (which so influenced our discussion in §6.4) that, for closed 3-manifolds, one is a nontrivial connected sum[3] iff the fundamental group is a nontrivial free product.

These can be found in any book on 3-manifolds, in particular, Hempel (1976) and Jaco (1980).

In higher dimensions, the Farrell fibering theorem gives an approach to the Borel conjecture for tori. Without using periodicity of structure sets, one has to argue indirectly and use periodicity of L-groups and $G/$Top and calculate. This was done via Wall (1968), Shaneson (1969), Hsiang and Shaneson (1970), and Farrell (1971b, 1996).[4] More generally, the splitting theorem of Cappell (1976a) can be thought of as a Mayer–Vietoris sequence for L-theory of amalgamated free products and HNN extensions (see Cappell, 1976b), except that in the non-square root closed situation one can run into UNil obstructions; or, phrased differently, there's an extra summand in one of the terms of the Mayer–Vietoris sequence.

The analogous situation in K-theory was perhaps not as immediately apparent to someone working on concrete questions, since frequently K_0 and Whitehead groups vanish.

Stallings (1965a) had shown early that $\mathrm{Wh}(G * H) = \mathrm{Wh}(G) \oplus \mathrm{Wh}(H)$. For polynomial extensions, Bass *et al.* (1964) gave the formula in the $\mathbb{Z}$ case of the K-theory Borel with coefficients. (This paper, strictly speaking, required

---

[3] Nontrivial excludes connected sum with homotopy spheres, but now that the Poincaré conjecture is a theorem, this does not need to be made explicit.

[4] Farrell (1971b) explains the close connection between fibering over a circle and the problem of putting a boundary on an open manifold, i.e. the problem studied in Siebenmann (1965).

coefficients to be a regular ring, which avoids Nil terms. However, Bass's 1968 book has a more complete discussion.) This formula led to Bass's definition of the negative K-groups and the desire for higher algebraic K-theory.

The twisted version of this formula was discovered by Farrell and Hsiang (1970) and was part and parcel of understanding the problem of fibering over the circle (monodromy requires allowing for twists – the problem always involves, as we've discussed, the Whitehead group). Waldhausen (1987) made a major advance when he proved the analogous statement for K-theory of amalgamated free products, using a new Nil functor to measure the lack of excision. His motivation was to understand why the homotopy equivalences between 3-manifolds discussed above were all simple! His answer was that the same structure that enables one to deform homotopy equivalences to homeomorphisms enables one to prove that the relevant Whitehead groups are trivial.

The story in algebraic K-theory is recapitulated in operator theory. The analogue of the Farrell and Hsiang (1970) formula is the Pimsner–Voiculescu sequence (Pimsner and Voiculescu, 1980) and the analogue of Cappell's theorem is Pimsner's theory of K-theory for "groups that act on trees" (Pimsner, 1986). (Interestingly, for the special problem of positive scalar curvature (see Chapter 5), the analogue of the key boundary map in the exact sequence, which above is produced via codimension-1 splitting, was constructed – at least in low dimensions – by Schoen and Yau (1979a), who showed how to use stable minimal hypersurfaces to use hierarchies to obstruct positive scalar curvature.[5])

It is surely worth observing that among the most developed methods for constructing strange groups is via amalgamated free products and HNN extensions (see e.g. Baumslag *et al.*, 1983). As a result, the Mayer–Vietoris sequences in K- and L-theory remain valuable for constructing examples.

Codimension-1 methods also are critical to the proofs that go through the controlled world. The basic results about controlled topology over finite-dimensional ANRs or bounded control over cones of polyhedra, etc., are all proved via appropriate codimension-1 splitting theorems (since, after all, the hard part in proving that something is a homology theory is almost always checking excision, also known as Mayer–Vietoris); see, for example, Quinn (1979, 1982b,c, 1986), Pedersen and Weibel (1989), and Ferry and Pederson (1995).

---

[5] In a recent preprint (Schoen and Yau, 2017), they remove the low-dimensional condition.

8.2 Induction and Control

Our next goal is to explain the ideas of Farrell and Hsiang (1970) that, for example, prove the Borel conjecture for flat manifolds. These are "merely" finite torsion-free extensions of free abelian groups, yet they are hard to understand directly. The arguments are a beautiful mix of algebra (induction theory) and controlled methods (one of the first applications of these to computing something) and have been very influential.

Actually the result of Farrell and Hsiang (1970) is more general: it gives a topological characterization of almost-flat manifolds in the sense of Gromov. A manifold M is almost flat if it has a sequence of Riemannian metrics with $|K| < 1$ and diam $(M) \to 0$ (or equivalently with bounded diameter and $K \to 0$). Such manifolds are exactly (see Buser and Karcher, 1981) infranilmanifolds.

Theorem 8.1 *A closed topological manifold has an almost-flat structure iff it is aspherical and its fundamental group is virtually nilpotent.*

Sketch The result boils down to showing that there is a unique aspherical manifold with the given fundamental group. We will ignore the low-dimensional cases, because they hold for the usual reasons: Perelman (geometricization) does dimension 3, and Freedman's work applies in dimension 4, because the relevant fundamental groups are all "small" (see Freedman and Quinn, 1990).

The nilpotent case is easy. Nilpotent groups are poly-$\mathbb{Z}$: there is a natural induction on the cohomological dimension, and such a group Γ always surjects to $\mathbb{Z}$ with a smaller such group as its kernel. Ultimately, one deduces the result from many applications of Farrell fibering and the Farrell and Hsiang (1970) variant of the Bass–Heller–Swan formula.

The remaining part of this argument is inspired by representation theory (see Serre, 1977, for everything we say). For finite groups, representations are determined by their characters, i.e. their restrictions to cyclic subgroups. However, not every representation is a sum of representations induced from their cyclic subgroups. (It is, if one allows rational coefficients, according to Artin's theorem.) To get these, one needs a larger class of groups (this is the content of Brauer induction).

These are proved by making use of $R(H)$ for all H in G. There are operations ind: $R(H) \to R(G)$ and res: $R(G) \to R(H)$ (and, of course, G can be replaced by any subgroup of G that contains H). It suffices to prove a formula in the algebra of operations of the form $1 = \sum a_H \mathrm{ind}_H^G(\mathrm{res}_H)$, where the a are some coefficients.   □

Suitable reciprocity would also give that, for any module over this algebra (i.e., suitable functors of groups that have appropriate behavior with respect to induction and restriction), restrictions to the family H will detect elements.

To make a long story short, the work of Dress (1975) on equivariant Witt rings can be used in a similar way to prove induction theorems for structure sets of manifolds, whenever one has a map $\pi_1 M \to G$ for a finite group G. Farrell and Hsiang (1983) use this for L-groups; Nicas (1982) gives a version for structure groups that is a bit more natural for our purposes:[6]

$$S(M) \to \bigoplus S(M_H)$$

is injective localized at a prime p, if H ranges over the p-hyperelementary subgroups (i.e. the groups containing normal cyclic subgroups with index a power of p), where M_H is the cover of M corresponding to the subgroup H (and the map is "transfer" to this cover). Rationally, it is injective when H ranges over the cyclic subgroups (just like in character theory!).

We now can sketch some of the ideas that go into the proof of the Farrell–Hsiang rigidity theorem. The actual proof for the general case involves more complex fibering rather than just over the circle (but from Chapter 6, this causes us no fear!). In the flat case where the holonomy is odd order, the proof is much simpler (Farrell and Hsiang, 1978b). In that case, they show that the following algebraic fact holds:

Proposition 8.2 *Assume Γ is the fundamental group of a flat manifold with holonomy G of odd order, so that there is an exact sequence*

$$1 \to A \to \Gamma \to G \to 1,$$

where A is (free) abelian. Then either there is a surjection $\Gamma \to \mathbb{Z}$, or for infinitely many $s = 1 \bmod \#(G)$, one has Γ / sA having the property that any hyperelementary subgroup that maps onto G, maps isomorphically onto G.

The conclusion of the proposition enables an induction on the cohomological dimension combined with the holonomy, except for the awkward case where one has an element that is transferred to the cover corresponding to a hyperelementary subgroup that is isomorphic to the manifold itself. Naively it looks like we're stuck with an impossible circular argument.

But we are not at all: the method of Example 4.17 in §4.6 now kicks in. When we identify the cover with the original M, one has used an expanding

---

[6] Of course, the upshot of modern surgery is that structure sets are essentially L-groups of a suitable object. As a result, precisely the same structure that gives induction for L-groups of groups gives it for structures. Induction arguments in smooth surgery are sometimes possible (see, for example, Madsen *et al.*, 1976).

map, so the point inverses are smaller. Ultimately, we are in the circumstances of Ferry's theorem and one gets that all the transfers are 0.

Consequently, by Dress induction, every element of the structure set vanishes,[7] and therefore the structure set does (and the assembly map is an isomorphism).

In §8.3, I will explain another example where transfers actually suffice for proving vanishing of structure sets. (Actually, I will focus on the easier case of Whitehead groups, for a reason that will be clearer in §8.3.)

Remark 8.3 Although the control ideas in the above proof are the most important piece for the Borel and Farrell–Jones conjectures,[8] the use of induction in *L*-theory and structure sets is fundamental throughout equivariant topology.

8.3 Dynamics and Foliated Control

To say that the subject was revolutionized by the work of Farrell and Jones through their long series of brilliant[9] and beautiful papers (Farrell and Jones, 1986, 1987, 1988, 1989, 1991a,b, 1993a,b, 1998a,b, 2003)[10] is an understatement.[11]

Here I will explain just one of the ideas, as an introduction to this body of work. Among their achievements, they proved the Borel conjecture for closed non-positively curved manifolds and for many manifolds with infinite volume, again under a curvature condition. But, also, they formulated (and proved many cases of) the "Farrell–Jones isomorphism conjecture" as a natural byproduct of what the method leads to. This conjecture is proved for lattices in Lie groups in Farrell and Jones (1993a) for *K*-theory and in Bartels *et al.* (2014a,b) for *L*-theory. We can see this already in their earliest result in this series,

Theorem 8.4 (Farrell and Jones, 1986, 1987) *If* Γ *is the fundamental group of a closed negatively curved manifold, then* Wh(Γ) = 0.

To get a feeling for this achievement, realize that until this point all previous groups that were understood had a clear algebraic nature, a place from which to

---

[7] Please see the cartoon in Figure 1.1 at the end of Chapter 1.

[8] However, see Bartels and Lück (2012b) for a use of their general inductive scheme.

[9] Difficult.

[10] It is also worth mentioning that there was follow-up work by Farrell and Jones and by Ontaneda that develops methods for proving results about negatively curved manifolds and spaces of negatively curved metrics building on these topological rigidity ideas.

[11] To say that this is an understatement, is an understatement; and so on, for another few iterations.

grab an inductive hold. Here we are impelled to use geometry, which seemed almost unheard-of in high dimensions.[12]

Here's the general idea. One starts with M the closed hyperbolic manifold with fundamental group Γ, which we assume is of dimension greater than 4. An element of the Whitehead group $\text{Wh}(\Gamma)$ is represented by an h-cobordism W, with $\partial W = M \cup M'$. The idea is to find a bundle X over W, so that this bundle has enough geometry that one can geometrically "flow" this new h-cobordism to "control it." Controlled h-cobordisms will be products, and if there is no information lost in the transfer map $\text{Wh}(M) \rightarrow \text{Wh}(X)$, then one will have proved that the original h-cobordism is trivial, and thus the theorem.

Of course, at the level of algebra there is a map $\text{Wh}(M) \rightarrow \text{Wh}(X)$, and the composite is just multiplying by $\chi(\text{F})$, the Euler characteristic of the fiber. (In L-theory, the monodromy plays a larger role, but it usually is multiplying by sign (F). This fairly straightforward topological step is a major obstacle in the Baum–Connes conjecture.[13])

None of these steps are straightforward.

For simplicity we will work on M rather than W. This means that we should work in a setting of geometric modules. The Whitehead group $\text{Wh}(M)$ is made out of free $Z\Gamma$ modules with automorphisms (or acyclic chain complexes of free based modules). Now imagine that each basis element is given a location on M, and that morphisms include paths that link generator to generator (see Quinn, 1982a, 1985a, 1987b). This enables defining controlled Whitehead group elements, and the "fundamental theorem of controlled topology" (i.e. the main theorem in Quinn, 1979, 1982b,c, 1986) asserts that controlled Whitehead groups are essentially the homology of the control space with coefficients in the Whitehead spectrum (so, for Z, they vanish). We will transfer up our geometric uncontrolled chain complex from M to X and flow it there. Hopefully, the curves that arise in the module will become smaller, and thus become controlled, and trivialize.[14]

---

[12] Of course, for most of a decade, Thurston had already been preaching the importance of geometry for low-dimensional topology. In the work of Thurston, the geometry is there because the manifold ends up being geometrizable. In the work of Farrell and Jones, it's because we can transport our problems to live over a geometric object, and study them there.

[13] Usually the signature of a fiber bundle is the product of the signature of base and fiber – at least if the monodromy is simple enough – but for other operators this is rarely the case. Part of the fascination with the elliptic genus is based on the magic that there *are* operators for which this is true for connected compact structural groups (see, for example Bott and Taubes, 1989) in smooth settings.

[14] This cannot actually happen – because of the Nil term in the Bass–Heller–Swan formula. Nothing in the geometry knows that the coefficients of the paths lie in Z, so we could do all this algebra with an arbitrary coefficient ring R; this proof, if it worked, would imply that $\text{Nil}(R) = 0$ for all R. However, as we will see, the kind of control that is gained is less than the control in Chapman and Ferry, or Quinn, and the Nil term will naturally come up in the end.

The first try for the bundle is the unit sphere bundle of M. This won't have the right transfer properties, so it will ultimately be modified. (It would suffice for M to be odd-dimensional to prove the vanishing of $\mathrm{Wh}(M) \otimes \mathbb{Z}[1/2]$.)

The unit sphere bundle of a Riemannian manifold has a natural flow on it, the geodesic flow: A point is a pair consisting of (m, v), with m a point of M and v a unit tangent vector at m. One then considers the geodesic going through m in the direction v for t seconds (transporting v to the other end).

The geodesic flow on a negatively curved manifold is Anosov. What this means is this. The tangent space to X breaks up into three pieces $F_s + F_u + \mathbb{R}$. In the F_s direction, the flow contracts (exponentially fast). In the F_u direction, the flow expands things (in negative time, it contracts). The $\mathbb{R}$ is the direction of the geodesic, where the flow leaves things alone.

A key element, then, is the asymptotic transfer, which lifts curves on M to ones on X.[15] It is like a connection, telling one to not just use the abstract bundle properties to lift homotopies, but how to use the geometry. The choice Farrell and Jones use is the "asymptotic transfer," defined to make curves shrink with respect to the geodesic flow, to at least end up close to the $\mathbb{R}$ directions, i.e. have no F_u parts.

The picture in Figure 8.1 describes the construction very clearly. A point on the universal cover of M and a vector determines a well-defined point at infinity.

---

[15] In transferring a torsion there is the base direction that we are in the midst of discussing in the body of the text, and also the fiber direction. The fiber direction, however, is the chain complex of the fiber, which is as controlled as one wishes (in this, and all the other known applications of this method), and we shall not discuss it.

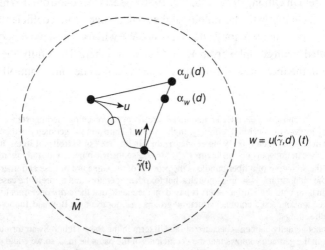

Figure 8.1 A picture of the asymptotic transfer. (Reproduced with thanks from Farrell's 2002 Trieste notes.)

Then, for any other point, there is a unique vector that points to that same point at infinity. Along a curve, one translates the vectors by this common asymptotic rule. (This rule is equivariant with respect to the action of the covering group, and is thus well defined on M as a way of transporting on curves.)

Note that if one transfers up a geodesic segment, it lifts to a geodesic segment (with its parallel translation of the initial vector) and nothing shrinks during the flow. But, at least nothing gets larger, and all of the other directions are exponentially shrinking. So after a while one has an h-cobordism (or acyclic geometric chain complex) that is "foliated-controlled over X," i.e. sizes can be made arbitrarily small in directions orthogonal to the leaves of the foliation.

More precisely, X (the unit sphere bundle) is foliated by the orbits of geodesic flow (i.e. the geodesics on M with their unit tangent vectors), i.e. it has a one-dimensional foliation. Most of the leaves are isomorphic to $\mathbb{R}$, but there is a countable number of exceptions – the closed geodesics[16] of M. After flowing, all the morphisms in the geometric module (i.e. the tracks of the homotopies in the h-cobordism[17]) end up lying as close as we want to leaves.

Essentially what happens is this. The $\mathbb{R}$ leaves contribute nothing – the neighborhoods of these can be rescaled shrinking the $\mathbb{R}$ direction, to be fully controlled. However, the S^1 are more serious. They can't be rescaled away, but they cause no trouble because $\mathrm{Wh}(\mathbb{Z}) = 0$.

It is here that, if we used another ring, the Nil terms in the Bass–Heller–Swan formula would enter. We get one for each closed geodesic (as had been mentioned in §5.5).

Now we have to deal with the issue that the transfer to the unit sphere bundle is not injective. What Farrell and Jones did was associate to an n-dimensional negatively curved manifold M, a negatively curved metric on $M \times \mathbb{R}$, and on this manifold there is an invariant upper and lower hemispherical tangent bundle. These are disk bundles. One uses h-cobordisms with compact supports and flows on this space, following the above pattern. Topologically, we know that nothing differs in the Whitehead theory of M from that of $M \times \mathbb{R}$ with compact supports, but metrically we have replaced spheres by disks.

In Bartels *et al.* (2008) a Rips complex, with Mineyev's version of geodesic flow on a hyperbolic group (Mineyev, 2005), replaces ordinary geodesic flow, and enables the proof of the Farrell–Jones conjecture in K-theory for hyperbolic groups.

Remark 8.5 The foliated control theory is critical to the Farrell–Jones pro-

---

[16] Remember that M is assumed negatively curved.

[17] For any homotopy equivalence, one considers the homotopy H from fg to the identity, and for each point p one has the tracks $H(t, p)$ for $0 < t < 1$.

gram (although it does not play as large an explicit role in some of the later work of Bartels, Lück, Reich, and others).

First of all, the fibered case of foliated control is essentially the same thing as controlling with respect to the quotient space. In many situations in foliated geometry, one wants to analyze algebraic topological invariants of the quotient space that does not exist in the conventional sense – see e.g. Connes (1985, 1994) and our discussion in Chapter 5. In all cases, the basic idea is to deal with the Hausdorff object that exists and never really take the quotient.

Remark 8.6 (Digression) Foliations also occur very naturally in the study of the asymptotics of topological phenomena if there is a bound on the local geometry. One can, for example, think about the Borel conjecture as a statement about vanishing of certain "periodic structure sets" – by passing to the universal cover, and then begin inquiring about aperiodic analogues – in ways that ape the theory of quasicrystals (see, for example, Bellissard, 1995). This would lead to a foliated Borel conjecture. Needless to say, Baum and Connes, in formulating their conjecture, also considered a foliated version.

Such foliations and their homology also naturally arise in topological data analysis (see, for example, Weinberger, 2014) because of their connection to "testable properties" or statistically "sampleable" invariants of manifolds (see, for example, Bergeron and Gaboriau, 2004; Elek, 2010; Abert *et al.*, 2017). Essentially one asks for invariants of manifolds that can be approximated by knowing the balls around a number of randomly chosen points. For this to be true, the invariant needs to be continuous in a suitable topology (a modification of the usual Gromov–Hausdorff metric that takes measure into account; see Benjamini and Schramm, 2001, for the case of graphs – also Lovasz, 2012. and Gromov, 1999). Limits of sequences of compact manifolds in this topology are actually foliated spaces (with a transverse measure) – where the leaves have the same dimension as the approximating manifold.

8.4 Tensor Square Trick

The results on L-theory and for the Borel package require other transfers.

Crossing with a sphere (of dimension greater than 1) is trivial in L-theory as is crossing with a disk, so it is necessary to find a new fibration (and transfer) $X \to M$. In Farrell and Jones (1989) the fiber is a modification of $F = S^{m-1} \times S^{m-1}/\mathbb{Z}_2$, the set of unordered pairs (s, s') on the sphere at ∞. When $s \neq s'$ there is a unique geodesic in the universal cover asymptoting to that pair. When $s = s'$ there is a unique geodesic going through a given a given point in $\tilde{M}$ at time 0 and asymptoting to (s, s). Thus, one considers the union of F with a $\mathcal{D}^m$.

It turns out that this stratified space has the property that crossing a manifold surgery problem with it does not lose any surgery obstruction. A similar approach to Siebenmann periodicity $S(M) \rightarrow S(M \times \mathcal{D}^4)$ via an "exotic product" with $\mathbb{CP}^2 \cup \mathcal{D}^3$ is given in Weinberger and Yan (2001).

In both cases the key feature is that the "main part" of the space is a homology manifold and so has a signature and that signature is 1. (Indeed, the space of Farrell and Jones is modified to give rise to an equivariant version of Siebenmann periodicity in Weinberger and Yan (2005) for compact group actions.[18])

As the program developed, more and more complicated transfers were constructed. A major problem for the situation of hyperbolic groups comes because their boundaries are almost always not even ANRs, let alone manifolds! Bestvina and Mess (1991), though, do show that the compactification of Eπ is an ANR.

The solution to this was a breakthrough in Bartels and Lück (2012a) and relies on a tensor square trick. The first point is that there is no reason that one has to "cross with a space" (perhaps in a twisted way) to induce a transfer. One can cross with a symmetric Poincaré complex, which should be geometric over a control space – so one can gain control to good effect – but it need not be the controlled symmetric signature of the control space (or some fancy variant thereof).

This is akin to the use of elliptic operator to set up an (equivariant) Thom isomorphism for complex bundles in Atiyah (1974). One does not need to write down a bundle – just a construction that leads to a suitable family of operators. If one thinks of K-theory as being related to, for example, normal invariants, then one sees an isomorphism that is associated to a nontopological construction on the fibers – as the Dirac operator is not topological and the signature operator causes difficulties at the prime 2.[19]

The basic point of the construction is that, if P is a projective module, then $P \otimes P^*$ naturally supports a symmetric bilinear form. More generally, if P is a chain complex, then $P \otimes P_{-*}$ supports a symmetric Poincaré structure, $(P \otimes P_{-*})^* \cong (P_{-*} \otimes P) \cong P \otimes P_{-*}$ interchanging factors. If P has Euler characteristic 1, then this tensor square has signature 1, and one has a formal

---

[18] The earlier paper only succeeding in doing this for abelian groups.

[19] And this made the periodicity theorems of Weinberger and Yan (2001) more difficult still. As we discussed in Chapter 7 regarding equivariant products, the equivariant signature operator for even-order groups does not give an orientation even rationally, because the localized contribution near 0 is a zero divisor in $R(G)$. Thus one is *forced* to do non-topologically invariant constructions.

process of turning the kind of transfer used in K-theory into one suitable for L-theory. This is a construction that is perfectly well controlled, as verified in Bartels and Lück (2012a), when one changes the control Z space of P to $Z \times Z/\mathbb{Z}_2$ (this being necessary because of the interchange of factors in the above).

Remark 8.7   Bartels and Lück (2012a) introduce another important technical innovation in that paper (necessary for their CAT(0) results) – namely the use of homotopy actions rather than actions.

This paper ushered in a sequence of important new advances on this problem (see, for example, Rüping, 2013; Bartels, 2014; Bartels *et al.*, 2014a,b; Bartels and Bestvina, 2019). It is too soon to be sure where the new "natural boundary" of the current technique is. One can hope that all linear groups over some field, and groups with some "non-positive curvature," will ultimately follow to extensions of these methods.

8.5 The Baum–Connes Conjecture

The serious reader should turn to the excellent survey of Higson and Guentner (2004)[20] for a very useful and insightful treatment.[21] While there have been developments since that paper was written, notably Lafforgue's work clarifying the obstacle of Property (T),[22] it remains, to my mind, the best single survey.

What follows is intended for the frivolous reader.

Remember playing the Novikov game (way back in Chapter 5)? The setting for the game involved improving the index of elliptic operators to lie in the K-theory of some appropriate C^*-algebra. We have focused on the analogy between the normal invariants of degree-1 normal maps, living in $L(e)$-homology theory of a group, and $K_i(B\pi)$ and correspondingly between $L(\pi)$ and $K(C^*\pi)$.

More precisely, associated to a group π, there are C^*-algebras $C_r^*\pi$ and $C_{max}^*\pi$ that are completions of $\mathbb{C}\pi$ thought of as an algebra of unitary operators on appropriate Hilbert spaces. Of the two, $C_{max}^*\pi$ is perhaps the more naive choice – it is the completion with respect to all unitary representations. It has the advantage of being a functorial construction on the category of groups. The

---

[20]   This seems like a good place to express my deep gratitude to Erik [Guentner] for spending a couple of very intensive weeks at Jerusalem cafes explaining this all to me (including assigning and critiqueing homework). And thanks to Nigel [Higson] for sending Erik.

[21]   Other recommended surveys are Valette (2002) and Gomez Aparicio *et al.* (2019) and, of course, Connes's 1994 book *Noncommutative Geometry*.

[22]   The two Bourbaki expositions by Skandalis (1999) and Puschnigg (2012) on Lafforgue's work are excellent next steps.

other, $C_r^* \pi$, is the completion with respect to the regular representation. It is not, in general, functorial, although it is functorial with respect to injections. We will soon return to the issue of functoriality.

What is important for us here is that, associated to any elliptic operator D on M^n with fundamental group π, there is an index $\text{ind}(D) \in K_n(C^* \pi)$ for either of these algebras. The symbol of D lies in $K_n(M)$ (as observed by Atiyah, 1975; see also Higson and Roe, 2000).

There is a natural[23] group homomorphism

$$K_n(M) \to K_n(C^* \pi)$$

that takes an elliptic operator to its index. Indeed, this factors through

$$K_n(B\pi) \to K_n(C^* \pi),$$

and, ultimately (using proper equivariant elliptic operators)

$$K_n^\pi(E\pi) \to K_n(C^* \pi),$$

analogous to our story about the (equivariant) controlled symmetric signature of manifolds, and their algebraic uncontrolled versions (or equivalently the surgery obstruction map in surgery).

Moreover, one can take "twisted coefficients," as we had done in K- and L-theory to accommodate problems of (block or approximate) fibration (and stratified spaces[24]). This leads to the following statement:

$$K_T(G, D) \to K(C_r^*(G, D)).$$

When D is just $\mathbb{C}$, G acting trivially, then the left-hand side is the equivariant K-homology group $K_G(EG)$, as above.

As mentioned in Chapter 5, knowing the injectivity of such a map is an analytic variant of the Novikov conjecture (and it's sometimes called the strong Novikov conjecture in the literature). It implies the usual Novikov conjecture when applied to the signature operator. When applied to an equivariant signature operator, it implies the pseudo-equivalence invariance statement discussed in Chapter 7. (As hinted at in §4.5, the topological invariance of the equivariant signature operator[25] can be proved – along the lines of Pedersen *et al.* (1995)

---

[23] For maps that preserve the fundamental group, using the pushforward of pseudodifferential operators that is introduced for the K-theoretic proof of the index theorem in Atiyah and Singer (1968a).

[24] Although there is much more that needs to be done to understand the natural elliptic operators on stratified spaces than arises for the signature operator (or in topology).

[25] Which implies, for example, that for odd-order groups nonlinear conjugacy of linear representations only occurs for linearly equivariant representations, the theorem of Hsiang–Pardon and Madsen–Rothenberg discussed in Chapter 6.

for the ordinary signature operator – using a metric space version of this kind of statement, which is indeed true for cones of G-ANRs by, for example, the reasoning in Roe, 1996).)

And applied to other operators it has further implications, e.g. for positive scalar curvature, and to higher Riemann–Roch kinds of theorems, etc. The first nontrivial case is when $G = \mathbb{Z}$, when this conjecture is verified by the Pimsner–Voiculescu exact sequence (Pimsner and Voiculescu, 1980). It is the result of applying Mayer–Vietoris to computing the left-hand side and combining it with the isomorphism above. It is the analogue of the Bass–Heller–Swan(–Farrell–Hsiang) theorem in algebraic K-theory. It is simpler in that there is no Nil. Indeed, there is no need for the Nil and UNils that arise in K- and L-theory Farrell–Jones conjectures. The analogue of the work of Waldhausen and Cappell described in the first section of this chapter is Pimsner (1986).

As in the previous paragraph, this implies all the cases of the Novikov conjecture discussed in §8.1.

Much of the immediately subsequent development took somewhat parallel turns in topology and operator theory. It is important to mention the work of Kasparov (1988) on non-positively curved complete manifolds as a highpoint (which inspired the work of Ferry and Weinberger (1991) that paralleled it, although it looked quite different at the time because of difference of emphasis[26]). For this work, Kasparov developed KK-theory, a bivariant version of K-theory that accepts a pair of C^*-algebras – which, together with variants such as E-theory, tend to be key technical tools in the area. The serious reader should study Blackadar (1998) and Higson (2000) to learn this tool.

One result on the operator algebra side that has no known analogue in topology is the theorem of Higson and Kasparov on the (idiosyncratically named[27]) a-T-menable groups.

Theorem 8.8 (Higson and Kasparov, 2001)   *If G acts metrically properly and isometrically on a Hilbert space (i.e. is a-T-menable), then the Baum–Connes map (with coefficients) is an isomorphism (with either completion)*

$$K^G(\underline{EG}, D) \to K(C^*(G, D)).$$

The condition of a-T-menability is somehow an opposite of Property (T): Property (T) groups always have fixed points for continuous isometric actions on Hilbert space.

---

[26] It was only later that Higson, Roe, and others elucidated the close parallels between these theories.

[27] By Gromov (1993).

A couple of hundred pages ago, amenability was also described as an opposite to Property (T). Amenable groups are, indeed, examples of a-T-menable groups.

This is not at all obvious. (Indeed, Gromov had asked the question in 1993, expecting the positive solution.) This was soon shown by Bekka *et al.* (1995). We will return to some of the relevant concepts in §8.6.

The groups $SO(n, 1)$ and $SU(n, 1)$ are also a-T-menable, as had been showed by Vershik, Gel'fand, and Graev almost 40 years ago (Vershik *et al.*, 1974; see also Cherix, 2001).

The Higson–Kasparov theorem has one aspect that cannot be improved upon – their ability to accommodate $C^*_{\max}$. At the opposite extreme, Property (T) groups, every finite-dimensional representation is isolated in the Fell topology;[28] these give rise to elements of infinite order in $K_0(C^*_{\max}\pi)$.

For instance, if π is, say, a lattice in a higher-rank Lie group or even $Sp(n, 1)$,[29] it has many finite-dimensional irreducible representations, so the right-hand side $K_0(C^*_{\max}\pi)$ is infinitely generated (while the domain of the assembly map is finitely generated, e.g. by Borel–Serre).

This is one of the difficulties with the conjecture. From its outset, one realized that, because of the general functoriality of the domain, one would want to use $C^*_{\max}$, but that Property (T) is an obstacle. In some sense the Higson–Kasparov theorem carves out the natural place to look where this difficulty will not arise.

We shall also see that that theorem has some extraordinary implications.

However, it underscores the extent to which Property (T) is an obstacle. It is thus very remarkable that Lafforgue (2002) was able to overcome this obstacle in some cases by including uniform lattices in $SL_3(\mathbb{R})$ and $SL_3(\mathbb{C})$, and hyperbolic groups (Lafforgue, 2012) – even with coefficients. These are based on making a variant of *KK*-theory for Banach algebras, which allow more deformations of representations that allow one to "pass through" the gaps that prevent deformations in *KK*-theory. However, Lafforgue (2008) describes a strengthening of Property (T) that obstructs all known techniques, and shows that $SL_3(\mathbb{Q}_p)$ has this property, so that it (and its lattices) definitely lie outside current technology.

In §8.6 we will discuss that, for general discrete groups, the Baum–Connes conjecture with coefficients fails – and the reason for this is because of expanders, a class of graphs that we have met in Chapter 3 as one of the first applications of Property (T).

---

[28] The reader might want to review some ideas from Chapter 3 to unravel this discussion.

[29] Since $Sp(n, 1)$ is a rank-1 group, its lattices are fundamental groups of negatively curved manifolds. As a result, the Higson–Kasparov theorem does not even extend to the situation of negative curvature.

8.6 A-T-menability, Uniform Embeddability, and Expanders

The hypothesis of the Higson–Kasparov theorem, a-T-menability, was first introduced by Haagerup in an equivalent form. A useful source on a-T-menability is Cherix (2001). The equivalent forms of a-T-menability have parallels among equivalent definitions of Property (T). These equivalences are generally useful for making constructions and in different applications.

(1) There is a proper function $\psi \colon \pi \to \mathbb{R}^+$ that is conditionally negative-definite (i.e. $\psi(g) = \psi(g^{-1})$) and, for any n-tuple of elements of π, the matrix $\psi(g_i g_j^{-1})$ is conditionally negative-definite (i.e. negative-definite on tuples $(a_1, \ldots, a_n)$ so that the sum of the a_i equals zero.)

(2) There is a sequence φ_n of continuous positive-definite functions on π with $\varphi_n(1) = 1$ that vanish at infinity but converge to 1 uniformly on compact subsets of π.

(3) π acts isometrically and metrically properly on a Hilbert space.

For Property (T) groups, every conditionally negative-definite function is bounded (i.e. not proper if π is noncompact). If a sequence of normalized positive-definite functions converges to 1 uniformly on compact sets, then it converges to 1 uniformly, so they can't vanish at infinity. And, finally, as noted before, every isometric action has a fixed point.

Many groups are a-T-menable (and, of course, many are not). As we mentioned, $SO(n, 1)$ and $SU(n, 1)$ and products of such, and amenable groups are. Also groups that act on CAT(0) cubical complexes or more generally "spaces with walls" are included in this class (Niblo and Reeves, 1997). So (as proved by Farley in his thesis – see Farley, 2003) the morally amenable (but not yet known to be (non-)amenable) Thompson group satisfies the Baum–Connes conjecture.

The Higson–Kasparov theorem is proved by an analogue of Atiyah's proof of Bott periodicity (Atiyah, 1968).

Rather than make any attempt at explaining the proof of the theorem, let's instead (for example, following the line of thought of §§4.8 and 4.9) think about the metric-space analogue of the theorem and then see what that buys us in terms of the Novikov conjecture. Of course, the metric-space version is of interest in its own right: there are many more bounded geometry metric spaces[30] than there are finitely generated groups!

---

[30] A metric space has bounded geometry if it is a path metric (i.e. distances are generated by a path geometry) and there are only "finitely many types" (or a compact space of types) of balls of a fixed radius. So all Cayley graphs of finitely generated groups have bounded geometry. As

Theorem 8.9 (Yu, 2000; Skandalis *et al.*, 2002) *If Γ is a (discrete) metric space of bounded geometry which uniformly embeds in a Hilbert space, then the bounded Baum–Connes assembly map is an isomorphism:*

$$K_n^{\mathrm{lf}}(|\Gamma|) = \lim K_n^{\mathrm{lf}}(N_k|\Gamma|) \to K_n^{\mathrm{lf}}(C^*|\Gamma|).$$

The left-hand side is the limit of K-homology of the nerve of covers of Γ by k-balls, as $k \to \infty$. (The discreteness of Γ is a convenience to make this cover locally finite: otherwise, one can replace a metric space of bounded geometry by a coarsely dense discrete subset.) The range is the K-theory of the Roe algebra: it is the closure of the bounded propagation speed operators on Γ. If one thinks of operators as being described by kernels (like matrices in a geometric module), then one is taking the limit of the operators where $k(x, y) = 0$ if $d(x, y) > R$ as $R \to \infty$. So an operator in this algebra can be well approximated by operators with finite propagation speed. We shall later see the implications of this approximation – which have no analogue in the purely topological world.

Note that if Γ acts properly and isometrically on a Hilbert space, then it uniformly embeds – indeed, the map $\Gamma \to H$ given by $\gamma \to \gamma(h)$ for any fixed h is a uniform embedding. Propriety means that one only returns finitely many times to any fixed neighborhood of h. The isometry condition then translates this into a condition for any group element (uniformly), i.e. that there is a proper non-negative increasing function f, so that

$$\|\gamma(h) - \gamma'(h)\| > f\big(d(\gamma, \gamma')\big).$$

Moreover, the map is Lipschitz with Lipschitz constant supremum sup $\|\gamma(h) - h\|$ as γ runs over generators of Γ.

The arguments given in §4.9 (i.e. the principle of descent) now enable one to show that, if in addition $B\Gamma$ is a finite complex, then the analytic Novikov conjecture with coefficient is true, i.e. the Baum–Connes assembly map with coefficients is split injective. However, things are better yet. Higson (2000) has shown that one can dispense with this finiteness and still get the result.

Theorem 8.10 (The Novikov conjecture for groups that uniformly embed; Skandalis *et al.*, 2002) *If Γ is a countable group which uniformly embeds in*

---

mentioned in §4.8 and explained in §5.3, without bounded geometry there are older counterexamples to the bounded Borel and Baum–Connes conjectures based on different principles than the examples we are about to explain regarding the Baum–Connes conjecture. I am not aware of any counterexamples to the bounded Borel conjecture.

Hilbert space,[31] *then for all coefficients D,*

$$K^{\Gamma}(\underline{E\Gamma}, D) \to K\big(C^*(\Gamma, D)\big)$$

is split injective.

Later work by Kasparov and Yu (2006) has weakened the hypothesis on which Banach space one needs to embed in for this result.

This theorem has the following corollary:

Corollary 8.11 (Guentner *et al.*, 2005)   *The Novikov conjecture holds for any countable* Γ *in* $GL_n(\mathbb{F})$ *for any field* $\mathbb{F}$.

Of course, this supplements the cases of non-positive curvature, amenable (a-T-menable) groups, hyperbolic groups, etc. that have been discussed before!

The proof uses a variant of condition (1) at the start of this section that describes a condition sufficient for uniformly embedding a discrete metric space into Hilbert space. Instead of a function from Γ to $\mathbb{R}$, one uses a function $\Gamma \times \Gamma \to \mathbb{R}$ that is a negative-type kernel, defined exactly the same as in condition (1). We now need the kernel to behave well with respect to the metric on Γ, i.e. that $\psi(g, h)$ can be bounded above and below in terms of $d(g, h)$.

If Γ were discretely embedded in $GL_n(\mathbb{C})$, one could use explicitly the geometry of $GL_n(\mathbb{C})/U(n)$ to construct the desired embedding in Hilbert space. $(GL_n(\mathbb{C})/U(n)$ is isomorphic to the parabolic group of upper triangular matrices, which is amenable, indeed solvable.) In general, the idea is to find enough valuations, so that Γ is discretely embedded in a product of GL_n of valuated rings and that each of these is embedded in a way appropriate to the geometry of building for that valuation.[32] Thus, the workhorse lemma is this:

Lemma 8.12   *For any finitely generated field, there is a countable number of valuations* d_i *(both archimedean and discrete) such that, for any finitely generated subring, R, and for any positive numbers* N_i, *the* $\{r \in R \mid d_i(r) < N_i\}$ *is finite.*

For $\mathbb{Q}$ one uses the usual valuations. The finitely generated subrings are of the form $\mathbb{Z}[1/N]$ for some N. Then one only needs finitely many valuations, namely the archimedean one and the ones corresponding to primes in N.

Remark 8.13   There is now a rather different approach to this corollary that works in the topological setting, at least with the hypothesis of finiteness of

---

[31] Assume that all of its finitely generated subgroups do, in order to avoid any questions about metrics.

[32] Needless to say, some care needs to be taken in combining a perhaps infinite number of embeddings to guarantee convergence and that one remains, for example, discrete.

BΓ. This is due to Guentner *et al.* (2012) and is based on clever limiting arguments and takes its start from the Novikov conjecture for groups of finite asymptotic dimension (Yu, 1998; Carlsson and Goldfarb, 2004; Chang *et al.*, 2008; Dranishnikov *et al.*, 2008; Bartels, 2014).

We close with a brief discussion of failure. First of all, not all discrete metric spaces of bounded geometry uniformly embed in Hilbert space. Although not the first examples, an example can be built from expanders, as observed by Gromov.

Proposition 8.14 *If X_i is a sequence of d-regular expander graphs, then their disjoint union cannot be uniformly embedded.*

Without loss of generality we can assume that X_i is embedded via f_i so that its mean value (in H) is trivial. We will now use the Laplacian characterization of expansion. Let's assume that neighbors in X_i are moved a distance at most 1. In that case

$$|(\Delta f_i, f_i)| = 1/d(\text{sum over neighbors } (v, w), \|f_i(v) - f_i(w)\|^2) < \|X_i\|.$$

But this gives an upper bound on $\|f_i\|$ by the expander property, which means that the average distance of f_i from the origin is uniformly bounded (in terms of d and the expansion constant), contradicting uniformity of the embedding (i.e. that far vertices are mapped far apart).

In fact, consider $e^{-t\Delta}$. It is a bounded propagation speed operator that converges to the projection to the locally constant functions on this disjoint union. Thus that projection is in $C^*|X|$. This is a bounded propagation speed operator whose definition requires expansion: one might expect that it does not lie in the image of the coarse Baum–Connes map. This is true. It is analogous to the fact that G-indices for free actions – i.e. ones that come from $K(X/G)$ – are multiples of the regular representation.

Gromov then showed that an expander family of large girth expanders (e.g. the ones that come from the Selberg theorem; see the appendix to §3.5) can be coarsely embedded in "random quotients" of hyperbolic groups (Gromov, 2003; Silberman, 2003),[33] which therefore do not uniformly embed in Hilbert space. Higson *et al.* (2002) converted such groups into counterexamples to Baum–Connes with coefficients.

Of course, this raises a number of questions. Do expanders obstruct the untwisted Baum–Connes conjecture? Can they be used to disprove the Novikov conjecture (in any form)? Or to disprove the Borel conjecture?

---

[33] See Sapir (2014) for a method of embedding these groups into geometrically finite groups, as the Gromov examples are only finitely generated.

However, now that we have seen that the original versions of these analytic analogues of the Borel conjecture fail, it seems, in the spirit of what we have argued throughout this book, that understanding what is true remains an important problem.

For example – it seems to me that understanding bounded propagation speed algebras could be useful in scientific situations far removed from manifold theory, and the hypotheses that the underlying metric space – network – is uniformly embeddable seems shockingly naive. Indeed, besides the issues caused by expanders, one frequently would want to dispense even with bounded geometry, bringing on many new issues.

References

Abert, M., N. Bergeron, I. Biringer, *et al.* (2017). On the growth of L^2-invariants for sequences of lattices in Lie groups. *Ann. of Math. (2)*, **185**, 711–790.

Abramenko, P. and K. Brown. (2008). *Buildings*. Berlin: Springer.

Abramovich, D., K. Karu, K. Matsuki, and J. Wlordczyk. (2002). Torification and factorization of birational maps. *J. Amer. Math. Soc.*, **15**, 531–572.

Agol, Ian. (2013). The virtual Haken conjecture. With an appendix by Agol, Daniel Groves, and Jason Manning. *Doc. Math.* **18**, 1045–1087.

Albin, Pierre, Eric Leichtnam, Rafe Mazzeo, and Paolo Piazza, Paolo. (2018). Hodge theory on Cheeger spaces. *Journal für die reine und angewandte Mathematik*, **744**, 29–102.

Alonso, J.M., T. Brady, D. Cooper, *et al.* (1991). Notes on word hyperbolic groups. In *Group Theory from a Geometrical Viewpoint (Trieste, 1990)*. River Edge: World Scientific, pp. 3–63.

Alperin, R. and P. Shalen. (1982). Linear groups of finite cohomological dimension. *Invent. Math.*, **66**, 89–98.

Ancel, R. and J. Cannon. (1979). The locally flat approximation of cell-like embedding relations. *Ann. of Math. (2)*, **109**(1), 61–86.

Anderson, D. and W.C. Hsiang. (1976). Extending combinatorial PL structures on stratified spaces. *Invent. Math.*, **32**(2), 179–204.

Anderson, D. and W.C. Hsiang. (1977). The functors K_{-i} and pseudo-isotopies of polyhedra. *Ann. of Math. (2)*, **105**(2), 201–223.

Anderson, D. and W.C. Hsiang. (1980). Extending combinatorial PL structures on stratified spaces II. *Trans. Amer. Math. Soc.*, **260**(1), 223–253.

Anderson, D., F. Connolly, S. Ferry, and E. Pedersen. (1994). Algebraic K-theory with continuous control at infinity. *J. Pure Appl. Algebra*, **94**(1), 25–47.

Ash, A. and A. Borel. (1990). Generalized modular symbols. In *Cohomology of Arithmetic Groups and Automorphic Forms*. Lecture Notes in Mathematics, 1447. Berlin: Springer, pp. 57–75.

Assadi, A. (1982). Finite group actions on simply-connected manifolds and CW complexes. *Mem. Amer. Math. Soc.*, **35**(257).

Assadi, A. and W. Browder. (1985). On the existence and classification of extensions of actions on submanifolds of disks and spheres. *Trans. Amer. Math. Soc.*, **291**(2), 487–502.

Assadi, A. and P. Vogel. (1987). Actions of finite groups on compact manifolds. *Topology*, **26**(2), 239–263.

Atiyah, M. (1961). Thom complexes. *Proc. London Math. Soc. (3)*, **11**, 291–310.

Atiyah, M. (1968). Bott periodicity and the index of elliptic operators. *Q. J. Math. Oxford Ser. (2)*, **19**, 113–140.

Atiyah, M. (1969). The signature of fibre-bundles. In *Global Analysis (Papers in Honor of K. Kodaira)*. Tokyo: University of Tokyo Press, pp. 73–84.

Atiyah, M. (1970). Global theory of elliptic operators. In *Proc. Int. Conf. on Functional Analysis and Related Topics (Tokyo, 1969)*. Tokyo: University of Tokyo Press, pp. 21–30.

Atiyah, M. (1974). Elliptic operators, discrete groups and von Neumann algebras. In *Colloque "Analyse et Topologie" en l'Honneur de Henri Cartan (Orsay, 1974)*. Astérisque, 32–33. Paris: Société Mathématique de France, pp. 43–72.

Atiyah, M. (1975). Eigenvalues and Riemannian geometry. In *Manifolds – Tokyo 1973 (Proc. Int. Conf.)*. Tokyo: University of Tokyo Press, pp. 5–9.

Atiyah, M. and R. Bott. (1967). A Lefschetz fixed point formula for elliptic complexes. I. *Ann. of Math. (2)*, **86**, 374–407.

Atiyah, M. and R. Bott. (1968). A Lefschetz fixed point formula for elliptic complexes. II. Applications. *Ann. of Math. (2)*, **88**, 451–491.

Atiyah, M. and F. Hirzebruch. (1970). Spin-manifolds and group actions. In *Essays on Topology and Related Topics (Mémoires dédiés à Georges de Rham)*. New York: Springer, pp. 18–28.

Atiyah, M. and W. Schmid. (1977). A geometric construction of the discrete series for semisimple Lie groups. *Invent. Math.*, **42**, 1–62.

Atiyah, M. and G. Segal. (1968). The index of elliptic operators. II. *Ann. of Math. (2)*, **87**, 531–545.

Atiyah, M. and I. Singer. (1968a). The index of elliptic operators. I. *Ann. of Math. (2)*, **87**, 484–530.

Atiyah, M. and I. Singer. (1968b). The index of elliptic operators. III. *Ann. of Math. (2)*, **87**, 546–604.

Atiyah, M. and I. Singer. (1971). The index of elliptic operators. IV. *Ann. of Math. (2)*, **93**, 119–138.

Atiyah, M., R. Bott, and A. Shapiro. (1964). Clifford modules. *Topology*, **3**(Suppl. 1), 3–38.

Atiyah, M., V. Patodi, and I. Singer. (1975a). Spectral asymmetry and Riemannian geometry. I. *Math. Proc. Cambridge Philos. Soc.*, **77**, 43–69.

Atiyah, M., V. Patodi, and I. Singer. (1975b). Spectral asymmetry and Riemannian geometry. II. *Math. Proc. Cambridge Philos. Soc.*, **78**(3), 405–432.

Aumann, R.J. (1956). Asphericity of alternating knots. *Ann. of Math. (2)*, **64**, 374–392.

Avramidi, G. (2013). Periodic flats and group actions on locally symmetric spaces. *Geom. Topol.*, **17**(1), 311–327.

Avramidi, G. (2014). Smith theory, L^2-cohomology, isometries of locally symmetric manifolds, and moduli spaces of curves. *Duke Math. J.*, **163**(1), 1–34.

Avramidi, G. (2018). Rational manifold models for duality groups. *Geom. Funct. Anal.*, **28**, 965–994.

Baily, W. and A. Borel. (1966). Compactification of arithmetic quotients of bounded symmetric domains. *Ann. of Math. (2)*, **84**, 442–528.

Ballman, W. and J. Siwatkowski. (1977). On L^2-cohomology and property (T) for automorphism groups of polyhedral cell complexes. *Geom. Funct. Anal.*, **7**, 615–645.

Banagl, M. and A. Ranicki. (2006). Generalized Arf invariants in algebraic L-theory. *Adv. Math.*, **199**(2), 542–668.

Bartels, A. (2003). Squeezing and higher algebraic K-theory. *K-Theory*, **28**(1), 19–37.

Bartels, A. (2014). On proofs of the Farrell–Jones conjecture. Preprint, arXiv:1210.1044.

Bartels, A. and M. Bestvina. (2019). The Farrell–Jones conjecture for mapping class groups. *Invent. Math.*, **215**, 651–712

Bartels, A. and W. Lück. (2012a). The Borel conjecture for hyperbolic and CAT(0)-groups. *Ann. of Math. (2)*, **175**(2), 631–689.

Bartels, A. and W. Lück. (2012b). The Farrell–Hsiang method revisited. *Math. Ann.*, **354**(1), 209–226.

Bartels, A. and H. Reich. (2007). Coefficients in the Farrell–Jones conjecture. *Adv. Math.*, **15**, 337–362.

Bartels, A., W. Lück, and H. Reich. (2008). The K-theoretic Farrell–Jones conjecture for hyperbolic groups. *Invent. Math.*, **172**(1), 29–70.

Bartels, A., W. Lück, and S. Weinberger. (2010). On hyperbolic groups with spheres as boundary. *J. Differential Geom.*, **86**(1), 1–16.

Bartels, A., F.T. Farrell, and W. Lück. (2014a). The Farrell–Jones conjecture for co-compact lattices in virtually connected Lie groups. *J. Amer. Math. Soc.*, **27**(2), 339–388.

Bartels, A., W. Lück, H. Reich, and H. Ruping. (2014b). K- and L-theory of group rings over $GL_n(\mathbb{Z})$. *Publ. Math. Inst. Hautes Études Sci.*, **119**, 97–125.

Bartholdi, L. and B. Virag. (2005). Amenability via random walks. *Duke Math. J.*, **130**, 39–56.

Bass, H. (1968). *Algebraic K-Theory* . New York: Benjamin.

Bass, H. (1976). Euler characteristics and characters of discrete groups. *Invent. Math.*, **35**, 155–196.

Bass, H. and M.P. Murthy. (1967). Grothendieck groups and Picard groups of abelian group rings. *Ann. of Math. (2)*, **86**, 16–73.

Bass, H., A. Heller, and R. Swan. (1964). The Whitehead group of a polynomial extension. *Publ. Math. Inst. Hautes Études Sci.*, **22**, 61–79.

Bass, H., J. Milnor, and J.-P. Serre. (1967). Solution of the congruence subgroup problem for $SL^n(n \geq 3)$ and $Sp^{2n}(n \geq 2)$. *Publ. Math. Inst. Hautes Études Sci.*, **33**, 59–137.

Bassalygo, L.A. and M.S. Pinsker (1973). The complexity of an optimal non-blocking commutation scheme without reorganization (in Russian). *Problemy Peredachi Informatsii*, **9**(1), 84–87. Translated into English in *Problems Inform. Transmiss*, **9**, 64–66 (1974).

Baum, P. and A. Connes. (2000). Geometric K-theory for Lie groups and foliations. *Enseign. Math. (2)*, **46**(1–2), 3–42.

Baum, P., W. Fulton, and R. MacPherson. (1975). Riemann–Roch for singular varieties. *Publ. Math. Inst. Hautes Études Sci.*, **45**, 101–145.

Baumslag, G., E. Dyer, and A. Heller. (1980). The topology of discrete groups. *J. Pure Appl. Algebra*, **16**(1), 1–47.

Baumslag, G., E. Dyer, and C. Miller III, (1983). On the integral homology of finitely presented groups. *Topology*, **22**(1), 27–46.

Bekka, B., P. Cherix, and A. Valette. (1995). Proper affine isometric actions of amenable groups. In *Novikov Conjectures, Index Theorems and Rigidity (Oberwolfach, 1993)*, Vol. 2. London Mathematical Society Lecture Note Series, 227. Cambridge: Cambridge University Press, pp. 1–4.

Bekka, B., P. de la Harpe, and A. Valette. (2008). *Kazhdan's Property (T)*. Cambridge: Cambridge University Press.

Belegradek, I. (2006). Aspherical manifolds, relative hyperbolicity, simplicial volume and assembly maps. *Algebr. Geom. Topol.*, **6**, 1341–1354.

Belegradek, I. (2007). Aspherical manifolds with relatively hyperbolic fundamental groups. *Geom. Dedicata*, **129**, 119–144.

Bellissard, J. (1995). Noncommutative geometry and quantum Hall effect. In *Proc. Int. Congress of Mathematicians (Zürich, 1994)*. Basel: Birkhäuser, pp. 1238–1246.

Benjamini, I. and O. Schramm. (2001). Recurrence of distributional limits of finite planar graphs. *Electron. J. Probab.*, **6**, 23.

Bergeron, N. and D. Gaboriau. (2004). Asymptotique des nombres de Betti, invariants l^2 et laminations. *Comment. Math. Helv.*, **79**, 362–395.

Bergeron, N. and A. Venkatesh. (2010). The asymptotic growth of torsion homology for arithmetic groups. Preprint, arXiv:1004.1083v1.

Berline, N., E. Getzler, and M. Vergne. (2004). *Heat Kernels and Dirac Operators*. Grundlehren Text Editions. Berlin: Springer. Corrected reprint of the 1992 original.

Bestvina, M. (2014). Geometric group theory and 3-manifolds hand in hand: the fulfillment of Thurston's vision for three-manifolds. *Bull. Amer. Math. Soc.*, **51**, 53–70.

Bestvina, M. and M. Feighn. (1992). A combination theorem for negatively curved groups. *J. Differential Geom.*, **35**, 85–101.

Bestvina, M. and G. Mess. (1991). The boundary of negatively curved groups. *J. Amer. Math. Soc.*, 4(3), 469–481.

Bieri, R. and B. Eckmann. (1973). Groups with homological duality generalizing Poincaré duality. *Invent. Math.*, **20**, 103–124.

Bing, R.H. (1959). The cartesian product of a certain nonmanifold and a line is E^4. *Ann. of Math. (2)*, **70**, 399–412.

Blackadar, B. (1998). *K-Theory for Operator Algebras*, 2nd edn. Mathematical Sciences Research Institute Publications, 5. Cambridge: University of Cambridge Press.

Bleecker, D. and B. Booss-Bavnbek. (2013). *Index Theory with Applications to Mathematics and Physics*. Boston: International Press.

Block, J. and S. Weinberger. (1992). Aperiodic tilings, positive scalar curvature and amenability of spaces. *J. Amer. Math. Soc.*, 5(4), 907–918.

Block, J. and S. Weinberger. (1997). Large scale homology theories and geometry. In *Geometric Topology (Athens, GA, 1993)*. AMS/IP Studies in Advanced Mathematics, 2. Providence: American Mathematical Society; Boston: International Press, pp. 522–569.

Block, J. and S. Weinberger. (1999). Arithmetic manifolds of positive scalar curvature. *J. Differential Geom.*, **52**(2), 375–406.

Block, J. and S. Weinberger. (2006). Todd classes and holomorphic group actions. *Pure Appl. Math. Q.*, 2(4), 1237–1253. Special issue in honor of Robert D. MacPherson, Part 2.

Block, J. and S. Weinberger. (2008). On the generalized Nielsen realization problem. *Comment. Math. Helv.*, **83**(1), 21–33.

Boileau, M., B. Leeb, and J. Porti. (2005). Geometrization of 3-dimensional orbifolds. *Ann. of Math. (2)*, **162**(1), 195–290.

Bökstedt, M., W.C. Hsiang, and I. Madsen. (1993). The cyclotomic trace and algebraic *K*-theory of spaces. *Invent. Math.*, **111**(3), 465–539.

Borel, A. (1960). *Seminar on Transformation Groups*, with contributions by G. Bredon, E. E. Floyd, D. Montgomery, and R. Palais. Annals of Mathematics Studies, 46. Princeton: Princeton University Press.

Borel, A. (1963). Compact Clifford–Klein forms of symmetric spaces. *Topology*, **2**, 111–122.

Borel, A. (1974). Stable real cohomology of arithmetic groups. *Ann. Sci. Éc. Norm. Supér. (4)*, **7**, 235–272.

Borel, A. (1983). On period maps of certain K(π, 1). In *Borel Collected Papers*, Vol. III. Berlin: Springer, pp. 57–60.

Borel, A. and Harish-Chandra. (1961). Arithmetic subgroups of algebraic groups. *Bull. Amer. Math. Soc.*, **67**, 579–583.

Borel, A. and F. Hirzebruch. (1959a). Characteristic classes and homogeneous manifolds, I. *Amer. J. Math.*, **80**, 458–538.

Borel, A. and F. Hirzebruch. (1959b). Characteristic classes and homogeneous manifolds, II. *Amer. J. Math.*, **81**, 315–382.

Borel, A. and F. Hirzebruch. (1960). Characteristic classes and homogeneous manifolds, III. *Amer. J. Math.*, **82**, 491–504.

Borel, A. and L. Ji. (2005). *Compactifications of Symmetric and Locally Symmetric Spaces*. Basel: Birkhäuser.

Borel, A. and J.-P. Serre. (1973). Corners and arithmetic groups. *Comment. Math Helv.*, **48**, 436–491.

Borisov, L. and A. Libgober. (2008). Higher elliptic genera. *Math. Res. Lett.*, **15**(3), 511–520.

Bott, R. and C. Taubes. (1989). On the rigidity theorems of Witten. *J. Amer. Math. Soc.*, **2**(1), 137–186.

Bousfield, A.K. and D. Kan. (1972). *Homotopy Limits, Completions and Localizations*. Lecture Notes in Mathematics, 304. Berlin: Springer.

Brasselet, J.-P., J. Schürmann, and S. Yokura. (2010). Hirzebruch classes and motivic Chern classes for singular spaces. *J. Topol. Anal.*, **2**(1), 1–55.

Bredon, G. (1967). *Sheaf Theory*. New York: McGraw-Hill.

Bredon, G. (1972). *Introduction to Compact Transformation Groups*. Pure and Applied Mathematics, 46. New York: Academic Press.

Bridson, M. and A. Haefliger. (1999). *Metric Spaces of Non-Positive Curvature*. Berlin: Springer.

Brooks, R. (1981). The fundamental group and the spectrum of the Laplacian. *Comment. Math. Helv.*, **56**, 581–598.

Browder, W. (1972). *Surgery on Simply Connected Manifolds*. Berlin: Springer.

Browder, W. and W.C. Hsiang. (1982). G-actions and the fundamental group. *Invent. Math.*, **65**(3), 411–424.

Browder, W. and J. Levine. (1966). Fibering manifolds over a circle. *Comment. Math. Helv.*, **40**, 153–160.

Browder, W. and R. Livesay. (1973). Fixed point free involutions on homotopy spheres. *Tôhoku Math. J. (2)*, **25**, 69–87.

Browder, W. and F. Quinn. (1973). A surgery theory for G-manifolds and stratified spaces. In *Manifolds – Tokyo 1973 (Proc. Int. Conf.)*. Tokyo: University of Tokyo Press, pp. 27–36.

Brown, K. (1982). *Cohomology of Groups*. Berlin: Springer.

Bryant, J., S. Ferry, W. Mio, and S. Weinberger. (1993). Topology of homology manifolds. *Bull. Amer. Math. Soc.*, **28**(2), 324–328.

Bryant, J., S. Ferry, W. Mio, and S. Weinberger. (1996). The topology of homology manifolds. *Ann. of Math. (2)*, **143**(3), 435–467.

Bryant, J., S. Ferry, W. Mio, and S. Weinberger. (2007). Desingularizing homology manifolds. *Geom. Topol.*, **11**, 1289–1314.

Burger, M. and S. Mozes. (2001). Lattices in products of trees. *Publ. Math. Inst. Hautes Études Sci.*, **92**, 151–194.

Burger, M., T. Gelander, A. Lubotzky, and S. Mozes (2002). Counting hyperbolic manifolds. *Geom. Funct. Anal.*, **12**(6), 1161–1173.

Buser, P. and H. Karcher. (1981). *Gromov's Almost Flat Manifolds*. Astérisque, 81. Paris: Société Mathématique de France.

Burghelea, D. (1985). The cyclic homology of the group algebras. *Comment. Math. Helv.*, **60**(3), 354–365.

Burghelea, D., R. Lashof, and M. Rothenberg. (1975). *Groups of Automorphisms of Manifolds*. Lecture Notes in Mathematics, 473. Berlin: Springer.

Calabi, E. (1961). On compact, Riemannian manifolds with constant curvature. I. In *Differential Geometry*. Proc. Symp. in Pure Mathematics, III. Providence: American Mathematical Society, pp. 155–180.

Calabi, E. and E. Vesentini. (1960). On compact, locally symmetric Kähler manifolds. *Ann. of Math. (2)*, **71**, 472–507.

Calegari, F. (2015). The stable homology of congruence subgroups. *Geom. Topol.*, **19**, 3149–3191.

Calegari, F. and M. Emerton. (2012). Completed cohomology—a survey. In *Non-Abelian Fundamental Groups and Iwasawa Theory*. London Mathematical Society Lecture Note Series, 393. Cambridge: Cambridge University. Press, pp. 239–257.

Calegari, F. and A. Venkatesh. (2019). *A Torsion Jacquet–Langlands Correspondence*. Astérisque, 409. Paris: Société Mathématique de France.

Cantat, S. (2017). Progrès récents concernant le programme de Zimmer [d'après A. Brown, D. Fisher, et S. Hurtado]. *Séminaire Bourbaki*, **2017–18**, 1136.

Cappell, S. (1973). Mayer–Vietoris sequences in Hermitian K-theory. In *Algebraic K-Theory, III: Hermitian K-Theory and Geometric Applications (Proc. Conf., Battelle Memorial Inst., Seattle, WA, 1972)*. Lecture Notes in Mathematics, 343. Berlin: Springer, pp. 478–512.

Cappell, S. (1974a). Unitary nilpotent groups and Hermitian K-theory. I. *Bull. Amer. Math. Soc.*, **80**, 1117–1122.

Cappell, S. (1974b). On connected sums of manifolds. *Topology*, **13**, 395–400.

Cappell, S. (1976a). A splitting theorem for manifolds. *Invent. Math.*, **33**(2), 69–170.

Cappell, S. (1976b). On homotopy invariance of higher signatures. *Invent. Math.*, **33**(2), 171–179.

Cappell, S. and J. Shaneson. (1974). The codimension-two placement problem and homology-equivalent manifolds. *Ann. of Math. (2)*, **99**, 277–348.

Cappell, S. and J. Shaneson. (1976). Piecewise linear embeddings and their singularities. *Ann. of Math. (2)*, **103**(1), 163–228.

Cappell, S. and J. Shaneson. (1978). An introduction to embeddings, immersions and singularities in codimension two. In *Algebraic and Geometric Topology (Stanford Univ., Stanford, CA, 1976)*, Part 2. Proc. Symp. in Pure Mathematics, 32. Providence: American Mathematical Society, pp. 129–149.

Cappell, S. and J. Shaneson. (1979). A counterexample on the oozing problem for closed manifolds. In *Algebraic Topology, Aarhus, 1978 (Proc. Symp., Univ. Aarhus, Aarhus, 1978)*. Lecture Notes in Mathematics, 763. Berlin: Springer, pp. 627–634.

Cappell, S. and J. Shaneson. (1981). Nonlinear similarity. *Ann. of Math. (2)*, **113**(2), 315–355.

Cappell, S. and J. Shaneson. (1982). The topological rationality of linear representations. *Publ. Math. Inst. Hautes Études Sci.*, **56**, 101–128.

Cappell, S. and S. Weinberger. (1987). A geometric interpretation of Siebenmann's periodicity phenomenon. In *Geometry and Topology (Athens, GA, 1985)*. New York: Marcel Dekker, pp. 47–52.

Cappell, S. and S. Weinberger. (1991a). A simple construction of Atiyah–Singer classes and piecewise linear transformation groups. *J. Differential Geom.*, **33**(3), 731–742.

Cappell, S. and S. Weinberger. (1991b). Classification de certains espaces stratifiés. *C. R. Acad. Sci. Paris, Sér. I Math.*, **313**(6), 399–401.

Cappell, S. and S. Weinberger. (1995). Replacement of fixed sets and of their normal representations in transformation groups of manifolds. In *Prospects in Topology (Princeton, NJ, 1994)*. Annals of Mathematics Studies, 138. Princeton: Princeton University Press, pp. 67–109.

Cappell, S., J. Shaneson, and S. Weinberger. (1991). Classes topologiques caractéristiques pour les actions de groupes sur les espaces singuliers. *C. R. Acad. Sci. Paris, Sér. I Math.*, **313**(5), 293–295.

Cappell, S., S. Weinberger, and M. Yan. (2013). Closed aspherical manifolds with center. *J. Topol.*, **6**(4), 1009–1018.

Cappell, S., A. Lubotzky, and S. Weinberger. (2015). A trichotomy theorem for transformation groups of locally symmetric manifolds and topological rigidity. Preprint, arXiv:1601.00262v1.

Cappell, S., S. Weinberger, and M. Yan. (2020). Fixed points of G-CW-complex with prescribed homotopy type. Preprint, arXiv:2010.14988

Carlsson, G. (1991). Equivariant stable homotopy and Sullivan's conjecture. *Invent. Math.*, **103**(3), 497–525.

Carlsson, G. (1995). Bounded K-theory and the assembly map in algebraic K-theory. In *Novikov Conjectures, Index Theorems and Rigidity (Oberwolfach, 1993)*, Vol. 2. London Mathematical Society Lecture Note Series, 227. Cambridge: Cambridge University Press, pp. 5–127.

Carlsson, G. and B. Goldfarb. (2004). The integral K-theoretic Novikov conjecture for groups with finite asymptotic dimension. *Invent. Math.*, **157**(2), 405–418.

Carlsson, G. and E. Pederson. (1995). Controlled algebra and the Novikov conjectures for K- and L-theory. *Topology*, **34**(3), 731–758.

Carter, D. (1980). Lower K-theory of finite groups. *Comm. Algebra*, **8**(20), 1927–1937.

318   *References*

Cartan, H. and S. Eilenberg. (1956). *Homological Algebra*. Princeton: Princeton University Press.

Cerf, J. (1970). La stratification naturelle des espaces de fonctions différentiables réelles et le théorème de la pseudo-isotopie. *Publ. Math. Inst. Hautes Études Sci.*, **39**, 5–173.

Casson, A. (1967). Fibrations over spheres. *Topology*, **6**, 489–499.

Chang, S. (2001). Coarse obstructions to positive scalar curvature in noncompact arithmetic manifolds. *J. Differential Geom.*, **57**(1), 1–21.

Chang, S. (2004). On conjectures of Mathai and Borel. *Geom. Dedicata*, **106**, 161–167.

Chang, S. and S. Weinberger. (2003). On invariants of Hirzebruch and Cheeger–Gromov. *Geom. Topol.*, **7**, 311–319.

Chang, S. and S. Weinberger. (2006). On Novikov-type conjectures. In *Surveys in Noncommutative Geometry*. Clay Mathematics Proceedings, 6. Providence: American Mathematical Society, pp. 43–70.

Chang, S. and S. Weinberger. (2007). Topological nonrigidity of nonuniform lattices. *Comm. Pure Appl. Math.*, **60**(2), 282–290.

Chang, S. and S. Weinberger. (2010). Taming 3-manifolds using scalar curvature. *Geom Dedicata*, **148**, 3–14.

Chang, S. and S. Weinberger. (2015). Modular symbols and the topological nonrigidity of arithmetic manifolds. *Comm. Pure Appl. Math.*, **68**(11), 2022–2051.

Chang, S. and S. Weinberger. (2020). *A Course in Surgery Theory*. Princeton University Press.

Chang, S., S. Ferry, and G. Yu. (2008). Bounded rigidity of manifolds and asymptotic dimension growth. *J. K-Theory*, **1**(1), 129–144.

Chang, S., S. Weinberger, and G. Yu. (2020). Positive scalar curvature and a new index theory for noncompact manifolds. *J. Geom. Phys.*, **149**, 103575.

Chapman, T. (1981). Approximation results in topological manifolds. *Mem. Amer. Math. Soc.*, **34**(251).

Chapman, T. (1983). *Controlled Simple Homotopy Theory and Applications*. Lecture Notes in Mathematics, 1009. Berlin: Springer.

Chapman, T. and S. Ferry. (1979). Approximating homotopy equivalences by homeomorphisms. *Amer. J. Math.*, **101**(3), 583–607.

Charney, R. (1984). On the problem of homology stability for congruence subgroups. *Comm. Algebra*, **12**, 2081–2123.

Charney, R. and M. Davis. (1995). Strict hyperbolization. *Topology*, **34**(2), 329–350.

Cheeger, J. (1970). A lower bound for the smallest eigenvalue of the Laplacian. In *Problems in Analysis (Papers dedicated to Salomon Bochner, 1969)*. Princeton: Princeton University Press, pp. 195–199.

Cheeger, J. (1980). On the Hodge theory of Riemannian pseudomanifolds. In *Geometry of the Laplace Operator (Univ. Hawaii, Honolulu, HI, 1979)*. Proc. Symp. in Pure Mathematics, 36. Providence: American Mathematical Society, pp. 91–146.

Cheeger, J. and D. Ebin. (2008). *Comparison Theorems in Differential Geometry*. Providence: American Mathematical Society; New York: Chelsea. Reprint of the 1975 North-Holland original.

Cheeger, J. and M. Gromov. (1985a). Bounds on the von Neumann dimension of L^2-cohomology and the Gauss–Bonnet theorem for open manifolds. *J. Differential Geom.*, **21**(1), 1–34.

Cheeger, J. and M. Gromov. (1985b). On the characteristic numbers of complete manifolds of bounded curvature and finite volume. In *Differential Geometry and Complex Analysis*. Berlin: Springer, pp. 115–154.

Cheeger, J. and M. Gromov. (1986). L^2-cohomology and group cohomology. *Topology*, 25(2), 189–215.

Cheeger, J., M. Gromov, and M. Taylor. (1982). Finite propagation speed, kernel estimates for functions of the Laplace operator, and the geometry of complete Riemannian manifolds. *J. Differential Geom.*, 17(1), 15–53.

Cherix, P.-A., M. Cowling, P. Jolissaint, P. Julg, and A. Valette. (2001). *Groups with the Haagerup Property: Gromov's a-T-menability*. Progress in Mathematics, 197. Basel: Birkhäuser.

Chern, S., F. Hirzebruch, and J.-P. Serre. (1957). On the index of a fibered manifold. *Proc. Amer. Math. Soc.*, 8, 587–596.

Clair, B. and K. Whyte. (2003). Growth of Betti numbers. *Topology*, 42, 1125–1142.

Cohen, M. (1970). Homeomorphisms between homotopy manifolds and their resolutions. *Invent. Math.*, 10, 239–250.

Cohen, M. (1973). *A Course in Simple-Homotopy Theory*. Graduate Texts in Mathematics, 10. New York: Springer.

Cohen, R. (1987). Pseudo-isotopies, K-theory, and homotopy theory. In *Homotopy Theory (Durham, 1985)*. London Mathematical Society Lecture Note Series, 117. Cambridge: Cambridge University Press, pp. 35–71.

Conner, P. and E. Floyd. (1959). On the construction of periodic maps without fixed points. *Proc. Amer. Math. Soc.*, 10, 354–360.

Conner, P. and F. Raymond. (1971). Injective operations of the toral groups. *Topology*, 10, 283–296.

Conner, P., F. Raymond, and P. Weinberger. (1972). Manifolds with no periodic maps. In *Proc. Second Conf. on Compact Transformation Groups (Univ. Massachusetts, Amherst, MA, 1971)*. Part II. Lecture Notes in Mathematics, 299. Berlin: Springer, pp. 81–108.

Connes, A. (1982). A survey of foliations and operator algebras. In *Operator Algebras and Applications (Kingston, Ont., 1980)*. Part I. Proc. Symp. in Pure Mathematics, 38. Providence: American Mathematical Society, pp. 521–628.

Connes, A. (1985). Noncommutative differential geometry. *Publ. Math. Inst. Hautes Études Sci.*, 62, 257–360.

Connes, A. (1994). *Noncommutative Geometry*. San Diego: Academic Press.

Connes, A. and M. Marcolli. (2008). A walk in the noncommutative garden. In *An Invitation to Noncommutative Geometry*. Singapore: World Scientific.

Connes, A., M. Gromov, and H. Moscovici. (1990). Conjecture de Novikov et fibrés presque plats. *C. R. Acad. Sci. Paris, Sér. I Math.*, 310(5), 273–277.

Connes, A., M. Gromov, and H. Moscovici. (1993). Group cohomology with Lipschitz control and higher signatures. *Geom. Funct. Anal.*, 3(1), 1–78.

Connolly, F. and J. Davis. (2004). The surgery obstruction groups of the infinite dihedral group. *Geom. Topol.*, 8, 1043–1078.

Connolly, F. and T. Kosniewski. (1990). Rigidity and crystallographic groups. I. *Invent. Math.*, 99(1), 25–48.

Connolly, F. and T. Kozniewski. (1991). Examples of lack of rigidity in crystallographic groups. In *Algebraic Topology (Poznań 1989)*. Lecture Notes in Mathematics, 1474. Berlin: Springer, pp. 139–145.

Connolly, F., J. Davis, and Q. Khan. (2014). Topological rigidity and H-negative involutions on tori. *Geom. Topol.*, **18**, 1719–1768.

Connolly, F., J. Davis, and Q. Khan. (2015). Topological rigidity of actions on contractible manifolds with discrete singular set. *Trans. Amer. Math. Soc. Ser. B*, **2**, 113–133.

Cooke, G. (1978). Replacing homotopy actions by topological actions. *Trans. Amer. Math. Soc.*, **237**, 391–406.

Coornaert, M., T. Delzant, and A. Papadopoulos. (1990). *Géométrie et Théorie des Groupes*. Lecture Notes in Mathematics, 1441. Berlin: Springer.

Dahamani, F. (2003). Combination of convergence groups. *Geom. Topol.*, **7**, 933–963.

Daverman, R. (2007). *Decompositions of Manifolds*. Providence: American Mathematical Society; New York: Chelsea. Reprint of the 1986 Academic Press original.

Daverman, R. and G. Venema. (2009). *Embeddings in Manifolds*. Graduate Studies in Mathematics, 106. Providence: American Mathematical Society.

Davis, J. (1983). The surgery semicharacteristic. *Proc. London Math. Soc. (3)*, **47**, 411–428.

Davis, J. and W. Lück. (1998). Spaces over a category and assembly maps in isomorphism conjectures in K- and L-theory. *K-Theory*, **15**(3), 201–252.

Davis, M. (1983). Groups generated by reflections and aspherical manifolds not covered by Euclidean space. *Ann. of Math. (2)*, **117**(2), 293–324.

Davis, M. (2000). Poincaré duality groups. In *Surveys on Surgery Theory*, Vol. 1. Annals of Mathematics Studies, 145. Princeton: Princeton University Press, pp. 167–193.

Davis, M. (2008). *The Geometry and Topology of Coxeter Groups*. London Mathematical Society Monograph Series, 32. Princeton: Princeton University Press.

Davis, M. and J.C. Hausmann. (1989). Aspherical manifolds without smooth or PL structure. In *Algebraic Topology (Arcata, CA, 1986)*. Lecture Notes in Mathematics, 1370. Berlin: Springer, pp. 135–142.

Davis, M. and T. Januszkiewicz. (1991). Hyperbolization of polyhedra. *J. Differential Geom.*, **34**(2), 347–388.

Davis, M., T. Januszkiewicz, and S. Weinberger. (2001). Relative hyperbolization and aspherical bordisms: an addendum to "Hyperbolization of polyhedra." *J. Differential Geom.*, **58**(3), 535–541.

DeGeorge, D.D. and N.R. Wallach. (1978). Limit formulas for multiplicities in $L^2(\Gamma \backslash G)$. I. *Ann. of Math. (2)*, **107**(1), 133–150.

DeGeorge, D.D. and N.R. Wallach. (1979). Limit formulas for multiplicities in $L^2(\Gamma \backslash G)$. II. *Ann. of Math. (2)*, **109**(3), 477–495.

Deligne, P. and G.D. Mostow. (1993). *Commensurabilities among Lattices in $PU(1, n)$*. Annals of Mathematics Studies, 132. Princeton: Princeton University Press.

Donaldson, S. (1983). An application of gauge theory to four-dimensional topology. *J. Differential Geom.*, **18**, 279–315.

Dovermann, K.H. and Schultz, R. (1990). *Equivariant Surgery Theories and their Periodicity Properties*. Berlin: Springer.

Dranishnikov, A. (1988). On a problem of P.S. Aleksandrov (in Russian). *Mat. Sb. (N. S.)*, **135**(177/4), 551–560. Transl. (1989). *Math. USSR-Sb.*, **63**(2), 539–545.

Dranishnikov, A., S. Ferry, and S. Weinberger. (2003). Large Riemannian manifolds which are flexible. *Ann. of Math. (2)*, **157**(3), 919–938.

Dranishnikov, A., S. Ferry, and S. Weinberger. (2008). An etale approach to the Novikov conjecture. *Comm. Pure Appl. Math.*, **61**(2), 139–155.

Dranishnikov, A., S. Ferry, and S. Weinberger. (2020). An infinite-dimensional phenomenon in finite-dimensional metric topology. *Cambridge Jour. Math.*, **8**(1), 95–147.

Dress, A. (1975). Induction and structure theorems for orthogonal representations of finite groups. *Ann. of Math. (2)*, **102**(2), 291–325.

Drutu, C. and M. Sapir. (2005). Tree-graded spaces and asymptotic cones of groups. *Topology*, **44**(5), 959–1058. With an appendix by D. Osin and M. Sapir.

Dundas, B., T. Goodwillie, and R. McCarthy. (2013). *The Local Structure of Algebraic K-Theory*. Berlin: Springer.

Dwyer, W.G. (1989). R-nilpotency in homotopy equivalences. *Israel J. Math.*, **66**(1–3), 154–159.

Dwyer, W.G. and E. Friedlander. (1986). Conjectural calculations of general linear group homology. In *Applications of Algebraic K-Theory to Algebraic Geometry and Number Theory, Part I, II (Boulder, CO, 1983)*. Contemporary Mathematics, 55. Providence: American Mathematical Society, pp. 135–147.

Dwyer, W.G. and C. Wilkerson. (1988). Smith theory revisited. *Ann. of Math. (2)*, **127**(1), 191–198.

Eberlein, P. (1997). *Geometry of Non-Positively Curved Manifolds*. Chicago: University of Chicago Press.

Eckmann, B. (1986). Cyclic homology of groups and the Bass conjecture. *Comment. Math. Helv.*, **61**, 193–202.

Elek, G. (2010). Betti numbers are testable. In *Fete of Computer Science and Combinatorics*. Berlin: Springer, pp. 139–149.

Evans, L.C. (2010). *Partial Differential Equations*, 2nd edn. Providence: American Mathematical Society.

Farb, B. (1998). Relatively hyperbolic groups. *Geom. Funct. Anal.*, **8**(5), 810–840.

Farb, B. and S. Weinberger. (2005). Hidden symmetries and arithmetic manifolds. In *Geometry, Spectral Theory, Groups, and Dynamics*. Contemporary Mathematics, 387. Providence: American Mathematical Society, pp. 111–119.

Farb, B. and S. Weinberger. (2008). Isometries, rigidity and universal covers. *Ann. of Math. (2)*, **168**(3), 915–940.

Farb, B. and S. Weinberger. (2010). The intrinsic asymmetry and inhomogeneity of Teichmüller space. *Duke Math. J.*, **155**(1), 91–103.

Farber, M. (1998). Geometry of growth: approximation theorems for L^2 invariants. *Math Ann.*, **311**, 335–375.

Farley, D. (2003). Proper isometric actions of Thompson's groups on Hilbert space. *Int. Math. Res. Not.*, **45**, 2409–2414.

Farrell, F.T. (1971a). The obstruction to fibering a manifold over a circle. In *Actes du Congrès International des Mathématiciens (Nice, 1970)*. Tome 2. Paris: Gauthier-Villars, pp. 69–72.

Farrell, F.T. (1971b). The obstruction to fibering a manifold over a circle. *Indiana Univ. Math. J.*, **21**, 315–346.

Farrell, F.T. (1996). *Lectures on Surgical Methods in Rigidity*. Tata Institute of Fundamental Research. Berlin: Springer.

Farrell, F.T. (2002). The Borel conjecture. In *Topology of High-Dimensional Manifolds (Trieste, 2001)*. ICTP Lecture Notes Series, IX. Trieste: ICTP.

Farrell, F.T. and W.C. Hsiang. (1970). A formula for $K_1 R_\alpha [T]$. In *Applications of Categorical Algebra (New York, 1968)*. Proc. Symp. in Pure Mathematics, XVII. Providence: American Mathematical Society, pp. 192–218.

Farrell, F.T. and W.C. Hsiang. (1978a). On the rational homotopy groups of the diffeomorphism groups of discs, spheres and aspherical manifolds. In *Algebraic and Geometric Topology (Stanford Univ., Stanford, CA, 1976)*. Part 1. Proc. Symp. in Pure Mathematics, 32. Providence: American Mathematical Society, pp. 325–337.

Farrell, F.T. and W.C. Hsiang. (1978b). The topological–Euclidean space form problem. *Invent. Math.*, **45**(2), 181–192.

Farrell, F.T. and W.C. Hsiang. (1981). On Novikov's conjecture for non-positively curved manifolds. I. *Ann. of Math. (2)*, **113**(1), 199–209.

Farrell, F.T. and W.C. Hsiang. (1982). The stable topological–hyperbolic space form problem for complete manifolds of finite volume. *Invent. Math.*, **69**, 155–170.

Farrell, F.T. and W.C. Hsiang. (1983). Topological characterization of flat and almost flat Riemannian manifolds M^n ($n \ne 3, 4$). *Amer. J. Math.*, **105**(3), 641–672.

Farrell, F.T. and L. Jones. (1986). K-theory and dynamics I. *Ann. of Math. (2)*, **124**, 531–569.

Farrell, F.T. and L. Jones. (1987). K-theory and dynamics II. *Ann. of Math. (2)*, **126**, 451–493.

Farrell, F.T. and L. Jones. (1988). Foliated control theory. I, II. *K-Theory*, **2**(3), 357–430.

Farrell, F.T. and L. Jones. (1989). A topological analogue of Mostow's rigidity theorem. *J. Amer. Math. Soc.*, **2**(2), 257–370.

Farrell, F.T. and L. Jones. (1991a). Stable pseudoisotopy spaces of compact non-positively curved manifolds. *J. Differential Geom.*, **34**(3), 769–834.

Farrell, F.T. and L. Jones. (1991b). Rigidity in geometry and topology. In *Proc. Int. Congress of Mathematicians (Kyoto, 1990)*, Vols. I, II. Tokyo: Mathematical Society of Japan, pp. 653–663.

Farrell, F.T. and L. Jones. (1993a). Isomorphism conjectures in algebraic K-theory. *J. Amer. Math. Soc.*, **6**(2), 249–297.

Farrell, F.T. and L. Jones. (1993b). Topological rigidity for compact non-positively curved manifolds. In *Differential Geometry: Riemannian Geometry (Los Angeles, CA, 1990)*. Part 3. Proc. Symp. in Pure Mathematics, 54. Providence: American Mathematical Society, pp. 229–274.

Farrell, F.T. and L. Jones. (1998a). Collapsing foliated Riemannian manifolds. *Asian J. Math.*, **2**(3), 443–494.

Farrell, F.T. and L. Jones. (1998b). Rigidity for aspherical manifolds with $\pi_1 \subset GL_m(\mathbf{R})$. *Asian J. Math.*, **2**(2), 215–262.

Farrell, F.T. and L. Jones. (2003). Local collapsing theory. *Pacific J. Math.*, **210**(1), 1–100.

Farrell, F.T. and J. Lafont. (2004). Finite automorphisms of negatively curved Poincaré duality groups. *Geom. Funct. Anal.*, **14**, 283–294.

Farrell, F.T., L. Göttsche, and W. Lück (eds). (2002). *Topology of High-Dimensional Manifolds (Trieste, 2001)*. ICTP Lecture Notes Series, IX. Trieste: ICTP.

Farrell, F.T., W. Lück, and W. Steimle. (2018). Approximately fibering a manifold over an aspherical one. *Math. Ann.*, **369**, 669–726.

Federer, H. (1956). A study of function spaces by spectral sequences. *Trans. Amer. Math. Soc.*, **82**, 340–361.

Ferry, S. (1979). Homotoping ε-maps to homeomorphisms. *Amer. J. Math.*, **101**(3), 567–582.

Ferry, S. (1981). A simple-homotopy approach to the finiteness obstruction. In *Shape Theory and Geometric Topology (Dubrovnik, 1981)*. Lecture Notes in Mathematics, 870. Berlin: Springer, pp. 73–81.

Ferry, S. (1992). On the Ancel–Cannon theorem. *Topology Proc.*, **17**, 41–58.

Ferry, S. (1994). Topological finiteness theorems for manifolds in Gromov–Hausdorff space. *Duke Math. J.*, **74**(1), 95–106.

Ferry, S. (2010). Epsilon–delta surgery over $\mathbb{Z}$. *Geom. Dedicata*, **148**, 71–101.

Ferry, S. and E. Pederson. (1995). Epsilon surgery theory. In *Novikov Conjectures, Index Theorems and Rigidity (Oberwolfach, 1993)*, Vol. 2. London Mathematical Society Lecture Note Series, 227. Cambridge: Cambridge University Press, pp. 167–226.

Ferry, S. and S. Weinberger. (1991). Curvature, tangentiality, and controlled topology. *Invent. Math.*, **105**(2), 401–414.

Ferry, S. and S. Weinberger. (1995). A coarse approach to the Novikov conjecture. In *Novikov Conjectures, Index Theorems and Rigidity (Oberwolfach, 1993)*, Vol. 1. London Mathematical Society Lecture Note Series, 226. Cambridge: Cambridge University Press, pp. 147–163.

Ferry, S., J. Rosenberg, and S. Weinberger. (1988). Phénomènes de rigidité topologique équivariante. *C. R. Acad. Sci. Paris, Ser. I Math.*, **306**(19), 777–782.

Ferry, S., A. Ranicki, and J. Rosenberg (eds). (1995). *Novikov Conjectures, Index Theorems and Rigidity (Oberwolfach, 1993)*. Two volumes. London Mathematical Society Lecture Note Series, 226 and 227. Cambridge: Cambridge University Press.

Ferry, S., W. Lück, and S. Weinberger. (2019). On the stable Cannon conjecture. *J. Topol.*, **12**(3), 799–832.

Floyd, E. and R. Richardson. (1959). An action of a finite group on an *n*-cell without stationary points. *Bull. Amer. Math. Soc.*, **65**, 73–76.

Folner, E. (1956). On groups with full Banach mean value. *Math. Scand.*, **3**(2), 243–254.

Fowler, J. (2009). Poincaré duality groups and homology manifolds. PhD thesis, University of Chicago.

Fowler, J. (2012). Finiteness properties for some rational Poincaré duality groups. *Illinois J. Math.*, **56**(2), 281–299.

Freedman, M. and F. Quinn. (1990). *Topology of 4-manifolds*. Princeton Mathematical Series, 39. Princeton: Princeton University Press.

Freitag, E. (1990). *Hilbert Modular Forms*. Berlin: Springer.

Gaboriau, D. (2002). Invariants l^2 de relations d'équivalence et de groupes. *Publ. Math. Inst. Hautes Études Sci.*, **95**, 93–150.

Gajer, P. (1987). Riemannian metrics of positive scalar curvature on compact manifolds with boundary. *Ann. Global Anal. Geom.*, **5**(3), 179–191.

Galewski, D. and R. Stern. (1980). Classification of simplicial triangulations. *Ann. of Math. (2)*, **111**, 1–34.

Garland, H. (1973). *P*-adic curvature and the cohomology of discrete subgroups of *p*-adic groups. *Ann. of Math. (2)*, **97**, 375–423.

Gersten, S. (1993). Quasi-isometry invariance of cohomological dimension. *C. R. Acad. Sci. Paris, Sér. I Math.*, **316**(5), 411–416.

Ghys, E. and P. de la Harpe (eds). (1990). *Sur les Groupes Hyperboliques d'après Mikhael Gromov.* Progress in Mathematics, 83. Boston: Birkhäuser.

Giffen, C. (1966). The generalized Smith conjecture. *Amer. J. Math.*, **88**, 187–198.

Gilkey, P. (1984). *Invariance Theory, the Heat Equation, and the Atiyah–Singer Index Theorem.* Mathematics Lecture Series, 11. Wilmington: Publish or Perish.

Gomez Aparicio, M., P. Julg, and A. Valette. (2019). The Baum–Connes conjecture: an extended survey. Preprint, arXiv:1905.10081v1.

Gong, D. (1998). Equivariant Novikov conjecture for groups acting on Euclidean buildings. *Trans. Amer. Math. Soc.*, **350**(6), 2141–2183.

Goodwillie, T. (1985). Cyclic homology, derivations, and the free loopspace. *Topology*, **24**(2), 187–215.

Goodwillie, T. and M. Weiss. (1999). Embeddings from the point of view of immersion theory. II. *Geom. Topol.*, **3**, 103–118.

Goodwillie, T., J. Klein, and M. Weiss. (2001). Spaces of smooth embeddings, disjunction and surgery. In *Surveys on Surgery Theory*, Vol. 2. Annals of Mathematics Studies, 149. Princeton: Princeton University Press, pp. 221–284.

Goresky, M. and R. MacPherson. (1980). Intersection homology I. *Topology*, **19**(2), 135–162.

Goresky, M. and R. MacPherson. (1983). Intersection homology II. *Invent. Math.*, **72**(1), 77–129.

Grabowski, L. (2014). On Turing dynamical systems and the Atiyah problem. *Invent. Math.*, **198**(1), 27–69.

Grigorchuk, R. (1984). Degrees of growth of finitely generated groups and the theory of invariant means. *Izv. Akad. Nauk SSSR Ser. Mat.*, **48**(5), 939–985. Transl. (1985). *Math. USSR-Izv.*, **25**(2), 259–300.

Gromov, M. (1978). Almost flat manifolds. *J. Differential Geom.*, **13**(2), 231–241.

Gromov, M. (1982). Volume and bounded cohomology. *Publ. Math. Inst. Hautes Études Sci.*, **56**, 5–99.

Gromov, M. (1986). *Partial Differential Relations.* Berlin: Springer.

Gromov, M. (1987). Hyperbolic groups. In *Essays in Group Theory.* Mathematical Sciences Research Institute Publications, 8. New York: Springer, pp. 75–263.

Gromov, M. (1991). Kähler hyperbolicity and L^2 Hodge theory. *J. Differential Geom.*, **33**, 263–292.

Gromov, M. (1993). Asymptotic invariants of infinite groups. In *Geometric Group Theory, (Sussex, 1991)*, Vol. 2. London Mathematical Society Lecture Note Series, 182. Cambridge: Cambridge University Press, pp. 1–295.

Gromov, M. (1996). Positive curvature, macroscopic dimension, spectral gaps, and higher signatures. In *Functional Analysis on the Eve of the 21st Century*, Vol. II. Progress in Mathematics, 132. Boston: Birkhäuser.

Gromov, M. (1999). *Metric Structures for Riemannian and Non-Riemannian Spaces.* Progress in Mathematics, 152. Boston: Birkhäuser. With appendices by M. Katz, P. Pansu and S. Semmes. Based on the 1981 French original. Transl. by S. M. Bates.

Gromov, M. (2003). Random walk in random groups. *Geom. Funct. Anal.*, **13**(1), 73–146.

Gromov, M. and L. Guth. (2012). Generalizations of the Kolmogorov–Barzdin embedding estimates (English summary). *Duke Math. J.*, **161**, 2549–2603.

Gromov, M. and H.B. Lawson. (1980a). The classification of simply connected manifolds of positive scalar curvature. *Ann. of Math. (2)*, **111**(3), 423–434.

Gromov, M. and H.B. Lawson. (1980b). Spin and scalar curvature in the presence of a fundamental group. I. *Ann. of Math. (2)*, **111**(2), 209–230.

Gromov, M. and H.B. Lawson. (1983). Positive scalar curvature and the Dirac operator on complete Riemannian manifolds. *Publ. Math. Inst. Hautes Études Sci.*, **58**, 83–196.

Gromov, M. and P. Pansu. (1991). Rigidity of lattices: an introduction. In *Geometric Topology: Recent Developments*. Lecture Notes in Mathematics, 1504. Berlin: Springer, pp. 39–137.

Gromov, M. and I. Piatetski-Shapiro. (1988). Nonarithmetic groups in Lobachevsky spaces. *Publ. Math. Inst. Hautes Études Sci.*, **66**, 93–103.

Gromov, M. and R. Schoen. (1992). Harmonic maps into singular spaces and p-adic superrigidity for lattices in groups of rank one. *Publ. Math. Inst. Hautes Études Sci.*, **76**, 165–246.

Gromov, M. and W. Thurston. (1987). Pinching constants for hyperbolic manifolds. *Invent. Math.*, **89**(1), 1–12.

Guentner, E., N. Higson, and S. Weinberger. (2005). The Novikov conjecture for linear groups. *Publ. Math. Inst. Hautes Études Sci.*, **101**, 243–268.

Guentner, E., R. Tesserra, and G. Yu. (2012). A notion of geometric complexity and its application to topological rigidity. *Invent. Math.*, **189**(2), 315–357.

Haefliger, A. (1964). Plongements de variétés dans le domaine stable. *Séminaire Bourbaki*, **1962–63**(1), 245.

Haefliger, A. (1966). Enlacements de sphères en codimension supérieure à 2. *Comment. Math. Helv.*, **41**, 51–72.

Haglund, Frédéric and Daniel Wise (2007). Special cube complexes. *Geom. Funct. Anal.*, **17**, 1–69.

Haglund, Frédéric and Daniel Wise (2012). A combination theorem for special cube complexes. *Ann. of Math. (2)* **176**93), 1427–1482.

Hambleton, I. and E. Pederson. (1991). Bounded surgery and dihedral group actions on spheres. *J. Amer. Math. Soc.*, **4**(1), 105–126.

Hambleton, I. and E. Pederson. (2005). Topological equivalence of linear representations of cyclic groups. I. *Ann. of Math. (2)*, **161**(1), 61–104.

Hambleton, I., R.J. Milgram, L. Taylor, and B. Williams. (1988). Surgery with finite fundamental group. *Proc. London Math. Soc. (3)*, **56**, 349–379.

Hanke, B. (2008). Positive scalar curvature with symmetry. *J. Reine Angew. Math.*, **614**, 73–115.

Hanke, B. and T. Schick. (2006). Enlargeability and index theory. *J. Differential Geom.*, **74**, 293–320.

Hanke, B., D. Kotschick, J. Roe, and T. Schick. (2008). Coarse topology, enlargeability and essentialness. *Ann. Sci. Éc. Norm. Supér. (4)*, **41**, 471–493.

Hatcher, A. (1973). The second obstruction for pseudo-isotopies. In Hatcher and Wagoner (1973), 239–275.

Hatcher, A. (1978). Concordance spaces, higher simple-homotopy theory, and applications. In *Algebraic and Geometric Topology (Stanford Univ., Stanford, CA, 1976).* Part 1. Proc. Symp. in Pure Mathematics, 32. Providence: American Mathematical Society, pp. 3–21.

Hatcher, A. (2017). Vector bundles & *K*-theory. Available at `https://www.math. cornell.edu/$\sim$hatcher/VBKT/VBpage.html`.

Hatcher, A. and J. Wagoner. (1973). *Pseudo-Isotopies of Compact Manifolds.* Astérisque, 6. Paris: Société Mathématique de France.

Hempel, J. (1976). *3-Manifolds.* Princeton: Princeton University Press.

Hesselholt, L. and I. Madsen. (2003). On the *K*-theory of local fields. *Ann. of Math. (2),* **158**(1), 1–113.

Higson, N. (2000). Bivariant *K*-theory and the Novikov conjecture. *Geom. Funct. Anal.,* **10**(3), 563–581.

Higson, N. and E. Guentner. (2004). Group C^*-algebras and *K*-theory. In *Noncommutative Geometry.* Lecture Notes in Mathematics, 1831. Berlin: Springer, pp. 137–251.

Higson, N. and G. Kasparov. (2001). *E*-theory and *KK*-theory for groups which act properly and isometrically on Hilbert space. *Invent. Math.,* **144**(1), 23–74.

Higson, N. and J. Roe. (2000). *Analytic K-homology.* Oxford Mathematical Monographs. Oxford: Oxford University Press.

Higson, N. and J. Roe. (2005a). Mapping surgery to analysis I. Analytic signatures. *K-Theory,* **33**(4), 277–299.

Higson, N. and J. Roe. (2005b). Mapping surgery to analysis II. Geometric signatures. *K-Theory,* **33**(4), 301–324.

Higson, N. and J. Roe. (2005c). Mapping surgery to analysis III. Exact sequences. *K-Theory,* **33**(4), 325 –346.

Higson, N. and J. Roe. (2010). K-homology, assembly and rigidity theorems for relative eta invariants. *Pure Appl. Math. Q.,* **6**(2), 555–601. Special issue in honor of Michael Atiyah and Isadore Singer.

Higson, N. and J. Roe. (in preparation). *Notes on the Index Theorem.*

Higson, N., V. Lafforgue, and G. Skandalis (2002). Counterexamples to the Baum–Connes conjecture. *Geom. Funct. Anal.,* **12**(2), 330–354.

Hilton, P. (1955). On the homotopy groups of unions of spheres. *J. London Math. Soc.,* **30**, 154–172.

Hilton, P., G. Mislin, and J. Roitberg. (1975). *Localization of Nilpotent Groups and Spaces.* North-Holland Studies in Mathematics. Amsterdam: North-Holland.

Hirsch, M. and S. Smale. (1959). Immersions of manifolds. *Trans. Amer. Math. Soc.,* **93**, 242–276.

Hirzebruch, F. (1971). The signature theorem: reminiscences and recreation. In *Prospects in Mathematics (Proc. Symp., Princeton Univ., Princeton, NJ, 1970).* Annals of Mathematics Studies, 70. Princeton: Princeton University Press, pp. 3–31.

Hirzebruch, F. (1995). *Topological Methods in Algebraic Geometry.* Berlin: Springer. With appendices by R.L.E. Schwarzenberger and A. Borel. Transl. from German by R.L.E. Schwarzenberger. Reprint of 1978 "Classics in Mathematics" edition.

Hitchin, N. (1974). Harmonic spinors. *Adv. Math.,* **14**, 1–55.

Hoory, S., N. Linial, and A. Wigserson. (2006). Expander graphs and their applications. *Bull. Amer. Math. Soc.,* **43**, 439–561.

Hrushovski, E., and Pillay, A. (1995). Definable subgroups of algebraic groups over finite fields. *J. Reine Angew. Math.*, **462**, 69–91.

Hsiang, W.C. and W. Pardon. (1982). When are topologically equivalent orthogonal transformations linearly equivalent? *Invent. Math.*, **68**(2), 275–316.

Hsiang, W.C. and H. Rees. (1982). Miscenko's work on Novikov's conjecture. In *Operator Algebras and K-theory (San Francisco, CA, 1981)*. Contemporary Mathematics, 10. Providence: American Mathematical Society, pp. 77–98.

Hsiang, W.C. and J. Shaneson. (1970). Fake tori. In *Topology of Manifolds (Proc. Topology of Manifolds Institute, Univ. of Georgia, Athens, GA, 1969)*. Chicago: Markham, pp. 18–51.

Hughes, B. (1996). Geometric topology of stratified spaces. *Electron. Res. Announc. Amer. Math. Soc.*, **2**(2), 73–81.

Hughes, B. and A. Ranicki. (1996). *Ends of Complexes*. Cambridge Tracts in Mathematics, 123. Cambridge: Cambridge University Press.

Hughes, B., L. Taylor, S. Weinberger, and B. Williams. (2000). Neighborhoods in stratified spaces with two strata. *Topology*, **39**(5), 873–919.

Hummel, C. and V. Schroeder. (1996). Cusp closing in rank one symmetric spaces. *Invent. Math.*, **123**, 283–307.

Ivanov, S.V. and A.Yu. Ol'shanskii. (1996). Hyperbolic groups and their quotients of bounded exponents. *Trans. Amer. Math. Soc.*, **348**, 2091–2138.

Jaco, W. (1980). *Lectures on Three-Manifold Topology*. CBMS Regional Conference Series in Mathematics, 43. Providence: American Mathematical Society.

Jones, L. (1971). The converse to the fixed point theorem of P. A. Smith. I. *Ann. of Math. (2)*, **94**, 52–68.

Jones, L. (1973). Patch spaces: a geometric representation for Poincaré spaces. *Ann. of Math. (2)*, **97**, 306–343. See also corrections: (1975). Erratum. *Ann. of Math. (2)*, **102**, 183–185.

Jones, L. (1986). Combinatorial symmetries of the *m*-dimensional ball. *Mem. Amer. Math. Soc.*, **62**(352).

Jost, J. and S.T. Yau. (1987). On the rigidity of certain discrete groups and algebraic varieties. *Math. Ann.*, **278**, 481–496.

Juschenko, K. and N. Monod. (2013). Cantor systems, piecewise translations, and simple amenable groups. *Ann. of Math. (2)*, **178**, 775–787.

Kahn, Jeremy and Vladimir Markovic. (2012). Immersing almost geodesic surfaces in a closed hyperbolic three manifold. *Ann. of Math. (2)*, **175**(3), 1127–1190.

Kapovich, M. (2001). *Hyperbolic Manifolds and Discrete Groups*. Basel: Birkhäuser.

Kasparov, G. (1988). Equivariant *KK*-theory and the Novikov conjecture. *Invent. Math.*, **91**(1), 147–201.

Kasparov, G. and G. Yu. (2006). The coarse geometric Novikov conjecture and uniform convexity. *Adv. Math.*, **206**, 1–56.

Kazdan, J. and F. Warner. (1975). Existence and conformal deformation of metrics with prescribed Gaussian and scalar curvatures. *Ann. of Math. (2)*, **101**, 317–331.

Kazhdan, D. (1967). On the connection of the dual space of a group with the structure of its closed subgroups (in Russian). *Funktsional. Anal. i Prilozhen.*, **1**, 71–74.

Kerckhoff, S. (1983). The Nielsen realization problem. *Ann. of Math. (2)*, **117**(2), 235–265.

Kervaire, M. (1969). Smooth homology spheres and their fundamental groups. *Trans. Amer. Math. Soc.*, **144**, 67–72.

Kervaire, M. and J. Milnor. (1963). Groups of homotopy spheres I. *Ann. of Math. (2)*, **77**, 504–537.

Kesten, H. (1959). Symmetric random walks on groups. *Trans. Amer. Math. Soc.*, **92**, 336–354.

Keswani, N. (2000). Relative eta-invariants and C^*-algebra K-theory. *Topology*, **39**(5), 957–983.

Kirby, R. (1969). Stable homeomorphisms and the annulus conjecture. *Ann. of Math. (2)*, **89**, 575–582.

Kirby, R. and M. Scharleman. (1979). Eight faces of the Poincaré homology 3-sphere. In *Geometric Topology (Proc. Georgia Topology Conf., Athens, GA, 1977)*. New York: Academic Press, pp. 113–146.

Kirby, R. and L. Siebenmann. (1977). *Foundational Essays on Topological Manifolds, Smoothings and Triangulations*. Princeton: Princeton University Press.

Kostant, B. (1975). On the existence and irreducibility of certain series of representations. In *Lie Groups and Their Representations (Proc. Summer School, Bolyai János Math. Soc., Budapest, 1971)*. New York: Halsted, pp. 231–329.

Kowalski, E. (2008). *The Large Sieve and Its Applications: Arithmetic Geometry, Random Walks and Discrete Groups*. Cambridge: Cambridge University Press.

Kreck, M. and W. Lück. (2005). *The Novikov Conjecture: Geometry and Algebra*. Basel: Birkhäuser.

Lafforgue, V. (2002). K-théorie bivariante pour les algèbres de Banach et conjecture de Baum–Connes. *Invent. Math.*, **149**(1), 1–95.

Lafforgue, V. (2008). Un renforcement de la propriété (T). *Duke Math. J.*, **143**(3), 559–602.

Lafforgue, V. (2012). La conjecture de Baum–Connes à coefficients pour les groupes hyperboliques. *J. Noncommut. Geom.*, **6**(1), 1–197.

Lannes, M. and L. Schwartz. (1986). À propos de conjectures de Serre et Sullivan (in French). *Invent. Math.*, **83**(3), 593–603.

Larsen, M. and R. Pink. (2011). Finite subgroups of algebraic groups. *J. Amer. Math. Soc.*, **24**(4), 1105–1158.

Lawson, H.B. and M.-L. Michelsohn. (1989). *Spin Geometry*. Princeton Mathematical Series, 38. Princeton: Princeton University Press.

Lawson, H.B. and S.T. Yau. (1972). Compact manifolds of non-positive curvature. *J. Differential Geom.*, **7**, 211–228.

Lawson, H.B. and S.T. Yau. (1974). Scalar curvature, non-abelian group actions, and the degree of symmetry of exotic spheres. *Comment. Math. Helv.*, **49**, 232–244.

Lee, K.B. and F. Raymond. (2010). *Seifert Fiberings*. Mathematical Surveys and Monographs, 166. Providence: American Mathematical Society.

Lee, R. and R. Szczarba. (1976). On the homology and cohomology of congruence subgroups. *Invent. Math.*, **33**, 15–53.

Lees, J.A. (1973). The surgery obstruction groups of C. T. C. Wall. *Adv. Math.*, **11**, 113–156.

Leichtnam, E., J. Lott, and P. Piazza. (2000). On the homotopy invariance of higher signatures for manifolds with boundary. *J. Differential Geom.*, **54**(3), 561–633.

Leichtnam, E., W. Lück and M. Kreck. (2002). On the cut-and-paste property of higher signatures of a closed oriented manifold. *Topology*, **41**(4), 725–744.

Levine, J. (1965). Unknotting spheres in codimension two. *Topology*, **4**, 9–16.

Levitt, N. and A. Ranicki. (1987). Intrinsic transversality structures. *Pacific J. Math.*, **129**(1), 85–144.

Liu, K. (1995). On modular invariance and rigidity theorems. *J. Differential Geom.*, **41**(2), 343–396.

Liu, K. and X. Ma. (2000). On family rigidity theorems. I. *Duke Math. J.*, **102**(3), 451–474.

Loday, J.-L. (1976) K-théorie algébrique et représentations de groupes. *Ann. Sci. Éc. Norm. Supér. (4)*, **9**(3), 309–377.

Loday, J.-L. (1998). *Cyclic Homology*, 2nd edn. Grundlehren der Mathematischen Wissenschaften, 301. Berlin: Springer. Appendix E by M. O. Ronco. Chapter 13 in collaboration with T. Pirashvili.

Lohkamp, J. (2006). Positive scalar curvature in dim≥8. *C. R. Acad. Sci. Paris, Sér. I Math.*, **343**(9), 585–588.

Lovasz, L. (2012). *Large Networks and Graph Limits*. Providence: American Mathematical Society.

Lubotzky, A. (1984). *Discrete Groups, Expanding Graphs and Invariant Measures*. Progress in Mathematics, 125. Basel: Birkhäuser.

Lubotzky, A. (1987). On finite index subgroups of linear groups. *Bull. London Math. Soc.*, **19**, 325–328.

Lubotzky, A. (1996). Free quotients and the first Betti number of some hyperbolic manifolds. *Transform. Groups*, **1**, 71–82.

Lubotzky, A. (2012). Expander graphs in pure and applied mathematics. *Bull. Amer. Math. Soc.*, **49**, 113–162.

Lubotzky, A. and D. Segal. (2003). *Subgroup Growth*. Basel: Birkhäuser.

Lubotzky, A., R. Phillips, and P. Sarnak. (1988). Ramanujan graphs. *Combinatorica*, **8**, 261–277.

Lück, W. (1987). Equivariant Eilenberg–Mac Lane spaces $K(G, \mu, 1)$ for possibly nonconnected or empty fixed point sets. *Manuscripta Math.*, **58**(1–2), 67–75.

Lück, W. (1994). Approximating L^2-invariants by their finite-dimensional analogues. *Geom. Funct. Anal.*, **4**, 455–481.

Lück, W. (2002a). A basic introduction to surgery theory. In *Topology of High-Dimensional Manifolds (Trieste, 2001)*. ICTP Lecture Notes Series, IX. Trieste: ICTP, pp. 1–224.

Lück, W. (2002b). L^2-*Invariants: Theory and Applications to Geometry and K-Theory*. Berlin: Springer.

Lück, W. (2022). *Isomorphism Conjectures in K- and L-Theory*. http://www.him.uni-bonn.de/lueck/data/ic200708.pdf

Lück, W. and I. Madsen. (1990a). Equivariant L-theory. I. *Math. Z.*, **203**(3), 503–526.

Lück, W. and I. Madsen. (1990b). Equivariant L-theory. II. *Math. Z.*, **2042**), 253–268.

Lück, W., H. Reich, J. Rognes, and M. Varisco. (2017). Algebraic K-theory of group rings and the cyclotomic trace map. *Adv. Math.*, **304**, 930–1020.

Lusztig, G. (1972). Novikov's higher signature and families of elliptic operators. *J. Differential Geom.*, **7**, 229–256.

Madsen, I. and R.J. Milgram. (1979). *The Classifying Spaces for Surgery and Cobordism of Manifolds*. Princeton: Princeton University Press.

Madsen, I. and M. Rothenberg. (1988a). On the classification of G-spheres. I. Equivariant transversality. *Acta Math.*, **160**(1–2), 65–104.

Madsen, I. and M. Rothenberg. (1988b). On the homotopy theory of equivariant automorphism groups. *Invent. Math.*, **94**(3), 623–637.

Madsen, I. and M. Rothenberg. (1989). On the classification of G-spheres. II. PL automorphism groups. *Math. Scand.*, **64**(2), 161–218.

Madsen, I., C. Thomas, and C.T.C. Wall. (1976). The topological spherical space form problem. II. Existence of free actions. *Topology*, **15**(4), 375–382.

Manolescu, C. (2016). Pin(2)-equivariant Seiberg–Witten Floer homology and the triangulation conjecture. *J. Amer. Math. Soc.*, **29**, 147–176.

Marcus, A., D. Spielman, and N. Srivastava. (2015). Interlacing families I: Bipartite Ramanujan graphs of all degrees. *Ann. of Math. (2)*, **182**(1), 307–325.

Margulis, G. (1988). Explicit group-theoretic constructions of combinatorial schemes and their applications in the construction of expanders and concentrators (in Russian). *Problemy Peredachi Informatsii*, **24**, 51–60. Translated into English in *Problems Inform. Transmiss.*, **24**, 39–46 (1988).

Margulis, G. (1991). *Discrete Subgroups of Semisimple Lie Groups*. Ergebnisse der Mathematik und ihrer Grenzgebiete, 17. Berlin: Springer.

Mathai, V. (1992). Spectral flow, eta invariants, and von Neumann algebras. *J. Funct. Anal.*, **109**(2), 442–456.

Matthews, C., L. Vasserstein, and B. Weisfeiler. (1984). Congruence properties of Zariski-dense subgroups. *Proc. London Math. Soc. (3)*, **48**, 514–532.

Matumoto, T. (1978). Triangulation of manifolds. In *Algebraic and Geometric Topology (Stanford Univ., Stanford, CA, 1976)*. Part 2. Proc. Symp. in Pure Mathematics, **32**. Providence: American Mathematical Society, pp. 3–6.

McCarthy, R. (1994). The cyclic homology of an exact category. *J. Pure Appl. Algebra*, **93**(3), 251–296.

McClure, J. (1986). Restriction maps in equivariant *K*-theory. *Topology*, **25**(4), 399–409.

Melnick, K. (2009). Compact Lorentz manifolds with local symmetry. *J. Differential Geom.*, **81**(2), 355–390.

Miller, C., III. (1971). *On Group-Theoretic Decision Problems and Their Classification*. Annals of Mathematics Studies, 68. Princeton: Princeton University Press.

Miller, H. (1987). The Sullivan conjecture and homotopical representation theory. In *Proc. Int. Congress of Mathematicians (Berkeley, CA, 1986)*, Vol. 1. Providence: American Mathematical Society, pp. 580–589.

Millson, J. (1976), On the first Betti number of a constant negatively curved manifold, *Ann. of Math* **104** pp. 235–247.

Milnor, J. (1954). Link groups. *Ann. of Math. (2)*, **59**, 177–195.

Milnor, J. (1957). Groups which act on S^n without fixed points. *Amer. J. Math.*, **79**, 623–630.

Milnor, J. (1961). Two complexes which are homeomorphic but combinatorially distinct. *Ann. of Math. (2)*, **74**, 575–590.

Milnor, J. (1963). *Morse Theory*. Princeton: Princeton University Press.

Milnor, J. (1966). Whitehead torsion. *Bull. Amer. Math. Soc.*, **72**, 358–426.

Milnor, J. (1971). *Introduction to Algebraic K-Theory*. Annals of Mathematics Studies, 72. Princeton: Princeton University Press; and Tokyo: University of Tokyo Press.

Milnor, J. and D. Husemoller. (1973). *Symmetric Bilinear Forms*. Ergebnisse der Mathematik und ihrer Grenzgebiete, 73. New York: Springer.

Mineyev, I. (2005). Flows and joins of metric spaces. *Geom. Topol.*, **9**, 403–482.

Mineyev, I. and G. Yu. (2002). The Baum–Connes conjecture for hyperbolic groups. *Invent. Math.*, **149**(1), 97–122.

Mischenko, A.S. (1974). Infinite dimensional representations of discrete groups and higher signatures (in Russian). *Izv. Akad. SSSR, Ser. Mat.*, **38**, 81–106.

Mischenko, A.S. (1976). Hermitian K-theory. Theory of characteristic classes, methods of functional analysis (in Russian). *Uspekhi Mat. Nauk*, **31**(2/188), 69–134.

Moore, C. and C. Schochet. (2006). *Global Analysis on Foliated Spaces*, 2nd edn. Mathematical Sciences Research Institute Publications, 9. New York: Cambridge University Press.

Morgan, J. (1984). On Thurston's uniformization theorem for three-dimensional manifolds. In *The Smith Conjecture*. Orlando: Academic Press, pp. 37–125.

Morgan, J. (1996). *The Seiberg–Witten Equations and Applications to the Topology of Smooth Four-Manifolds*. Princeton: Princeton University Press.

Morgan, J. and D. Sullivan. (1974). The transversality characteristic class and linking cycles in surgery theory. *Ann. of Math. (2)*, **99**, 463–544.

Morgan, J. and G. Tian. (2014). *The Geometrization Conjecture*. Clay Mathematics Monographs, 5. Providence: American Mathematical Society.

Mostow, G.D. (1954). Factor spaces of solvable groups. *Ann. of Math. (2)*, **60**(1), 1–27.

Mostow, G.D. (1968). Quasi-conformal mappings in n-space and the rigidity of the hyperbolic space forms. *Publ. Math. Inst. Hautes Études Sci.*, **34**, 53–104.

Mostow, G.D. (1973). *Strong Rigidity of Locally Symmetric Spaces*. Annals of Mathematics Studies, 78. Princeton: Princeton University Press.

Munkholm, H.J. (1979). Simplices of maximal volume in hyperbolic space, Gromov's norm, and Gromov's proof of Mostow's rigidity theorem (following Thurston). In *Topology Symposium, Siegen 1979*. Lecture Notes in Mathematics, 788. Berlin: Springer, pp. 109–124.

Nabutovsky, A. and S. Weinberger. (1999). Algorithmic aspects on homeomorphism problems. *Contemp. Math.*, **231**, 245–250.

Neumann, W. (1979). Signature related invariants of manifolds. I. Monodromy and γ-invariants. *Topology*, **18**(2), 147–172.

Niblo, G. and L. Reeves. (1997). Groups acting on CAT(0) cube complexes. *Geom. Topol.*, **1**, 1–7.

Nicas, A. (1982). Induction theorems for groups of homotopy manifold structures. *Mem. Amer. Math. Soc.*, **39**(267).

Nori, Madhav V. (1987). On subgroups of $GL_n(\mathbf{F}_p)$. *Invent. Math.*. **88**(2), 257–275.

Novikov, S.P. (1966). On manifolds with free abelian fundamental group and their application (in Russian). *Izv. Akad. Nauk SSSR Ser. Mat.*, **30**, 207–246.

Olbrich, M. (2002). L^2-invariants of locally symmetric spaces. *Doc. Math.*, **7**, 219–237.

Oliver, R. (1975). Fixed-point sets of group actions on finite acyclic complexes. *Comment. Math. Helv.*, **50**, 155–177.

Oliver, R. (1976a). A proof of the Conner conjecture. *Ann. of Math. (2)*, **103**, 637–644.

Oliver, R. (1976b). Smooth compact Lie group actions on disks. *Math. Z.*, **149**(1), 79–96.

Oliver, R. and T. Petrie. (1982). G-CW surgery and $K_0(\mathbf{Z}G)$. *Math. Z.*, **179**, 11–42.

Ol'shanskii, A.Ju. (1980). On the question of the existence of an invariant mean on a group (in Russian). *Uspekhi Mat. Nauk*, **35** (4/214), 199–200.

Ontaneda, P. (2011). Pinched smooth hyperbolization. Preprint, arXiv:1110.6374v1.

C. Rourke and D.P. Sullivan (1971). On the Kervaire obstruction. *Annals of Math.*, **94** 393–413.

Palais, R.S. (1965). *Seminar on the Atiyah–Singer Index Theorem*, with contributions by M. F. Atiyah, A. Borel, E. E. Floyd, R. T. Seeley, W. Shih and R. Solovay. Annals of Mathematics Studies, 57. Princeton: Princeton University Press.

Papakyriakopoulos, C.D. (1957). On Dehn's lemma and the asphericity of knots. *Ann. of Math. (2)*, **66**, 1–26.

Paterson, A. (1988). *Amenability*. Providence: American Mathematical Society.

Pedersen, E. (2000). Continuously controlled surgery theory. In *Surveys on Surgery Theory*, Vol. 1. Annals of Mathematics Studies, 145. Princeton: Princeton University Press, pp. 307–321.

Pedersen, E. and C. Weibel. (1989). K-theory homology of spaces. In *Algebraic Topology (Arcata, CA, 1986)*. Lecture Notes in Mathematics, 1370. Berlin: Springer, pp. 346–361.

Pedersen, E., J. Roe, and S. Weinberger. (1995). On the homotopy invariance of the boundedly controlled analytic signature of a manifold over an open cone. In *Novikov Conjectures, Index Theorems and Rigidity (Oberwolfach, 1993)*, Vol. 2. London Mathematical Society Lecture Note Series, 227. Cambridge: Cambridge University Press, pp. 285–300.

Perelman, G. (2002). The entropy formula for the Ricci flow and its geometric applications. Preprint, arXiv:math/0211159v1.

Petrie, T. (1978). Pseudoequivalences of G-manifolds. In *Algebraic and Geometric Topology (Stanford Univ., Stanford, CA, 1976)*. Proc. Symp. in Pure Mathematics, 32. Providence: American Mathematical Society, pp. 119–163.

Piazza, P. and V. Zenobi. (2016). Additivity of the rho map on the topological structure group. Preprint, arXiv:1607.07075v2.

Pimsner, M. (1986). KK-groups of crossed products by groups acting on trees. *Invent. Math.*, **86**(3), 603–634.

Pimsner, M. and D. Voiculescu. (1980). Exact sequences for K-groups and Ext-groups of certain cross-product C^*-algebras. *J. Operator Theory*, **4**(1), 93–118.

Prasad, G. (1973). Strong rigidity of Q-rank 1 lattices. *Invent. Math.*, **21**, 255–286.

Puschnigg, M. (2012). The Baum–Connes conjecture with coefficients for word-hyperbolic groups, following V. Lafforgue. *Séminaire Bourbaki*, **2012–13**, 1062.

Quinn, F. (1970). A geometric formulation of surgery. In *Topology of Manifolds (Proc. Topology of Manifolds Institute, Univ. of Georgia, Athens, GA, 1969)*. Chicago: Markham, pp. 500–511.

Quinn, F. (1972). Surgery on Poincaré and normal spaces. *Bull. Amer. Math. Soc.*, **78**, 262–267.

Quinn, F. (1979). Ends of maps. I. *Ann. of Math. (2)*, **110**(2), 275–331.

Quinn, F. (1982a). Resolutions of homology manifolds, and the topological characterization of manifolds. *Invent. Math.*, **72**, 267–284. See also corrections: (1986). Erratum. *Invent. Math.*, **85**, 653.

Quinn, F. (1982b). Ends of maps. II. *Invent. Math.*, **68**(3), 353–424.

Quinn, F. (1982c). Ends of maps. III. Dimensions 4 and 5. *J. Differential Geom.*, **17**(3), 503–521.

Quinn, F. (1985a). Geometric algebra. In *Algebraic and Geometric Topology (New Brunswick, NJ, 1983)*. Lecture Notes in Mathematics, 1126. Berlin: Springer, pp. 182–198.

Quinn, F. (1985b). Algebraic K-groups of poly(finite-or-cyclic) groups. *Bull. Amer. Math. Soc.*, **12**, 221–226.

Quinn, F. (1986). Ends of maps. IV. Controlled pseudoisotopy. *Amer. J. Math.*, **108**(5), 1139–1161.

Quinn, F. (1987a). Applications of topology with control. In *Proc. Int. Congress of Mathematicians (Berkeley, CA, 1986)*, Vol. 1. Providence: American Mathematical Society, pp. 598–606.

Quinn, F. (1987b). An obstruction to the resolution of homology manifolds. *Michigan Math. J.*, **34**(2), 285–291.

Quinn, F. (1988). Homotopically stratified sets. *J. Amer. Math. Soc.*, 1(2), 441–499.

Quinn, F. (2010). A controlled-topology proof of the product structure theorem. *Geom. Dedicata*, **148**, 303–308.

Raghunathan, M.S. (1972). *Discrete Subgroups of Lie Groups*. Berlin: Springer.

Ranicki, A. (1979a). Localization in quadratic *L*-theory. In *Algebraic Topology, Waterloo, 1978 (Proc. Conf., Univ. Waterloo, Waterloo, ON, 1978)*. Lecture Notes in Mathematics, 741. Berlin: Springer, pp. 102–157.

Ranicki, A. (1979b). The total surgery obstruction. In *Algebraic Topology, Aarhus, 1978 (Proc. Symp., Univ. Aarhus, Aarhus, 1978)*. Lecture Notes in Mathematics, 763. Berlin: Springer, pp. 275–316.

Ranicki, A. (1980a). Algebraic theory of surgery I. Foundations. *Proc. London Math. Soc. (3)*, **40**, 87–192.

Ranicki, A. (1980b). Algebraic theory of surgery II. Applications to topology. *Proc. London Math. Soc. (3)*, **40**, 193–283.

Ranicki, A. (1992). *Algebraic L-theory and Topological Manifolds*. Cambridge: Cambridge University Press.

Ranicki, A. (2002). *Algebraic and Geometric Surgery*. Oxford: Oxford University Press.

Raymond, F. and L. Scott. (1977). Failure of Nielsen's theorem in higher dimensions. *Arch. Math. (Basel)*, **29**(6), 643–654.

Roe, J. (1988a). An index theorem for open manifolds I. *J. Differential Geom.*, **27**, 87–113.

Roe, J. (1988b). An index theorem for open manifolds II. *J. Differential Geom.*, **27**, 115–136.

Roe, J. (1988c). Partitioning noncompact manifolds and the dual Toeplitz problem. In *Operator Algebras and Applications*, Vol. 1. London Mathematical Society Lecture Note Series, 135. Cambridge: Cambridge University Press, pp. 187–228.

Roe, J. (1993). Coarse cohomology and index theory on complete Riemannian manifolds. *Mem. Amer. Math. Soc.*, **104**(497).

Roe, J. (1996). *Index Theory, Coarse Geometry, and Topology of Manifolds*. CBMS Regional Conference Series in Mathematics, 90. Providence: American Mathematical Society.

Roe, J. (1998). *Elliptic Operators, Topology and Asymptotic Methods*, 2nd edn. Pitman Research Notes in Mathematics Series, 395. Harlow: Longman.

Roe, J. (2003). *Lectures on Coarse Geometry*. University Lecture Series, 31. Providence: American Mathematical Society.

Rognes, J. and F. Waldhausen. (2013). *Spaces of PL Manifolds and Categories of Simple Maps*. Annals of Mathematics Studies, 186. Princeton: Princeton University Press.

Rosenberg, J. (1983). *C*-algebras, positive scalar curvature, and the Novikov conjecture. Publ. Math. Inst. Hautes Études Sci.*, **58**, 197–212.

Rosenberg, J. (1986a). *C*-algebras, positive scalar curvature, and the Novikov conjecture, II. In Geometric Methods in Operator Algebras (Kyoto, 1983)*. Pitman Research Notes in Mathematics Series, 123. Harlow: Longman, pp. 341–374.

Rosenberg, J. (1986b). *C*-algebras, positive scalar curvature, and the Novikov conjecture—III. Topology*, **25**(3), 319–336.

Rosenberg, J. (1991). The KO-assembly map and positive scalar curvature. In *Algebraic Topology Poznań1989*. Lecture Notes in Mathematics, 1474. Berlin: Springer, pp. 170–182.

Rosenberg, J. (1996). *Algebraic K-Theory and Its Applications*. Graduate Texts in Mathematics. Berlin: Springer.

Rosenberg, J. (2008). An analogue of the Novikov conjecture in complex algebraic geometry. *Trans. Amer. Math. Soc.*, **360**(1), 383–394.

Rosenberg, J. and S. Stoltz. (2001). Metrics of positive scalar curvature and connections with surgery. In *Surveys on Surgery Theory*, Vol. 2. Annals of Mathematics Studies, 149. Princeton: Princeton University Press, pp. 353–386.

Rosenberg, J. and S. Weinberger. (1988). Higher G-indices and applications. *Ann. Sci. Éc. Norm. Supér. (4)*, **21**(4), 479–495.

Rosenberg, J. and S. Weinberger. (1990). An equivariant Novikov conjecture. *K-Theory*, **4**(1), 29–53. With an appendix by J. P. May.

Rosenberg, J. and S. Weinberger. (2006). The signature operator at 2. *Topology*, **45**, 47–63.

Rothenberg, M. (1978). Torsion invariants and finite transformation groups. In *Algebraic and Geometric Topology (Stanford Univ., Stanford, CA, 1976)*, Part 1. Proc. Symp. in Pure Mathematics, 32. Providence: American Mathematical Society, pp. 267–311.

Rothenberg, M. and J. Sondow. (1979). Nonlinear smooth representations of compact Lie groups. *Pacific J. Math.*, **84**(2), 427–444.

Rourke, C. and B. Sanderson. (1968a). Block bundles I. *Ann. of Math. (2)*, **87**, 1–28.

Rourke, C. and B. Sanderson. (1968b). Block bundles. II. Transversality. *Ann. of Math. (2)*, **87**, 256–278.

Rourke, C. and B. Sanderson. (1968c). Block bundles. III. Homotopy theory. *Ann. of Math. (2)*, **87**, 431–483.

Rourke, C. and B. Sanderson. (1982). *Introduction to PL Topology*. Berlin: Springer.

Rourke, C. and D. Sullivan. (1971). On the Kervaire obstruction. *Ann. of Math. (2)*, **94**, 397–413.

Rueping, H. (2013). The Farrell–Jones conjecture for some general linear groups. Thesis, Bonn.

Sadun, L. (1998). Some generalizations of the pinwheel tiling. *Discrete Comput. Geom.*, **20**, 79–110.

Sapir, M. (2014). A Higman embedding preserving asphericity. *J. Amer. Math. Soc.*, **27**(1), 1–42.

Sarnak, P. and X. Xue. (1991). Bounds for multiplicities of automorphic representations. *Duke Math. J.*, **64**(1), 207–227.

Schoen, R. and S.-T. Yau. (1979a). Existence of incompressible minimal surfaces and the topology of three-dimensional manifolds with nonnegative scalar curvature. *Ann. of Math. (2)*, **110**(1), 127–142.

Schoen, R. and S.-T. Yau. (1979b). Compact group actions and the topology of manifolds of non-positive curvature. *Topology*, **18**, 361–380.

Schoen, R. and S.-T. Yau. (2017). Positive scalar curvature and minimal hypersurface singularities. Preprint, arXiv:1704.05490v1.

Schultz, R. (1975). Circle actions on homotopy spheres not bounding spin manifolds. *Trans. AMS*, **213**, 89–98.

Schultz, R. (1977). On the topological classification of linear representations. *Topology*, **16**(3), 263–269.

Schultz, R. (1985). Transformation groups and exotic spheres. In *Group Actions on Manifolds (Boulder, CO, 1983)*. Contemporary Mathematics, 36. Providence: American Mathematical Society, pp. 243–267.

Schultz, R. (1987). Homology spheres as stationary sets of circle actions. *Michigan Math. J.*, **34**(2), 183–200.

Schwermer, J. (2010). Geometric cycles, arithmetic groups and their cohomology. *Bull. Amer. Math. Soc.*, **47**(2), 187–279.

Scott, P. (1983). The geometries of 3-manifolds. *Bull. London Math. Soc.*, **15**, 401–487.

Selberg, A. (1960). On discontinuous groups in higher-dimensional symmetric spaces. In *Contributions to Function Theory (Int. Colloq. on Function Theory, Bombay, 1960)*. Bombay: Tata Institute of Fundamental Research, pp. 147–164.

Selberg, A. (1965). On the estimation of Fourier coefficients of modular forms. In *Theory of Numbers*. Proc. Symp. in Pure Mathematics, VIII. Providence: American Mathematical Society, pp. 1–15.

Serre, J.-P. (1951). Homologie singulière des espaces fibrés. *Ann. of Math. (2)*, **54**(3), 425–505.

Serre, J.-P. (1973). *A Course in Arithmetic*. Graduate Texts in Mathematics, 7. New York: Springer.

Serre, J.-P. (1977). *Linear Representations of Finite Groups*. Graduate Texts in Mathematics, 42. New York: Springer.

Serre, J.-P. (2003). *Trees*. Springer Monographs in Mathematics. Berlin: Springer. Transl. from French original by J. Stillwell. Corrected 2nd printing of 1980 English translation.

Shalom, Y. (2000). Rigidity of commensurators and irreducible lattices. *Invent. Math.*, **141**(1), 1–54.

Shalom, Y. (2006). The algebraization of Kazhdan's property (T). In *Proc. Int. Congress of Mathematicians (Madrid)*. Available at http://www.icm2006.org/proceedings/Vol_II/contents/ICM_Vol_2_60.pdf.

Shaneson, J. (1969). Wall's surgery obstruction groups for $G \times \mathbb{Z}$. *Ann. of Math. (2)*, **90**, 296–334.

Siebenmann, L. (1965). The obstruction to finding a boundary for an open manifold of dimension greater than five. Thesis, Princeton.

Siebenmann, L. (1970a). A total Whitehead torsion obstruction to fibering over the circle. *Comment. Math. Helv.*, **45**, 1–48.

Siebenmann, L. (1970b). Infinite simple homotopy types. *Indag. Math. (Proc.)*, **73**, 479–495.

Siebenmann, L. (1972). Approximating cellular maps by homeomorphisms. *Topology*, **11**, 271–294.

Siegel, C.L. (1988). *Lectures on the Geometry of Numbers*. Berlin: Springer. Rewritten by K. Chandrasekharan with the assistance of R. Suter.

Siegel, P. (1983). Witt spaces: a geometric cycle theory for KO-homology at odd primes. *Amer. J. Math.*, **105**, 1067–1105.

Silberman, L. (2003). Addendum to: "Random walk in random groups" by M. Gromov. *Geom. Funct. Anal.*, **13**(1), 147–177.

Singer, I. (1977). Some remarks on operator theory and index theory. In *K-Theory and Operator Algebras (Proc. Conf., Univ. of Georgia, Athens, GA, 1975)*. Lecture Notes in Mathematics, 575. Berlin: Springer, pp. 128–138.

Skandalis, G. (1999). Progrès récents sur la conjecture de Baum–Connes. Contribution de Vincent Lafforgue. *Séminaire Bourbaki*, **1999**, 869.

Skandalis, G., J. Tu, and G. Yu. (2002). The coarse Baum–Connes conjecture and groupoids. *Topology*, **41**(4), 807–834.

Spanier, E. (1981). *Algebraic Topology*. Berlin: Springer. Corrected reprint.

Spivak, M. (1967). Spaces satisfying Poincaré duality. *Topology*, **6**, 77–101.

Stallings, J. (1962). The piecewise-linear structure of Euclidean space. *Proc. Cambridge Philos. Soc.*, **58**, 481–488.

Stallings, J. (1965a). Whitehead torsion of free products. *Ann. of Math. (2)*, **82**, 354–363.

Stallings, J. (1965b). On infinite processes leading to differentiability in the complement of a point. In *Differential and Combinatorial Topology (A Symposium in Honor of Marston Morse)*. Princeton: Princeton University Press, pp. 245–254.

Steinberger, M. (1988). The equivariant topological *s*-cobordism theorem. *Invent. Math.*, **91**(1), 61–104.

Steinberger, M. and J. West. (1987). Approximation by equivariant homeomorphisms. I. *Trans. Amer. Math. Soc.*, **302**(1), 297–317.

Stoltz, S. (1992). Simply connected manifolds of positive scalar curvature. *Ann. of Math. (2)*, **136**, 511–540.

Stoltz, S. (1995). Positive scalar curvature metrics – existence and classification questions. In *Proc. Int. Congress of Mathematicians (Zürich, 1994)*, Vol. 2. Basel: Birkhäuser, pp. 625–636.

Stoltz, S. (1996). A conjecture concerning positive Ricci curvature and the Witten genus. *Mathematische Annalen*, **304**, 785–800.

Sullivan, D. (2005). Geometric topology: localization, periodicity and Galois symmetry. In *The 1970 MIT Notes*. K-Monographs in Mathematics, 8. Dordrecht: Springer. Edited and with a preface by A. Ranicki.

Takeuchi, K. (1977). Arithmetic triangle groups. *J. Math. Soc. Japan*, **29**, 91–106.

Taylor, L. and B. Williams. (1979a). Surgery spaces: formulae and structure. In *Algebraic Topology, Waterloo, 1978 (Proc. Conf., Univ. Waterloo, Waterloo, ON, 1978)*. Lecture Notes in Mathematics, 741. Berlin: Springer, pp. 170–195.

Taylor, L. and B. Williams. (1979b). Local surgery: foundations and applications. In *Algebraic Topology, Aarhus, 1978 (Proc. Symp., Univ. Aarhus, Aarhus, 1978)*. Lecture Notes in Mathematics, 763. Berlin: Springer, pp. 673–695.

Taylor, M. (2011a). *Partial Differential Equations I. Basic Theory*, 2nd edn. Applied Mathematical Sciences, 115. New York: Springer.

Taylor, M. (2011b). *Partial Differential Equations II. Qualitative Studies of Linear Equations*, 2nd edn. Applied Mathematical Sciences, 116. New York: Springer.

Taylor, M. (2011c). *Partial Differential Equations III. Nonlinear Equations*, 2nd edn. Applied Mathematical Sciences, 117. New York: Springer.

Teleman, N. (1980). Combinatorial Hodge theory and signature operator. *Invent. Math.* **61**, 227–249.

Thom, R. (1954). Quelques propriétés globales des variétés différentiables. *Comment. Math, Helv.*, **28**, 18–88.

Thurston, W. (1982). Three-dimensional hyperbolic geometry. *Bull. Amer. Math. Soc.*, **6**, 357–381.

Thurston, W. (2002). *The Geometry and Topology of Three-Manifolds*. Classical Princeton 1980 notes retypeset by S. Newbery. Available at http://library.msri.org/books/gt3m/PDF/1.pdf.

Tits, J. (1974). On buildings and their applications. In *Proc. Int. Congress of Mathematicians (Vancouver, BC, 1974)*, Vol. 1. Ottawa: Canadian Mathematical Congress, pp. 209–220.

tom Dieck, T. (1979). *Transformation Groups and Representation Theory*. Lecture Notes in Mathematics, 766. Berlin: Springer.

Valette, A. (2002). *Introduction to the Baum–Connes Conjecture*. Basel: Birkäuser.

Van Limbeek, W. (2014). Riemannian manifolds with local symmetry. *J. Topol. Anal.*, **6**(2), 211–236.

Vershik, A.M., I.M. Gel'fand, and M.I. Graev. (1974). Irreducible representations of the group G^X and cohomology (in Russian). *Funktsional. Anal. i Prilozhen.*, **8**(2), 67–69.

Volic, I. (2006). Finite type knot invariants and the calculus of functors. *Comp. Math.*, **142**, 222–250.

Wagon, S. (1993). *The Banach–Tarski Paradox*. Cambridge: Cambridge University Press. Corrected reprint of the 1985 original.

Waldhausen, F. (1968). On irreducible 3-manifolds which are sufficiently large. *Ann. of Math. (2)*, **87**, 56–88.

Waldhausen, F. (1978). Algebraic K-theory of generalized free products. I, II, III, IV. *Ann. of Math. (2)*, **108**(1), 135–256.

Waldhausen, F. (1987). An outline of how manifolds relate to algebraic K-theory. In *Homotopy Theory (Durham, 1985)*. London Mathematical Society Lecture Note Series, 117. Cambridge: Cambridge University Press, pp. 239–247.

Wall, C.T.C. (1965). Finiteness conditions for CW complexes. *Ann. of Math. (2)*, **81**, 56–69.

Wall, C.T.C. (1968). *Surgery on Compact Manifolds*. New York: Academic Press.

Wall, C.T.C. (1974). On the classification of Hermitian forms. V. Global rings. *Invent. Math.*, **23**, 261–288.

Wall, C.T.C. (1976a). On the classification of Hermitian forms. VI. Group rings. *Ann. of Math. (2)*, **103**(1), 1–80.

Wall, C.T.C. (1976b). Formulae for surgery obstructions. *Topology*, **15**, 189–210. See also corrections: Erratum. (1977). *Topology*, **16**, 495–496.

Wall, C.T.C. (1979). List of problems. In *Homological Group Theory*. London Math. Soc. Lecture Note Series, 36. Cambridge: Cambridge University Press, pp. 369–394.

Wang, H.C. (1972). Topics on totally discontinuous groups. In *Symmetric Spaces (Short Courses Presented at Washington University)*. New York: Dekker, pp. 459–487.

Wegge-Olsen, N. (1993). *K-Theory and C*-algebras: A Friendly Approach*. New York: Oxford University Press.

Wehrfritz, B. (1973). *Infinite Linear Groups: an Account of the Group-Theoretic Properties of Infinite Groups of Matrices*. Berlin: Springer.

Weibel, C. (2005). Algebraic K-theory of rings of integers in local and global fields. In *Handbook of K-Theory*, Vol. 1. Berlin: Springer, pp. 139–190.

Weil, A. (1960). On discrete subgroups of Lie groups. *Ann. of Math. (2)*, **72**, 369–384.

Weil, A. (1962). On discrete subgroups of Lie groups. II. *Ann. of Math. (2)*, **75**, 578–602.

Weil, A. (1964). Remarks on the cohomology of groups. *Ann. of Math. (2)*, **80**, 149–157.

Weinberger, S., (1985a). Constructions of group actions: a survey of some recent developments. In *Group Actions on Manifolds (Boulder, CO, 1983)*. Contemporary Mathematics, 36. Providence: American Mathematical Society, pp. 269–298.

Weinberger, S. (1985b). Group actions and higher signatures. *Proc. Natl. Acad. Sci. USA*, **82**(5), 1297–1298.

Weinberger, S. (1986). Homologically trivial group actions. I. Simply connected manifolds. *Amer. J. Math.*, **108**(5), 1005–1021.

Weinberger, S. (1987). Group actions and higher signatures. II. *Comm. Pure Appl. Math.*, **40**(2), 179–187.

Weinberger, S. (1988a). Class numbers, the Novikov conjecture, and transformation groups. *Topology*, **27**(3), 353–365.

Weinberger, S. (1988b). Homotopy invariance of η-invariants. *Proc. Natl. Acad. Sci. USA*, **85**(15), 5362–5363.

Weinberger, S. (1989). Semifree locally linear PL actions on the sphere. *Israel J. Math.*, **66**(1–3), 351–363.

Weinberger, S. (1994). *The Topological Classification of Stratified Spaces*. Chicago Lectures in Mathematics. Chicago: University of Chicago Press.

Weinberger, S. (1995). Nonlocally linear manifolds and orbifolds. In *Proc. Int. Congress of Mathematicians (Zürich, 1994)*, Vol. 2. Basel: Birkhäuser, pp. 637–647.

Weinberger, S. (1996). Rationality of ρ-invariants. Appendix to Farber, M. and J. Levine. (1996). Jumps of the eta-invariant. *Math. Z.*, **223**(2), 197–246.

Weinberger, S. (1999a). Nonlinear averaging, embeddings, and group actions. In *Tel Aviv Topology Conf.: Rothenberg Festschrift (1998)*. Contemporary Mathematics, 231. Providence: American Mathematical Society, pp. 307–314.

Weinberger, S. (1999b). Higher ρ-invariants. In *Tel Aviv Topology Conf.: Rothenberg Festschrift (1998)*. Contemporary Mathematics, 231. Providence: American Mathematical Society, pp. 315–320.

Weinberger, S. (2014). The complexity of certain topological inference problems. *Found. Comput. Math.*, **14**, 1277–1285.

Weinberger, S. and M. Yan. (2001). Equivariant periodicity for abelian group actions. *Adv. Geom.*, **1**, 49–70.

Weinberger, S. and M. Yan. (2005). Equivariant periodicity for compact group actions. *Adv. Geom.*, **5**, 363–376.

Weinberger, S. and G. Yu. (2015). Finite part of operator K-theory for groups finitely embeddable into Hilbert space and the degree of nonrigidity of manifolds. *Geom. Topol.*, **19**(5), 2767–2799.

Weinberger, S., Z. Xie, and G. Yu. (2020). Additivity of higher rho invariants and nonrigidity of topological manifolds. Preprint, arXiv:1608.03661v5.

Weisfeiler, B. (1984). Strong approximation for Zariski-dense subgroups of semi-simple algebraic groups. *Ann. of Math. (2)*, **120**, 271–315.

Weiss, M. (1996). Calculus of embeddings. *Bull. Amer. Math. Soc.*, **33**(2), 177–187.

Weiss, M. (1999). Embeddings from the point of view of immersion theory. I. *Geom. Topol.*, **3**, 67–101. See also corrections: (2011). Erratum. *Geom. Topol.*, **15**(1), 407–409.

Weiss, M. and B. Williams. (2001). Automorphisms of manifolds. In *Surveys on Surgery Theory*, Vol. 2. Annals of Mathematics Studies, 149. Princeton: Princeton University Press, pp. 165–220.

Witte-Morris, D. (2015). *Introduction to Arithmetic Groups*. Lethbridge: Deductive Press.

Whyte, K. (1999). Amenability, bilipschitz equivalence, and the von Neumann conjecture. *Duke Math J.*, **99**, 93–112.

Xue, X. (1992). On the Betti numbers of a hyperbolic manifold. *Geom. Funct. Anal.*, **2**(1), 126–136.

Yamasaki, M. (1987). L-groups of crystallographic groups. *Invent. Math.*, **88**(3), 571–602.

Yan, M. (1993). The periodicity in stable equivariant surgery. *Comm. Pure Appl. Math.*, **46**(7), 1013–1040.

Young, R. (2005). Counting hyperbolic manifolds with bounded diameter. *Geom. Dedicata*, **116**, 61–65.

Yu, G. (1998). The Novikov conjecture for groups with finite asymptotic dimension. *Ann. of Math. (2)*, **147**(2), 325–355.

Yu, G. (2000). The coarse Baum–Connes conjecture for spaces which admit a uniform embedding into Hilbert space. *Invent. Math.*, **139**(1), 201–240.

Zeeman, E.C. (1962). A note on an example of Mazur. *Ann. of Math. (2)*, **76**(2), 235–236.

Zimmer, R. (1984). *Ergodic Theory and Semisimple Groups*. Basel: Birkhäuser.

Zuk, A. (2003). Property (T) and Kazhdan constants for discrete groups. *Geom. Funct. Anal.*, **13**(3), 643–670.

Zuk, A. (2011). Automata groups. In *Topics in Noncommutative Geometry*. Clay Mathematics Proceedings, 16. Providence: American Mathematical Society.

Subject Index

Printed in the United States
by Baker & Taylor Publisher Services